A NEW HISTORICAL GEOGRAPHY OF
ENGLAND BEFORE 1600

A NEW HISTORICAL GEOGRAPHY OF ENGLAND BEFORE 1600

Edited by
H. C. DARBY

CAMBRIDGE UNIVERSITY PRESS

CAMBRIDGE

LONDON   NEW YORK   MELBOURNE

CAMBRIDGE UNIVERSITY PRESS
Cambridge, New York, Melbourne, Madrid, Cape Town,
Singapore, São Paulo, Delhi, Tokyo, Mexico City

Cambridge University Press
The Edinburgh Building, Cambridge CB2 8RU, UK

Published in the United States of America by Cambridge University Press, New York

www.cambridge.org
Information on this title: www.cambridge.org/9780521291446

First published as part of *A New Historical Geography of England*, 1973
First published separately, 1976
Re-issued 2011

A catalogue record for this publication is available from the British Library

Library of Congress Cataloguing in Publication data
Darby, Henry Clifford, 1909–
A new historical geography of England before 1600.
Comprises the first six chapters of the editor's
A new historical geography of England published in 1973.
Includes bibliographical references and index.
1. England – Historical geography. 1. Title.
DA610.D274 1976 942 76–26141

ISBN 978-0-521-22122-1 Hardback
ISBN 978-0-521-29144-6 Paperback

CONTENTS

MAPS AND DIAGRAMS

PREFACE

In 1936 the Cambridge University Press published *An historical geography of England before A.D. 1800*. It has been reprinted a number of times but clearly the moment has come not for a further reprint nor for a new edition but for an entirely new volume based upon the enormous amount of work that has been done since 1936, and especially since 1945. Much of the new work has appeared in the pages of three journals, *The Economic History Review*, *The Agricultural History Review* and the *Transactions and Papers of the Institute of British Geographers*. Moreover, the English Place-Name Society has continued to produce its scholarly volumes year by year; the Domesday Book has been analysed geographically, so have the Lay Subsidies of 1334 and 1524–5. Our views of agriculture and industry in later times have also been modified by a variety of monographs.

Not only has there been much exploration of sources, but also much discussion about the method of historical geography. In particular, a contrast has often been drawn between the reconstructions of past geographies and the study of geographical changes through time, between the so-called 'horizontal' and 'vertical' approaches. This present volume seeks to combine both, and the combination was suggested by J. O. M. Broek's *The Santa Clara Valley, California: a study in landscape changes* (Utrecht, 1932).

The new work begins with the coming of the Anglo-Saxons in the belief that so far as there ever is a new beginning in history, that event was such a beginning. It continues beyond the eighteenth century, up to 1900 or so, but this has not been taken as a rigid date because 1914, rather than 1900, marks the effective end of the nineteenth century. In contemplating the result, one can only be very conscious of what remains to be done. As far as sources are concerned, although much recent work has been done on the Tithe Returns of the 1840s, a comprehensive treatment has yet to appear. There have also been interesting studies on enclosure, but no large-scale attack on a geographical basis. Or, to take another example, even so obvious a source as the Census Returns, from 1801 onwards, still awaits comprehensive analysis and interpretation. As far as method is concerned, much further enquiry is needed to see to what extent statistical techniques and locational analysis can be applied to historical data of varying quality and coverage.

In another generation or so the materials for an historical geography of England will not be as we know them now. A wider range of sources will have been explored and evaluated. Fresh ideas about method will have prepared the way for a more sophisticated presentation. And by that time we and our landscape will have become yet one more chapter in some other *Historical Geography of England*.

I am greatly indebted to my fellow contributors for their co-operation and patience. All of us must thank the staff of the Cambridge University Press for their skill and care. We must also thank Mr G. R. Versey. He has not only drawn all the maps and diagrams but has compiled many of them and has also given much general assistance at all stages of the work.

H. C. DARBY

KING'S COLLEGE
CAMBRIDGE

Candlemas, 1973

NOTE TO TWO-VOLUME EDITION

This edition consists of the first six chapters of *A new historical geography of England* (1973). The text remains the same but the opportunity has been taken to make a few minor corrections, and the Index has been adjusted to meet the needs of the volume. An Introduction has been added to explain the thinking behind the plan of the work as a whole.

H.C.D.

KING'S COLLEGE
CAMBRIDGE
St Basil's Day, 1976

INTRODUCTION

An account of the geography of a past age that aims to explain as well as to describe should, like one of the present-day, take into consideration the relevant circumstances of earlier times. But when we contemplate a chronological series of cross-sections, we are at once faced with more complicated considerations. Two very different methods of treatment can be envisaged. If, on the one hand, each cross-section in a sequence aims at being a balanced geographical account, compounded of description and explanation, there will, of necessity, be much repetition and varying degrees of overlap as each cross-section ranges backwards to satisfy its own needs. If, on the other hand, each cross-section is limited strictly to its own contemporary materials, a valid criticism might be that the sequence constitutes a series of static pictures that ignore the process of becoming.

It is possible to compromise between these two extremes in a way suggested by J. O. M. Broek in *The Santa Clara Valley, California: a study in landscape changes* (Utrecht, 1932). Here, four cross-sections are separated by three studies of the economic and social forces that led to successive changes in the landscape. The device of separating the explanatory narrative from the description of a landscape at each period serves not only to explain each landscape but also to provide connecting links between the successive views. For *A new historical geography of England* the choice consists of six narratives and six cross-sections as follows:

1 *The Anglo-Scandinavian foundations*
2 Domesday in England
   3 *Changes in the early Middle Ages*
4 England *circa* 1334
   5 *Changes in the later Middle Ages*
6 England *circa* 1600
   7 *The age of the improver: 1600–1800*
8 England *circa* 1800
   9 *Changes in the early railway age: 1800–1850*
10 England *circa* 1850
   11 *The changing face of England: 1850–circa 1900*
12 England *circa* 1900

The choice of dates for the various cross-sections depends partly on the march of events, and also (let us admit it) upon the availability of sources of information. Even so, such a choice can only be an individual one, and the thinking that lies behind it must now be described.

The work as a whole begins with the coming of the Anglo-Saxons 'in the belief that so far as there ever is a new beginning in history, that event was such a beginning'. This is not to imply that pre-Saxon contributions were unimportant. Clearly, continuity between Roman Britain and Anglo-Saxon England was much greater than was at one time believed, and further work may strengthen this belief. But this does not affect the fact that the coming of the English to England was a new beginning – with all it meant in the peopling, the language and the institutions of much of these islands. When the Normans arrived in 1066, the villages they encountered bore names that were certainly not Celtic – except in Cornwall. Part of Britain had become England.

The period between the end of Roman rule and the year of the Norman Conquest was one of great fluidity, and the sources are so limited that it is difficult for the pen to catch the scene. Yet it was a formative period of the greatest importance for the landscapes of later times. Anglo-Saxons and Scandinavians covered Roman Britain with their new villages, and proceeded rapidly to clear its woodland. No later invasion of peoples significantly modified the Anglo-Scandinavian pattern; the Norman Conquest was the transposition of an aristocracy and not a folk movement of new settlers on the land. Twenty years after their coming, the Normans instituted the enquiry that resulted in Domesday Book. Its unique character as a source enables us to survey the results of the centuries of migration and settlement and to present a view of England in 1086.

In the years after the Conquest, the countryside continued to be cleared and towns grew and prospered. But this expansive movement did not continue uninterruptedly throughout the Middle Ages. In places it slowed down; in others it ceased; and in yet other places the frontiers of cultivation even retreated, and the populations of towns may have declined. Certainly in England, as in most of Western and Central Europe, medieval agrarian and commercial effort had reached a peak by 1300, and the great age of expanding arable and trade was succeeded in the fourteenth and fifteenth centuries by one of stagnation and recession. The decline was especially marked during the hundred years between 1350 and 1450. The early fourteenth century seemed therefore suitable

for another view of England. Moreover for this critical period a convenient and unique source is at hand – the lay subsidy of 1334. It is
convenient coming as it did before the full impact of the recession of the
later Middle Ages was felt. It is unique because the assessment agreed
upon in that year continued to be the basis for later subsidies until 1623.
This means that although the rolls of 1334 (as for other years) have not
entirely survived, the missing figures can be recovered from later rolls.
It is possible, therefore, to achieve a reasonably complete cover from
which to construct a general framework for delineating the geographical
and economic condition of the country as a whole.

A possible date for a cross-section at the end of the Middle Ages might
be 1500. But although Henry Tudor had won the throne of England in
1485, the years around 1500 had yet to see the recession of the later
Middle Ages merge into the full flowering of the sixteenth century, a
flowering reflected in the writings of such men as Leland and Camden,
in the maps of Saxton and Norden, and in Speed's *Theatre of the Empire of
Great Britaine* (1611) which aimed at presenting 'an exact geography' of
the realm. The Tudors, in the words of Charles Whibley, 'recognized
that the most brilliant discovery of a brilliant age was the discovery of
their own country', and in so doing helped to provide material that makes
1600 rather than 1500 a more suitable moment for another cross-section.
That the Tudor dynasty came to an end with the death of Queen Elizabeth
in 1603 did not determine the choice, but, at any rate, made it not
inappropriate.

The choice of dates for cross-sectional views in the modern period
can also be argued. Macaulay chose 1685 as the basis for the famous
third chapter in his *History of England* (1848), but this is too near 1600 in
the present scheme. Another date that might have been chosen is some
year in the 1720s. Scotland had been united with England in 1707, the
War of Spanish Succession was over in 1714, and Daniel Defoe's *Tour*
appeared in 1724. This latter date was used by G. M. Trevelyan in his
England Under Queen Anne (1930), and, like Macaulay's choice, was
appropriate in the context of his own work. Other conceivable dates
might have been 1760, at the beginning of the canal age, or even 1780
when the annual rate of industrial growth was first greater than 2%, but
both dates are too near 1800 which for many reasons appeared as an ideal
date for a cross-section in the full work. The first Census was taken in
1801; this was also the year of the so-called Acreage Returns for different crops; furthermore, between 1793 and 1815 appeared the *General*

Views of each county issued by the Board of Agriculture. Taken together they yield a remarkable body of information which reinforces the choice of the year 1800.

Thus it was that the span of years from 1600 to 1800 was covered in one chapter. The clearing of the wood may have been the great epic of the Middle Ages but now, after 1600 came other epics – the draining of the marsh, the reclamation of the heath, the enclosure of the arable, the spread of landscape gardens and the beginning of the later seats of industry. And so the changes of these two centuries were brought together under the title 'The Age of the Improver'.

After 1800 it is clear that the time-intervals between the cross-sections need to be shorter. Not only was the pace of change accelerating, but the written evidence about it was increasing prodigiously. J. H. Clapham chose 1820 and 1886–7 as dates for describing what he called 'the face of the country'; both were suitable for the development of the themes of his *Economic history of modern Britain* (1926–38). But for us 1820 is too near 1800; and 1886–7 is too near 1900. We chose the year 1850 because it may be said to mark the end of 'the early railway age', some of the results of which were indicated in the census of 1851. As for agriculture, the tithe surveys of the 1840s, and the county reports in the *Journal of the Royal Agricultural Society* (1845–69) together provided the only detailed picture since the *General Views* of the Board of Agriculture. Moreover, mid-century, on the eve of the Great Exhibition of 1851, seemed a suitable moment at which to pause.

The years after 1850 saw the full development of Britain as an industrial state. Towns became the birthplaces of the major part of the population, and agriculture declined to a subordinate position in face of overseas competition. The final cross-section is tied to the year 1900. We might have chosen 1910 or 1914 which marked the effective end of the nineteenth century; the last two chapters certainly do not hesitate on occasions to reach towards these dates. Rightly or wrongly, we chose 1900 for the title of the last chapter. At any rate when Queen Victoria died on 22 January 1901 many felt that a great epoch had closed.

H. C. D

Chapter 1

THE ANGLO-SCANDINAVIAN FOUNDATIONS

H. C. DARBY

The Anglo-Saxon settlement was a new beginning in the history of Britain. It made a decisive contribution to the peopling of the south-east; it determined the language of the area; and it brought institutions that formed the basis of all later development. In addition, it laid the foundations of the later geography of England. With the coming of the Anglo-Saxons, a new chapter in the history of settlement and land utilisation was begun, and, with but little interruption, the story has been continuous up to the present day.

It is true that the Anglo-Saxons did not come into an empty land, and that many contributions from pre-Saxon days have entered into the making of England. The various strains in the present population may be traced far back into prehistoric times, and the early peoples of those times left remains that are still visible today. These remains are particularly numerous on the downlands of the southern counties. Megalithic monuments (long barrows, tumuli and stone circles) are prominent in the chalk country of Wessex; here is Stonehenge, the most famous of all megalithic monuments, and not far away is the stone circle of Avebury and the mysterious mound of Silbury. Earthworks, or 'camps', like Maiden Castle in Dorset and Windmill Hill in Wiltshire, with their ramparts and ditches, are also prominent local features that remain to excite our imagination. Then again, old trackways can still be traced on the present surface of the ground. The Icknield Way below the Chilterns, and the Pilgrims' Way along the North Downs, may have been already old when the Roman legionaries tramped the country, and there are others like them. Finally, there are traces of primitive cultivation that have also survived the changes of later time. Many have been destroyed by later cultivation, but others, now grass-grown, can still be seen in the present landscape. These early monuments and features are especially characteristic of the chalk downlands, but they are also to be found elsewhere. Thus there are hut circles and primitive corn plots on Dartmoor, and megalithic remains on the Yorkshire Wolds and along many stretches of the western seaboard.

The Romans who came in A.D. 43 bequeathed, in turn, a substantial legacy to the geography of succeeding ages. The lines of many Roman roads are still in use as arterial ways. The strategical locations of many cities were recognised in Roman as in later times. In the north, the remains of the Roman Walls still cross the landscape in a distinctive fashion; and in the south many of the traces of early cultivation date from Romano-British times. The legacy of Rome to the geography of England is no mean one. But, even so, as far as there ever is a new beginning in history, the coming of the Angles, Saxons and Jutes was such a beginning.

The circumstances of this new beginning, however, are far from clear. The period that separates the end of Roman rule in Britain from the emergence of the earliest English states is obscure and baffling. The prelude to the settlement of the newcomers from across the North Sea consisted of plundering raids that began as early as A.D. 300, and even before. In the west and north, too, the raids of the Picts and the Scots left trails of devastation. But it was not until about A.D. 410 that the Roman legions were withdrawn and the cities of Britain told to look to their own defence. Whether this injunction marked the end of Roman rule in Britain is a disputed matter. At any rate, within a generation or so, it is clear that the Romano-British were incapable of withstanding the raiders from overseas. At what date this raiding passed into settlement is uncertain, but it would seem that the so-called 'Adventus Saxonum' took place in the years around the middle of the fifth century.

THE ANGLO-SAXON SETTLEMENT

The arrival of the English

The evidence that relates to the coming of the English is of three kinds – literary, archaeological and place-name.[1] Apart from a few brief and scattered references, the main body of the literary evidence is derived

---

[1] General sources for the account that follows include: (1) R. G. Collingwood and J. N. L. Myres, *Roman Britain and the English settlements* (Oxford, 2nd ed. 1937); (2) F. M. Stenton, *Anglo-Saxon England* (Oxford, 2nd ed. 1947); (3) P. H. Blair, *An introduction to Anglo-Saxon England* (Cambridge, 1956); (4) R. H. Hodgkin, *A history of the Anglo-Saxons*, 2 vols. (Oxford, 3rd ed. 1952); (5) K. Jackson, *Language and history in early Britain* (Edinburgh, 1953); (6) H. R. Loyn, *Anglo-Saxon England and the Norman Conquest* (London, 1962).

from the writings of Gildas (*circa* 550), Bede (*circa* 730), Nennius (*circa* 800) and from the earlier parts of the Anglo-Saxon Chronicle (*circa* 900). Each of these four sources is unsatisfactory, but taken together, they indicate a sequence of events that may well bear some approximation to the truth. In about A.D. 449 a certain British chieftain named Vortigern, so runs the tradition, enlisted the help, from across the sea, of the Saxons, under two brothers named Hengest and Horsa, to repel the raids of the Picts and the Scots. The mercenaries were then joined by others of their kind, and, after some dispute about their provisions, the newcomers turned against their employers and raided through the land from sea to sea. In the years that followed, others continued to come from across the North Sea. Thus did Nemesis fall upon the Britons, and, in the words of Gildas, the fire of the Saxons burned across the island until 'it licked the western ocean with its red and savage tongue'. The result of the calamity was a devastated countryside, ruined cities, and the complete collapse of British resistance. The newcomers were divided by Bede into Angles, Saxons and Jutes, but recent work has shown that Bede's generalisation was too simple and that the newcomers may also have included some Frisians.[1]

The first collapse was followed by a rally of the British forces and by a period of indecisive warfare, victory alternating with defeat, until a great British victory was won at a place called Mons Badonicus. Its date seems to have been about 500. Its site has been variously identified: one possible location is Badbury Rings to the north-west of Wimborne Minster in Dorset.[2] There is no doubt about its decisive nature for it was followed by a respite that continued for some fifty years, that is up to the time that Gildas wrote. Nennius tells us that among those who had fought against the Saxons was a man named Arthur; 'he fought against them with the kings of Britain, but he himself was the military commander'. Then follow the names of his twelve battles, and the twelfth was none other than that of Mons Badonicus. Later ages were to add legend after legend to the name of Arthur. It is interesting to think that at the core of all the romance of the Arthurian cycle, which has so fascinated generation after generation, lies something of the story of the defence of Britain against the English.

[1] F. M. Stenton, 'The historical bearing of place-name studies: the English occupation of South Britain', *Trans. Roy. Hist. Soc.*, 4th ser., XXII (1940), 5–6; H. R. Loyn, 27.

[2] K. Jackson, 'The site of Mount Badon', *Jour. Celtic Studies*, II (Temple Univ., Baltimore, Md., 1958), 152–5.

Its hero has been called the last of the Romans, the last to use Roman ideas of warfare for the benefit of the British people, and, wrote Professor Collingwood, 'the story of Roman Britain ends with him'.[1]

The last of the four sources, the Anglo-Saxon Chronicle, adds some further details, and describes how the Britons of Kent 'fled from the English like fire'; but, in common with the other sources, it tells us little about the shifting frontier between the advancing English and the Britons. The little it does tell us raises many difficulties. Under the year 571 it records an English victory over the Britons, and the capture of the four settlements of Limbury, Aylesbury, Benson and Eynsham, to the west of the Chilterns. 'No annal in the early sections of the Chronicle', wrote Sir Frank Stenton, 'is more important than this, and there is none of which the interpretation is more difficult.'[2] It is hard to see how the Britons could have been here at this date. Was their territory an enclave surrounded by the English? Or might it have been territory at first over-run by the English, then lost after Mons Badonicus, and then finally re-covered once more in 571? Whatever the interpretation, the entry does suggest that the settlement of the English was no easy appropriation of south-east Britain. The presence of Britons here, well over a hundred years after the traditional date of Hengest's landing, may indicate that the triumph of the Anglo-Saxons was by no means a foregone conclusion. The English occupation of south Britain may well have comprised two phases separated by the fifty years or so that followed their defeat at Mons Badonicus, i.e. by the earlier half of the sixth century.

Revealing as the literary evidence is, it consists only of the fragments of a story that is far from complete. It makes specific mention of the coming of the invaders only to Kent, Sussex and Wessex. It tells us nothing, for example, of the arrival of the East Anglians and the East Saxons, nor any-thing about that of the Mercians, the Middle Anglians and the North Anglians. For an idea of the early spread of the Anglo-Saxons as a whole we must turn to the evidence of archaeology and of place-names. The bulk of the archaeological evidence consists of cemeteries and burial sites belonging to the age before the Anglo-Saxons became Christian, and it dates therefore roughly from between 450 and 650. To the north of the Thames, as Fig. 1 shows, was a widespread occupation both of the Midlands and of southern England by about 570. The finds are strikingly concentrated around the entrance of the Wash, and it is clear that many

[1] R. G. Collingwood and J. N. L. Myres, 34.
[2] F. M. Stenton (1947), 27.

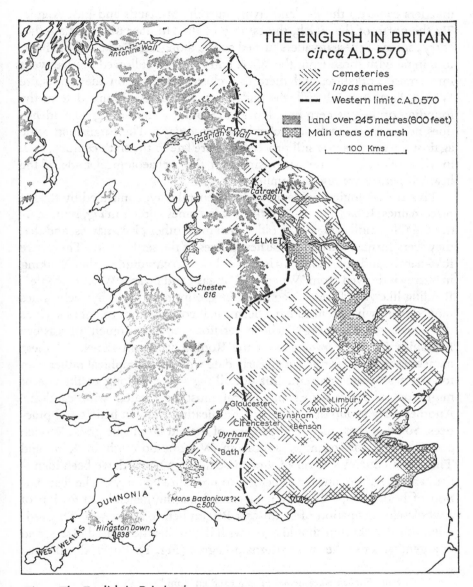

THE ENGLISH IN BRITAIN
circa A.D. 570

⦚⦚⦚	Cemeteries
⫻⫻	*ingas* names
– – –	Western limit *c.*A.D.570
▓	Land over 245 metres (800 feet)
▒	Main areas of marsh

100 Kms

Antonine Wall

Hadrian's Wall

Catraeth c.600

ÉLMET

Chester 616

Gloucester

Limbury

Aylesbury

Eynsham

Cirencester

Benson

Dyrham 577

Bath

Mons Badonicus? c.500

DUMNONIA

WEST WEALAS

Hingston Down 838

Fig. 1 The English in Britain *circa* A.D. 570
  Based on: (1) E. Ekwall, *English place-names in -ing* (Lund, 1935); (2) *Map of Britain in the Dark Ages* (Ordnance Survey, 2nd ed., 1966).

invaders came up the Fenland rivers into the Midlands and East Anglia.
To the north of the Wash, the estuary of the Humber provided another
entry; some of these invaders turned south, along the Trent and the Soar,
to mingle with those from the Wash entry; others followed the Derwent
into south Yorkshire, and there is a cluster of finds on the Yorkshire
Wolds. Farther north still, the finds are few. The traditional date of the
founding of the northern kingdom is 547, but the archaeological evidence
does not carry us back very clearly to those days. The English struggle
against the Britons was still proceeding here late in the sixth century, and,
in view of this, it is not surprising that the archaeological finds of the
heathen period are few in number.

This archaeological evidence is, in a general way, confirmed by that of
place-names. It has been shown that place-names ending in *ing*, representing the Old English *ingas*, are older than most other place-names, and that
they were formed during the earliest phases of the settlement.[1] They were
folk-names, and referred not to localities but to communities; thus Woking
in Surrey is derived from 'Wocc' and 'ingas', and means 'Wocc's people';
Reading likewise 'Reada's people'. The groups designated by such names
ranged from what seem to have been small communities to tribes such as
the *Sunningas* (Sonning), whose territory covered much of eastern
Berkshire and the *Hrothingas* of the Roding valley in Essex. All these
names ceased to be formed at an early date when topographical rather than
folk-names became current. But not all *ing* names belong to the age of
migration, and *ing* names in general must be treated with caution.[2]
Another early group of names is that indicating heathen beliefs and practices. Some of these include the names of the old Germanic gods; Woden
and Thunor, for example, appear in Woodnesborough in Kent and
Thursley in Surrey.[3] Over fifty of these heathen names have been identified, and they, too, appear to belong to pre-Christian days. The distribution of both groups of early names confirms the general impression of
a wholesale occupation of south-east Britain before about 570. Theoreti-
cally, this distribution should agree with that of heathen burial places, and
in a general sense the two patterns do agree (Fig. 1). But in detail some

---

[1] E. Ekwall, *English place-names in -ing* (2nd ed. Lund, 1962).
[2] See A. H. Smith (1) *English place-name elements* (Cambridge, 1956), 282–303;
(2) 'Place-names and the Anglo-Saxon settlement', *Proc. British Academy*, XLII (1956),
73–80.
[3] F. M. Stenton, 'The historical bearing of place-name studies: Anglo-Saxon
heathenism', *Trans. Roy. Hist. Soc.*, 4th ser., XXIII (1941), 1–24.

differences are at once apparent. There are, for example, few pagan cemeteries in Essex but many *ing* names. On the other hand, there are few *ing* names in the Upper Thames area where there are early cemeteries.

It is impossible to assess the significance of these discrepancies, but in view of the element of chance involved both in the survival of cemeteries and of place-names, it would be remarkable if they showed a detailed correspondence in every locality. Moreover, some anomalies may be accounted for by our inability to distinguish other types of archaic place-names 'which may have had a greater frequency in some districts'.[1] Many *ing* names, moreover, may belong to a period 'later than, but soon after, the immigration-settlement that is recorded in the early pagan-burials'.[2] Furthermore, many names ending in 'ham' may be older than was at one time supposed, and may belong to the earliest days of the settlement.[3]

The Anglo-Saxon kingdoms

Whatever the obscurity of the two centuries following 450, we know that in the course of the sixth century the Anglo-Saxons became organised into states. The process by which the historic kingdoms emerged is lost for ever from our sight. By the seventh century the names of as many as eleven kingdoms are to be found, and there may have been more. Some of these were joined to others until the result was the seven kingdoms of the Heptarchy, the boundaries of which often reflected local features of wood and marsh and heath (Fig. 2). To this Anglo-Saxon age belong some of the numerous linear bank-and-ditch earthworks that are found in many parts of the country, and that commonly have such names as Devil's Dyke or Grim's Dyke. Some were constructed primarily as boundaries of, say, cattle ranches or large sheep walks; such is Grim's Ditch in the north-east of Cranborne Chase. Others were for defensive purposes, and to these belong the striking Cambridgeshire system of dykes that cross the open chalk country between fen and wood and that were probably built by the East Angles against the Mercians (Fig. 3).

The earliest kingdoms to emerge seem to have been those of the south-east – East Anglia, Essex, Kent and Sussex; but they were soon out of

[1] A. H. Smith, 'Place-names and the Anglo-Saxon settlement', 84.

[2] J. McN. Dodgson, 'The significance of the distribution of the English place-name in -*ingas*, *inga* in south-east England', *Med. Archaeol.*, x (1966), 19. See also J. N. L. Myres, 'Britain in the Dark Ages', *Antiquity*, IX (1935), 455–64.

[3] B. Cox, 'The significance of the distribution of English place-names in *hām* in the Midlands and East Anglia', *Jour. English Place-Name Soc.*, v (1973), 15–73.

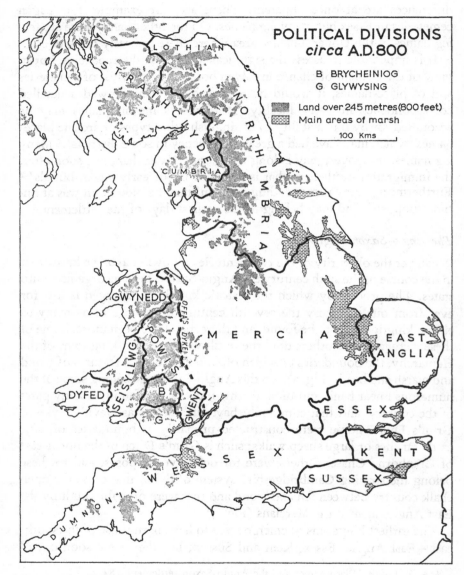

Fig. 2 Political divisions *circa* A.D. 800
    Based on: (1) R. L. Poole, *Historical atlas of modern Europe* (Oxford, 1902),
    plate 16; (2) W. Rees, *An historical atlas of Wales* (2nd ed., Cardiff, 1959),
    plate 22.

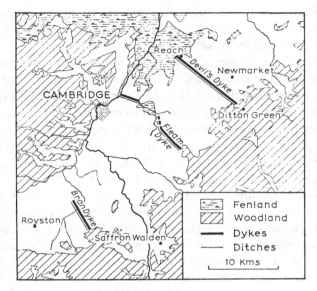

Fig. 3 The Cambridgeshire dykes
Based on C. Fox, 'Dykes', *Antiquity*, III (1929), 137.

touch with the advancing frontier against the Britons, and the great powers of the later Anglo-Saxon period came to be those of Wessex, Mercia and Northumbria. Their rivalry features prominently in the records of English history during the seventh and eighth centuries, but the complexities of their mutual relations did not prevent continued advance to the west.

The evidence for the origin of the kingdom of Wessex has been much debated. On the one hand, the Anglo-Saxon Chronicle, in entries for the years between 495 and 519, speaks of West Saxon chieftains landing on the Hampshire coast and fighting and killing many Britons. On the other hand, the archaeological evidence seems to show that the West Saxons arrived from the east, maybe from the Wash along the Icknield Way to the Upper Thames area and so southwards. The two versions have been reconciled on the assumption that the archaeological evidence bears witness to a mass movement of people, while the literary evidence records the exploits of individuals who were to become the ancestors of the royal house of Wessex. Whatever be the truth, the frontier against the Britons was being vigorously pressed long before the end of the sixth century. Under the year 577 the Chronicle records a victory over the Britons at Dyrham, and it tells how the Saxons took Gloucester, Cirencester and

Bath. Dyrham is six miles north of Bath, and the victory opened up the lower Severn valley to Saxon colonists, thus separating the Britons of the south-west peninsula from those of Wales.

The details of the further advance of the Saxons into the south-west are far from clear. The place-names of Dorset and Somerset suggest Saxon occupation in the seventh century.[1] To the west lay the Welsh kingdom of Dumnonia, the memory of which survives in the name of Devon; and it would seem that the eastern part of this passed into Saxon hands before the end of the seventh century, and that the northern and western parts were occupied after 710 when Ine of Wessex defeated Geraint, king of Dumnonia.[2] The Saxon occupation of Devon was thorough. Its place-names are overwhelmingly English, both in the west as well as in the east of the county; and Welsh names are, surprisingly, fewer than in Somerset or Dorset.[3] The reason may possibly lie in the fact that the Welsh population of Devon was sparse when the Saxons arrived, partly because, during the fifth and sixth centuries and earlier, there had been a large migration across the sea to transform Armorica into what later became Brittany.[4] Even so, it has been suggested that the Welsh constituted 'a far from negligible element in the population of Devon' in the Dark Ages.[5]

Farther west still, were the Welsh of Cornwall into which Dumnonia had contracted, and there are echoes of warfare between Welsh and Saxon. They were fighting in 710, in 722 and again in 753, and in 815 the Saxon king 'laid waste West Wales from east to west'. Ten years later the Welsh raided into Devon, and in 838 they joined a Danish army against Wessex only to be defeated at Hingston Down, to the west of the Tamar in Cornwall. Henceforward, Cornwall came under Saxon overlordship, but a native dynasty of Welsh kings seems to have survived probably into the early years of the tenth century. By this time, the force of Saxon colonisation had spent itself, and Cornwall, except in the extreme east, did not lose its Celtic character with its independence.

The details of the expansion westward across the Midland Plain are lost from our sight. But in the early seventh century various groups in the area emerged as the kingdom of Mercia. A limit to the westward expansion

[1] F. M. Stenton (1947), 63.

[2] W. G. Hoskins, *The westward expansion of Wessex* (Leicester, 1960).

[3] J. E. B. Gover et al., *The place-names of Devon*, pt 1 (Cambridge, 1931), xix.

[4] H. R. Loyn, 47; Norah K. Chadwick, 'The colonization of Brittany from Celtic Britain', *Proc. British Academy*, LI (1965), 235–99.

[5] H. P. R. Finberg, *The early charters of Devon and Cornwall* (Leicester, 1953), 31.

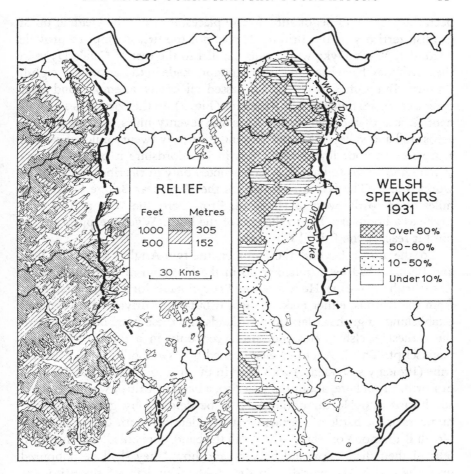

Fig. 4 Offa's Dyke
  Based on: (1) C. Fox, *Offa's Dyke* (London, 1955), 166; (2) D. T. Williams,
  'A linguistic map of Wales according to the 1931 Census, with some observations
  on its historical and geographical setting', *Geog. Jour.*, LXXXIX (1937), 148.
  Modern county boundaries are shown.

of the Mercian kings was set by the Welsh foothills. Little is known of the
process by which the boundary between the English and the Welsh
evolved, and only a few echoes of the English advance have come to us
from this period. It was about 790, during the reign of Offa, that the
famous dyke that now bears his name was constructed.[1] It seems to have

  [1] C. Fox, *Offa's Dyke* (British Academy, London, 1955).

been the product of negotiation for, in places, it meant the yielding up of English territory to the Britons. It was not the first attempt to provide a frontier. Wat's Dyke which runs parallel to the east of Offa's Dyke in the north may have been an earlier attempt, made in the seventh or eighth century. But Offa's Dyke far surpassed all earlier attempts, and after twelve centuries much of it still remains (Fig. 4). In the north it is a prominent feature of the countryside for some seventy miles, cutting straight across hill and valley alike, and running for nearly twenty miles above the 1,000 feet contour line. Southwards in Herefordshire it becomes intermittent and seems to have been constructed only in clearings made in the wooded area. The last stretch reaches the Wye near Bridge Sollers, six miles or so to the west of Hereford. From here, the Wye replaced the Dyke as a boundary, but to the east of the lower Wye it reappears again until it reaches the Bristol Channel.

To the north, beyond the Humber, the two Anglian states of Deira and Bernicia had been united to form the kingdom of Northumbria by Aethelfrith (593–617). He was, wrote Bede, notable for acquiring territory from the Britons, either making them tributary or driving them out and establishing English settlers on their lands. To these years belongs one of the earliest Welsh poems, but preserved only in a thirteenth-century manuscript. This is Aneirin's elegy called the 'Gododdin' after a British tribe (Ptolemy's Otadenoi) who lived in the area to the south of the Firth of Forth. From here, about the year 600, a band of British warriors raided southward into Northumbria, only to be defeated by greatly superior numbers at the battle of Catraeth, usually identified with Catterick in the North Riding of Yorkshire. Almost all the band were killed, and the poem tells of their distinction and bravery and how 'they shall be honoured until the end of the world'.[1] In 616, Aethelfrith defeated the Britons at Chester. It is not clear whether this was an isolated raid or an important event that brought the English to the western sea as had the battle of Dyrham in the south. In any case, it is clear that by the middle of the seventh century the Britons of the west had been separated into three groups – those of Cornwall and Devon, those of Wales itself and those of the north-west. One obstacle to the westward advance of the Northumbrians had been the presence of the British kingdom of Elmet, between the Aire and the Wharfe roughly north of what is now Leeds. This was

---

[1] I. Williams, 'The poems of Llywarch Hên', *Proc. British Academy*, XVIII (1932), 270–1; K. Jackson, 'The "Gododdin" of Aneirin', *Antiquity*, XIII (1939), 25–34; C. A. Gresham, 'The book of Aneirin', *Antiquity*, XVI (1942), 237–57.

conquered by Edwin, Aethelfrith's successor, sometime between 616 and 632. Before the end of the century, Anglian settlement and control extended in the north-east over the lowlands of Lothian to the Firth of Forth, and in the north-west into the plain of Carlisle and the lands around the Solway Firth. Affairs in the latter area were complicated by the survival of the Welsh kingdom of Strathclyde with its own royal line. At its greatest extent this seems to have stretched northward from the Ribble well into south-west Scotland. Its political relations and its extent at various times are very confused, but amid much that is obscure it is clear that Britons survived in the Lakeland mountains, leaving the name Cumberland to remind us that this was once the land of the Cymry, as the Welsh still call themselves today.

The survival of the Romano-Britons

To what extent was there continuity between the new Anglo-Saxon villages and those of the Romano-Britons? For some areas, at any rate, the answer to this question is clear, and has been given with the aid of air photography. On the chalk country of Wessex and Sussex the Anglo-Saxon settlement seems to have involved a complete break with the past, and the Romano-British villages ceased to be inhabited. 'Their sites were abandoned,' wrote O. G. S. Crawford, and are now 'a maze of grass-covered mounds. We do not know what happened during that century and a half of darkness and confusion; but when the dawn broke through once more, we find a totally new system, Teutonic in character, and different in every way from its predecessor. New villages with new Saxon names have sprung up along the valleys; the once populous uplands are deserted.'[1]

It does not follow from this that the Romano-Britons had not lived in the valleys. Evidence of early settlement has less chance to survive in the valleys because it is removed by the ploughs of later occupiers or buried beneath their houses, but the gravel terraces of the Thames and of other rivers show that in places the Britons were there before the Saxons. On several sites the association of Roman and Saxon remains seems to indicate 'a virtual continuity of occupation from one age into the other'.[2] And for

---

[1] O. G. S. Crawford, 'Air survey and archaeology', *Geog. Jour.*, LXI (1923), 353. See also C. C. Taylor, 'The study of settlement patterns in pre-Saxon Britain' in P. J. Ucko *et al.*, *Man, settlement and urbanism* (London, 1972), 109–13.

[2] F. M. Stenton (1947), 25. See also S. Applebaum's chapter entitled 'Continuity?' (pp. 250–65) in H. P. R. Finberg (ed.), *The agrarian history of England and Wales*, vol. I, pt 2, A.D. 43–1042 (Cambridge, 1972).

Withington, in the Cotswolds, H. P. R. Finberg has also argued the case for continuity.[1]

One of the questions that inevitably comes to mind is to what extent did the Britons survive the changes of the time. There are various references, in the Chronicle and elsewhere, to the slaughter of Britons, but these can hardly refer to the whole period of the Anglo-Saxon advance. Even so, whatever the degree of survival the fact remains that the English language is almost entirely free of British influence. Celtic words relating to agriculture and to domestic life are virtually absent from English speech.[2] What is more, the place-names of villages and hamlets are almost entirely non-Celtic, except in Cornwall. As might be expected, the counties with the greatest number of Celtic names are those of the western border. In Herefordshire, names like Dinedor and Moccas are anglicised Celtic names; so are Cumwhiton, Blencarn and the like in Cumberland; and Eccles and Haydock in Lancashire. Scattered Celtic place-names are also found in many other counties. Among those of West Riding, there is the name of the village of Wales (i.e. the *Wealas* or Welshmen) to the south-east of Sheffield. There are a few Celtic place-names in the Midlands and some even in the south and east. Wiltshire is a county with a relatively large number. They support the view that the Saxon occupation of Wiltshire did not take place until after the first destructive phase of invasion was over, and they point to a period of peaceful relations between Britons and newcomers.[3]

But the main indication of British survival comes not from names of places but from those of natural features, from the names of hills and woods and, above all, rivers. Celtic names of hills and woods include Cannock, Morfe and Kinver in Staffordshire, Chute and Savernake in Wiltshire, Chiltern in Hertfordshire and Brill and Chetwode in Buckinghamshire. More numerous still are the Celtic names of streams, both large and small. It is true that they are rare in Sussex and in the eastern counties generally, but they become frequent towards the west. The Aire, the Trent, the Derwent and the numerous rivers called Avon or Ouse, are only a few out of a large number. Eilert Ekwall summed up this evidence by saying that 'the old theory of a wholesale extermination or displacement of the British population is no doubt erroneous or exaggerated, but it may come near the truth in the districts first conquered'.[4]

[1] H. P. R. Finberg, *Roman and Saxon Withington* (Leicester, 1955).
[2] Max Forster, *Keltisches Wortgut im Englischen* (Halle, 1921).
[3] J. E. B. Gover *et al.*, *The place-names of Wiltshire* (Cambridge, 1939), xv–xvi.
[4] E. Ekwall, *English river-names* (Oxford, 1928), lxxxix.

There is also other evidence for the survival of the British. The laws of Ine (*circa* 690) show that within a Wessex that did not yet include Devon and Cornwall, there were not only Welsh slaves but Welsh freemen of some standing. There was even a Welsh strain in the West Saxon royal house.[1] But, in general, the term 'Welsh', derived from the English word 'walh' or 'wealh', was commonly taken to imply a foreigner or a slave; the Welsh called, and still call, themselves 'Cymry'.

At the other end of the country, the persistence of a strong Celtic element in the institutions of Northumbria must surely point to a substantial survival of the British among the Anglian settlers.[2] Furthermore, Glanville Jones has postulated the survival of British arrangements in the system of central manors with satellite hamlets that is a feature of northern England in the early Middle Ages and that meets our eye in the sokelands and berewicks of Domesday Book. He also thinks that the complex manors with dependencies to be found in southern England likewise result from the survival of Romano-British institutions.[3] Here is an hypothesis that needs further investigation.

The conclusion must be that there is not one but many answers to that debated question 'Did the native population survive?'. In some areas the British may have been exterminated. In many other areas they must have survived. In yet other areas they may have been absorbed into the economy and social structure of the newcomers at an early date. No single generalisation will cover the truth.

English place-names

Whatever the degree of British survival, and whatever the precise relationship between the two groups of peoples, the new villages established by the invaders were the villages of later times. Where, for example, Waels and his men settled in Norfolk, there is Walsingham today. Where Babba made a 'stoc' or settlement in Wiltshire, there Baverstock still stands. It is true that some Anglo-Saxon sites have been deserted, and that in Nor-

---

[1] F. M. Stenton, 'The historical bearing of place-name studies: England in the sixth century', *Trans. Roy. Hist. Soc.*, 4th ser., XXI (1939), 13–14; L. F. R. Williams, 'The status of the Welsh in the laws of Ine', *Eng. Hist. Rev.*, XXX (1915), 271–7.

[2] J. E. A. Jolliffe, 'Northumbrian institutions', *Eng. Hist. Rev.*, XLI (1926), 1–42.

[3] G. R. J. Jones, (1) 'Basic patterns of settlement distribution in northern England', *The Advancement of Science*, XVIII (1961), 192–200; (2) 'Early territorial organisation in England and Wales', *Geofriska Annaler*, XLIII (1961), 174–81; (3) 'Settlement patterns in Anglo-Saxon England', *Antiquity*, XXXV (1961), 221–32. See also F. M. Stenton (1947), 311.

man and later times some new villages came into being. But in general, the village geography of the sixth to the ninth century has formed the basic element in the pattern of much of the English countryside up to the present day. The rise and fall of the kingdoms of the Heptarchy occupy a prominent place in the annals of Anglo-Saxon England, but behind the clash and thunder of Anglo-Saxon warfare, there went on an almost silent, and for the most part unrecorded, process by which the English occupied and cultivated the land that was now theirs.

Two very frequent Anglo-Saxon elements in village-names are 'ham' and 'ton'. As the English advanced westwards, the word 'ham' seems to have become obsolete and, at the same time, the meaning of 'tun' was changing from 'enclosure' to 'farmstead' and then to 'village'. The result is that 'ham' names are most frequent in the south-east and in East Anglia, and that 'ton' or 'tun' names are most frequent in the Midlands and the west. As the settlement was consolidated, there came into being the great mass of terminations that give interest and variety to our place-names – 'cote', 'ford', 'stede', 'worth' and the like.[1] But while the Anglo-Saxons were thus laying the foundations of one village geography they were interrupted by a new group of invaders, who in turn were to continue and intensify the spread of settlement in the north and east. These were the Scandinavians.

THE SCANDINAVIAN SETTLEMENT

The Scandinavian invasions

Under the year 787 the Anglo-Saxon Chronicle records the first coming of Scandinavian raiders to the shores of England: 'then the reeve rode to the place, and would have driven them to the king's town, because he knew not who they were, and they there slew him'. From other sources we know that the place was Portland in Dorset, and that the reeve was the reeve of Dorchester. The Chronicle records the plundering of Lindisfarne in 793 and of Jarrow in 794. Other raids followed upon the coasts of western Scotland and Ireland, and to the litany of the Church in the ninth century was added the prayer: 'From the fury of the Northmen good Lord deliver us.' In 835, raiders appeared at the mouth of the Thames, and plundered the Isle of Sheppey. Others followed, and we hear of their activities in East Anglia, Lindsey, Devon and elsewhere. The Chronicle

[1] The authoritative survey is A. H. Smith, *English place-name elements*, 2 vols. (Cambridge, 1956).

entry for 851 says that the raiders 'for the first time remained over winter in Thanet'; and in 855 they likewise wintered for the first time in Sheppey.

It was not until 865 that invasion really started with the arrival of 'a great heathen army' in East Anglia. Here, in the following year, the Chronicle tells us, they obtained horses; and, thus equipped, they extended their activities during the next three years into southern Northumbria and eastern Mercia. Under the year 876 we read that they shared out the land of Northumbria and that 'they began to plough and make a livelihood'. In the following year they 'shared out' part of Mercia, and gave the rest to a puppet English king. We are not told whether this 'sharing out' was a division into small units for the purpose of cultivation, or one into large districts such as those connected with the fortified centres of what later came to be known as the 'Five Boroughs' of Lincoln, Nottingham, Derby, Leicester and Stamford. In 879, the land of East Anglia was also 'shared out'.

In the meantime, the Danes had advanced westward, to attack Wessex, and the defence of the English devolved upon King Alfred from his base on the Isle of Athelney in the Somerset marshes. There followed a period of confused warfare that culminated in a treaty traditionally ascribed to 886, the year in which Alfred captured London. The treaty defined the boundary between the political power of the English and that of the Danes. The line ran along the Thames and then along the Lea to its source, whence it continued in a straight line to Bedford and so along the Ouse to Watling Street; how far it continued north-westward we cannot say. To the east lay the territory that became generally known as the Danelaw, with its three divisions of East Anglia, Mercia and southern Northumbria.

The treaty of 886 marked not the end of conflict but a very temporary balance of forces, and before his death in 899 Alfred had begun a systematic fortification of southern England by building a series of 'burhs' or forts, a list of which survives in the so-called Burghal Hidage.[1] The work of recovery was carried on by Alfred's son Edward the Elder of Wessex and by his daughter Aethelflaed, 'the Lady of the Mercians'.[2] Acting in concert during the years 907–16, they continued the practice of building burhs until a zone of fortified centres stretched from the Mersey to Essex

---

[1] See pp. 36–7, below.

[2] F. T. Wainwright, 'Æthelflaed, Lady of the Mercians', in P. Clemoes (ed.), *The Anglo-Saxons* (London, 1959), 53–69.

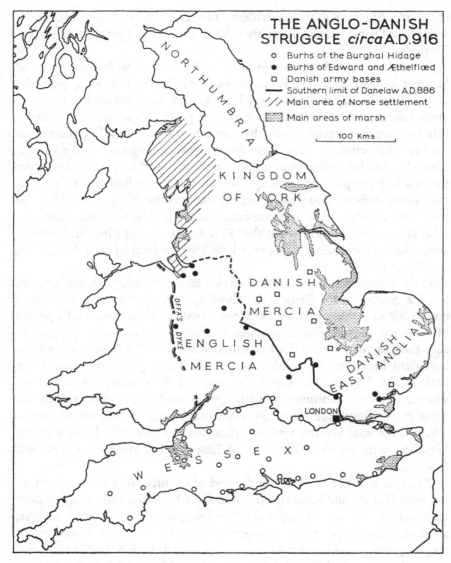

Fig. 5 The Anglo-Danish struggle *circa* A.D. 916
      Based on: (1) F. T. Wainwright, 'Æthelflaed, Lady of the Mercians', in
      P. Clemoes (ed.), *The Anglo-Saxons* (London, 1959), 59; (2) J. F. Benton,
      Town origins: the evidence from medieval England (Boston, Mass., 1968), 49.
      The possible continuation of the English–Danish boundary to the north-west
      is marked.

(Fig. 5). The campaign of the years 917 and 918 resulted in the recovery of East Anglia and Danish Mercia by the English; and, with the death of Aethelflaed in the latter year, English Mercia was joined to Wessex. When Edward the Elder died in 924, the northern boundary of his kingdom stretched from the Humber to the Mersey.

Moreover, to the north the Scandinavian kingdom of York (southern Northumbria), the English relic of Northumbria beyond the Tees, and the Celtic kingdom of Strathclyde had acknowledged the king of Wessex as their overlord. But this success was followed by rebellion and much confusion, increased, as we shall see, by the arrival of the Norse from Ireland. It was not until 954 that an independent Scandinavian kingdom of York ceased to exist. Any clear chronology of events is made difficult by the extreme obscurity of the sources.

While the Danes had thus been advancing from the east coast, another, and rather different, Scandinavian immigration was taking place from the west. Norsemen from south-western Norway had established settlements in the Shetlands and the Orkneys, and from there one stream of migration went northward to the Faroes, to Iceland and beyond, while another stream came to the northern mainland of Scotland, to the Western Isles and southward into the Irish Sea. The monastery of Iona was plundered in 795, and during the next thirty years or so the Irish annals made many references to Viking raiders. In the tenth century a line of Norse kings reigned in the Isle of Man, and powerful Norse kingdoms arose around the coastal towns of Ireland, around Wexford, Waterford, Cork, Limerick and, above all, around Dublin.

It was from these older colonies in Ireland, the Isle of Man and the Western Isles that the Norse came to Cumberland, Westmorland, Lancashire, Cheshire and to the coasts of North and South Wales. The only literary record (although much later in date) of this movement concerns the expedition of a certain Ingimund from Ireland to establish a Norse colony in the Wirral peninsula of Cheshire about the year 902.[1] The settlement seems to have been very largely a peaceful one, and this may also have been true generally of the Norse settlement to the north, in Lancashire and elsewhere. It would seem that in the tenth century there was still much uninhabited land in north-west England, and that the Norse were content to occupy less attractive districts neglected by the English or

[1] F. T. Wainright, 'Ingimund's invasion', *Eng. Hist. Rev.*, LXIII (1948), 145–69. See also F. T. Wainright, 'North-west Mercia', *Trans. Hist. Soc. Lancs. and Cheshire*, XCIV (1942), 3–55.

Welsh.[1] They spread eastward across the Pennines, so that the areas of Norse and Danish settlement overlapped. Norsemen from Ireland also played a part in the political complications of northern England. They even invaded Northumbria and, in 919, founded a new kingdom of York which continued with intervals until 954. During this period the Norse kings of Dublin at times also ruled in York. To the Danish settlement in Yorkshire was thus added a Norwegian–Irish element.

Before the end of the tenth century there began a second episode of Danish power in England. From the year 980 onwards there was a fresh series of Danish raids, especially along the south and south-east coasts; and recurrent Danegelds from 991 onwards failed to satisfy the raiders. Under the year 1004 the Anglo-Saxon Chronicle tells of the plundering of Norwich and Thetford. Under the year 1006 we hear of 'every shire in Wessex sadly marked by burning and plundering'. In 1009 a large army appeared off the coast of Kent. Repulsed by the city of London, a Danish force made its way through the Chilterns to Oxford, which was burned. After returning to Kent to repair their ships, the Danes appeared in 1010 off East Anglia which they 'ravaged and burned'. Here, 'they were horsed' and they raided through the countryside, destroying men and cattle; they burned Thetford and Cambridge, and continued south to the Thames and westward into Oxfordshire and Buckinghamshire and Bedfordshire, before returning 'to their ships with their booty'. By 1011, says the Chronicle, they had ravaged the whole or part of eighteen counties. Although it was sporadic, local and relatively ephemeral, this repeated devastation was an element of no little importance in the changing economic geography of eleventh-century England.

The invasions culminated in 1013 with the submission of London and the conquest of the realm by Swein who was followed by his son Canute of Denmark in 1016. With Canute's death in 1035 England ceased to be part of the Scandinavian empire, but a Danish dynasty did not cease to rule in England until 1042 when Edward the Confessor of the royal house of Wessex was chosen as king. In the meantime, the Anglo-Scottish frontier was emerging. Strathclyde was united with Scotland about 945, but Carlisle and the district to the south became part of England in 1092. To the east, Lothian was ceded to Scotland in 975 or shortly before. Thus did the boundary between the two kingdoms become the Tweed–Cheviot–Solway line, at any rate in name.

[1] F. T. Wainright, 'The Scandinavians in Lancashire', *Trans. Lancs. and Cheshire Antiq. Soc.*, LVIII (1946), 71–116.

Scandinavian place-names

Although Scandinavian political power vanished from England, the Scandinavians themselves remained. They left little archaeological evidence as compared with the Anglo-Saxons, and this has sometimes been attributed to the fact that the Scandinavians soon adopted the Christian faith and so discarded heathen burial practices. Whatever be the truth, the main evidence for their settlement is derived from place-names. One characteristic Scandinavian place-name element is 'by' implying a village, and there are about 250 Yorkshire village names that end in 'by', and nearly that number in Lincolnshire. Well over a half of these are compounded with personal names and may belong to a period 'within at most a generation or two' of the original settlement; some may even be the names of Danes who had taken part in the conquest.[1] A detailed examination of these 'by' names in relation to geology and soil led Kenneth Cameron to the view that the Danes settled in the less attractive localities left empty by the English and that they came 'predominantly as colonisers, occupying new sites'.[2]

Next in importance is the element 'thorpe' meaning 'hamlet' or outlying settlement from an older village. Many of these are compounded with personal names that suggest they were made by individuals from parent villages; generally they were later than 'by' names.[3] There is also a wide variety of other Scandinavian elements – *toft* (homestead), *garth* (enclosure), *lathe* (barn) and many others. The greater number of these words were common to both Danes and Norse, but there were a few words peculiar to each; thus *thorpe*, *both* (booth) and *hulm* (holme) were Danish, while *gill* (valley), *skali* (hut) and *erg* (shieling or hill pasture) were Norse or Norse borrowings from Irish. One further point must be made. While many Scandinavian names go back to the early days of the settlement, 'there is no safe way of distinguishing those that arose in the ninth century from those that may have arisen at a later date'.[4] The Scandinavian language continued to be spoken until at least the eleventh

[1] F. M. Stenton (1947), 516.
[2] K. Cameron, *Scandinavian settlement in the territory of the five boroughs: The place-name evidence* (University of Nottingham, 1965), 13.
[3] A. H. Smith, *English place-name elements*, pt 2, 209.
[4] F. T. Wainright, 'Early Scandinavian settlement in Derbyshire', *Jour. Derbyshire Archaeol. and Nat. Hist. Soc.*, n.s., XX (1947), 100.

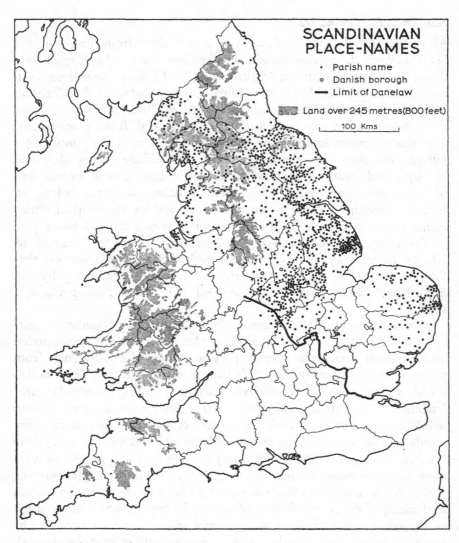

Fig. 6 Scandinavian place-names
   Based on A. H. Smith, *English place-name elements*, pt 1 (Cambridge, 1956).
   The five boroughs are these of Derby, Leicester, Lincoln, Nottingham and
   Stamford.

century,[1] and, moreover, the English language itself became highly Scandinavianised. Many 'Scandinavian' place-names date from the Middle English period, and cannot be regarded 'as affording in themselves evidence of a Scandinavian settlement'.[2]

The distribution of Scandinavian parish names as plotted on Fig. 6 shows that the area of settlement was not as extensive as the Danelaw itself; political control had been wider than actual settlement.[3] But a map such as Fig. 6 can give only a very inadequate idea of the Scandinavian element in place nomenclature. In the first place, not all Scandinavian names are of equal value in indicating the spread of settlement. Some names are Scandinavian in the strictest sense, that is words that were formed by people who spoke a Scandinavian language, names like Laceby and Litherland. Other names are only Scandinavianised versions of English ones, such as Keswick and Louth, which in their pure Anglian forms would have been Cheswick and Loud. Then there are hybrid names such as Grimston which is a combination of the Scandinavian personal name 'Grim' with the English 'tun'. In the second place, a parish with an English name often contains many minor names that are Scandinavian, names of topographical features and of hills. The hilly districts of the western parts of the North Riding and the West Riding, for example, are rich in Scandinavian names of minor features; so is north Lancashire; and the Lake District, especially in its remote parts, is even more rich. Names ending in *erg*, *saetr* and *scale* denote the former presence of Norse shielings. Thousands of such minor names supplement the evidence of village names, but it must be said at once that many, or perhaps most of them, date from a period long after the first settlements of the ninth century. In the present state of our knowledge, it would be impossible to construct a map that

---

[1] E. Ekwall, 'How long did the Scandinavian language survive in England?' in *A grammatical miscellany offered to Otto Jespersen on his seventieth birthday* (Copenhagen and London, 1930), 17–30.

[2] H. Lindkvist, *Middle English place-names of Scandinavian origin*, pt 1 (Uppsala, 1912), liii.

[3] Convenient summaries of the Scandinavian place-name evidence may be found in the following: (1) E. Ekwall, 'The Scandinavian element', in A. Mawer and F. M. Stenton (eds.) *Introduction to the survey of English place-names* (Cambridge, 1929), 55–92; (2) E. Ekwall, 'The Scandinavian settlement' in H. C. Darby (ed.), *An historical geography of England before A.D. 1800* (Cambridge, 1936), 133–64; (3) F. M. Stenton, 'The historical bearing of place-name studies: the Danish settlement of Eastern England', *Trans. Roy. Hist. Soc.*, 4th ser., XXIV (1942), 1–24. Individual elements are discussed in A. H. Smith (1956). See also the relevant county volumes of the English Place-Name Society.

took both these groups of considerations into account. Fig. 6 must stand as giving a general impression.

The density of the Scandinavian settlement

The general impression given by Fig. 6 leaves at least one major question unsettled. How numerous were the newcomers in relation to the existing English population? We can never answer that question with any certainty, but some scholars have seen clues in two types of evidence. One is the indication given by the personal names that occur in charters of the twelfth and thirteenth centuries. Personal names are always subject to fashion, and some English people may well have adopted Scandinavian names and vice versa. Yet, bearing this in mind, the percentage of Scandinavian personal names in many districts is very high. Sir Frank Stenton declared that in the northern Danelaw more than one half the native personal names that survived the Norman Conquest were Scandinavian, and that they were borne by men and women of widely different social condition.[1] They suggest, he wrote, that 'something like a genuine migration may have taken place in the ninth century, that, in particular, the armies may have sent for their womenkind when they turned from war to agriculture'.[2] The twelfth-century Scandinavian personal names on a series of Norfolk manors amount to about 10%,[3] and on a series of Suffolk manors to about 8½%.[4]

Another indication of the density of the Scandinavian settlement has been seen in the proportion of sokemen to total population as recorded in the Domesday Book. This is on the assumption that they represented 'the rank and file of the Scandinavian armies' of the ninth century. The percentage of sokemen in Lincolnshire wapentakes ranged from 20 to 73%, and it averaged 50%. In Leicestershire it ranged from 27 to 50% and it averaged about 30%. In Nottinghamshire the range was from 11 to 53% and the average was again about 30%.[5] The position in East Anglia was complicated by the presence of freemen as well as sokemen, but taken together they averaged over 40% in Norfolk and in Suffolk, and they, or

[1] F. M. Stenton, *Documents illustrative of the social and economic history of the Danelaw* (British Academy, London, 1920), cxiv–cxvi.

[2] F. M. Stenton, 'The Danes in England', *Proc. British Academy*, XIII (1927), 32.

[3] J. R. West, *St Benet of Holme, 1020–1210* (Norfolk Record Society, 1932), 258–9.

[4] D. C. Douglas, *Feudal documents from the abbey of Bury St Edmunds* (British Academy, London, 1932), cxx.

[5] F. M. Stenton, 'The free peasantry of the northern Danelaw', *Bulletin de la Société Royale des Lettres de Lund, 1925–19 26*, XXVI (1926), 73–185.

many of them, have been thought to be 'descendants of the Danish settlers of the ninth century'.[1]

This belief in an intensive Scandinavian colonisation of much of the north and east of England has not gone without criticism. H. W. C. Davis has pointed to the fact that the distribution of the free peasantry of East Anglia was far from coincident with that of Scandinavian place-names, and that there already was an English free peasantry here before 890. The 'by' names compounded with Scandinavian personal names might well refer 'not to peasants, but to the lords or owners of the villages'. There is no reason why the villagers themselves should not have been English. And as for personal names, he believes it is 'a complete delusion to think that every man with a Scandinavian name was of Scandinavian origin'. Personal names change with fashion as may be seen from the very large number of Norman names in East Anglia at the end of the twelfth century in spite of the fact that the Normans contributed little or nothing to the settlement of the area.[2]

Similar views have been expressed by P. H. Sawyer. He believes that the Danish armies were to be numbered in hundreds, not in thousands, and that 'the Danish settlements were the work of far fewer men than has often been supposed'. The Danish soldiers who settled here 'were probably joined by others about whom we hear as little as we do of Norse settlers in the north-west'. These later arrivals did not overwhelm the English; 'they settled where they could, most often on land left that the English had not yet occupied'. Even these do not account for all the Scandinavian and Scandinavianised place-names, many of which were not formed until after the Norman Conquest and so are 'clearly irrelevant for the study of the original Scandinavian settlements'.[3]

While there may be much in these arguments, particularly in so far as they concern East Anglia, the very large number of Scandinavian place-names and the evidence of Scandinavian influence upon law, language and administrative divisions, have made it difficult for some scholars to believe that an intensive colonisation did not take place.[4] The armies themselves

[1] B. Dodwell, 'The free peasantry of East Anglia in Domesday', *Trans. Norfolk and Norwich Archaeol. Soc.* XXVII (1939), 153.
[2] H. W. C. Davis, 'East Anglia and the Danelaw', *Trans. Roy. Hist. Soc.*, 5th ser., V (1955), 23–29.
[3] P. H. Sawyer, (1) 'The density of the Danish settlement in England', *Univ. o, Birmingham Hist. Jour.*, VI (1958), 1–17; (2) *The age of the Vikings* (London, 1962), 145–67.
[4] H. R. Loyn, 55. See also K. Cameron, 10–11.

may well have been small in size, but behind them, and protected by them, considerable immigration may well have taken place. One can only conclude that the Scandinavian settlement may have been a much more complicated process than was at one time thought.

THE CHANGING COUNTRYSIDE

The location of Anglo-Saxon and Scandinavian villages indicates a rational selection of sites – maybe along the juxtaposition of two geographical formations where springs gave water, or upon gravel terraces above the flood-levels of streams, or on some other 'dry-point' sites. The territories attached to these villages show how the needs of largely self-sufficing communities were met by including a variety of soils within the jurisdiction of each community. In changing the geography of Britain, the newcomers were themselves influenced by the varieties of soil and vegetation that confronted them. The sizes and shapes of the parishes of later times frequently provide an indication of the conditions under which early settlement took place. Arable, meadow, wood and pasture were the main ingredients of the usual village territory.

Exactly how that arable was cultivated we do not know, and there has been much discussion about the beginnings of the open-field system in England. To what extent did it exist in Britain before the coming of the Anglo-Saxons? Did the Anglo-Saxons introduce it from the Continent? And, in any case, what was the relation of this early system to the fully developed common-field system of later times? Or again, what was the influence, if any, of the Scandinavian settlement upon the agricultural arrangements they found? Any certain answers to these questions must await further investigation. A well-known passage in the laws of Ine has often been taken to imply the existence of some kind of open-field agriculture in Wessex about A.D. 690. It is not, however, until the tenth century that charters provide fairly numerous glimpses of unmistakable intermixed strips; but, even so, doubt has again been expressed as to whether, at this early date, they were grazed in common by the stock of the villagers after harvest and in fallow seasons. Not until post-Domesday times does a relatively clear picture emerge.[1]

The existence of open fields must not be taken to imply any uniformity

[1] H. P. R. Finberg (1972), 261–3, 416–20, 487–93; R. G. Collingwood and J. N. L. Myres, 210–13, 442–3; H. L. Gray, *English field systems* (Cambridge, Mass., 1915), 50–61.

in their organisation. If we may judge from later evidence, the open fields of the south-east and of East Anglia were organised in a different way from those of midland England. The Danes in East Anglia,[1] and either the Romans[2] or the Jutes[3] in Kent, and the Frisians in both areas[4] have been held responsible for the differences but these are very disputable matters. In some areas, natural features reflected themselves in individual agrarian arrangements, in, for example, the Fenland, the Weald and the wooded Chilterns. Then again, not only open fields but other forms of agrarian arrangements were to be found in the south-west and in the upland areas of the north and the north-west of England, where relief, soil and climate made pastoral activities important and where isolated farms and hamlets may have been more important than villages. The Northumbrian charters point to 'the use of summer and winter grazings remote from the agricultural settlements'.[5] In the north-west, Norse and Irish – Norse terminations in *erg*, *booth* and *saetr* also indicate shielings to which cattle were sent in summer,[6] but one must hasten to add that the earliest place-names incorporating these elements belong to a much later period. Although we have to rely upon the uncertainty of retrospective deduction for hints about agrarian arrangements in Anglo-Scandinavian times, there can be no doubt about the variety of those arrangements and about the fact that there were many different types of settlements.

The establishment of these Anglo-Saxon and Scandinavian settlements had profound consequences for the English countryside. In the earlier charters of the Anglo-Saxon period, grants of land are only vaguely defined as lying near this stream or as bounded by that wood. They reflect a time when settlement had not become intensive and when village rights were not precisely defined in the waste and wood around. But the charters of the tenth and eleventh centuries show a remarkable precision of topographical detail. Boundaries of estates are defined with great exactness and the charters abound in the local names of minor natural and other features. This contrast between the earlier and later charters 'indicates the nature of the unrecorded changes which had come over English country

[1] D. C. Douglas, *The social structure of medieval East Anglia* (Oxford, 1927), 50.

[2] H. L. Gray, 50 and 415–16.

[3] J. E. A. Jolliffe, *Pre-feudal England: the Jutes* (Oxford, 1933), 73 *et seq.*

[4] G. C. Homans, 'The rural sociology of medieval England', *Past and Present*, no. 4 (1953), 32–43.

[5] J. E. A. Jolliffe, 'Northumbrian institutions', *Eng. Hist. Rev.*, XLI (1926), 12.

[6] E. Ekwall in A. Mawer and F. M. Stenton (1929), 89.

life between the eighth and the tenth centuries'.[1] Some later villages bear names which had appeared merely as those of boundary marks in earlier charters. A charter of 1002 defines the boundaries of Little Haseley in Oxfordshire as beginning 'at Roppanford' and as passing 'over against Stangedelf'. The first of these boundary marks gave its name to the later village of Rofford, and the second to the later hamlets of Upper and Little Standhill. Such examples suggest the filling out of an earlier pattern of settlement, and it is clear that new settlements were being formed throughout a long period.[2]

It was F. W. Maitland who drew our attention to 'the numerous hints that our map gives us of village colonisation'.[3] A settlement with a wide territory often came to have another settlement within its limits. Such names as Great and Little Shelford or Guilden Morden and Steeple Morden in Cambridgeshire point to a time when a single territory 'by some process of colonisation or subdivision' became two territories. Not only two, but sometimes three or four or more settlements bear the same basic name, e.g. the three Rissingtons in Gloucestershire, the four Ilketshalls in Suffolk, the five Deverills in Wiltshire, the six South Elmhams in Suffolk, the seven Burnhams in Norfolk and the eight Rodings in Essex. Sometimes it is clear that the division had formally taken place before 1086. The Domesday folios for Hertfordshire refer to 'Hadham' and 'Parva Hadham', i.e. Much and Little Hadham; and those for Northamptonshire refer to 'Heiforde' and 'altera Haiford', i.e. Upper and Nether Heyford; those for Lincolnshire already speak of 'Nortstoches' and 'Sudstoches', i.e. North and South Stoke. Many such divided territories were to acquire the names of landholders of importance in the twelfth and thirteenth centuries. Thus the Domesday settlements of 'Emingeforde' and 'Alia Emingeforde', in Huntingdonshire, became known as Hemingford Abbots and Hemingford Grey after the Abbot of Ramsey and the Grey family. Usually, however, the Domesday text does not reveal whether the division had already taken place. The three Buckinghamshire villages of Bow Brickhill, Great Brickhill and Little Brickhill, separately distinguished by 1200 or so, are represented in the Domesday text only by the single name of 'Brichella'.

The process of colonisation went steadily on behind the clash of warfare

---

[1] F. M. Stenton (1947), 283.
[2] F. M. Stenton, in A. Mawer and F. M. Stenton (1929), 40 and 48.
[3] F. W. Maitland, *Domesday Book and beyond* (Cambridge, 1897), 365.

and the complications of the Anglo-Danish conflict. The dwellings of both groups of people were so many pioneer settlements battling to reduce the wood and waste around. The frontiers of expansion came to lie not to the west and north against the Welsh, but everywhere against the woods of the heavy clay soils. An echo of this activity comes to us in a remarkable passage in a treatise by King Alfred (871–901):

We wonder not that men should work in timber-felling and in carrying and building, for a man hopes that if he has built a cottage on the *laenland* of his lord, with his lord's help, he may be allowed to lie there awhile, and hunt and fish and fowl, and occupy the *laenland* as he likes, until through his lord's grace he may perhaps obtain some day boc-land and permanent inheritance.[1]

In time the log-hut might become one of a cluster, and this, in turn, might become a group of homesteads. The lumberman with his axe and his pick became the ploughman with his oxen. The cleared land was divided and tilled by the common plough, and thus there grew up a new hamlet or village in the outlying wood of some 'ham' or 'ton'. It is not surprising that the Anglo-Saxon poet described the ploughman as the 'enemy of the wood'.[2]

In spite of four centuries of Roman civilisation, Britain was very largely a wooded land when the Anglo-Saxons arrived. Perhaps the greatest single physical characteristic of most of the Anglo-Saxon countryside was its wooded aspect; the heavier clays in particular still carried great woodlands, which provided pasturage for swine feeding on acorns and beechmast. Of the group of descriptive place-names that reflect the primitive landscape of England, those denoting the presence of wood, and the clearing of wood, are the most frequent and the most important. In recalling these names we are, in a sense, looking at the countryside through the eyes of those who saw it in an unreclaimed condition.

Among the later Anglo-Saxon place-names belong those with such terminations as *ley*, *hurst*, *holt* and *hey*. There are also those numerous place-names that make specific references to species of tree, e.g. Oak-hanger, Ashton, Elmstead, Buckholt (beechwood) and Berkhamsted (homestead among the birches). Sometimes a place-name tells its story

[1] 'Blossom Gatherings out of St Augustine', Brit. Museum, Vit A, xv, f. 1; quoted in F. Seebohm, *The English village community* (Cambridge, 1883), 169.

[2] W. S. Mackie (ed.), *The Exeter Book*, pt 2 (Early English Text Soc., London, 1934), 111; R. K. Gordon, *Anglo-Saxon poetry* (Everyman's Library, London, 1954), 295.

by embodying a personal name. The village names of Knowsley and Winstanley in Lancashire are derived from men named Cynewulf and Wynstan respectively, while Chorley, found in Lancashire and elsewhere, was originally the clearing of the ceorls or peasants.[1] These examples happen to be the names of parishes, but the names of hamlets, farms and topographical features within a parish are no less significant. The parish of Chiddingford in Surrey contains the names of Frillinghurst, Killinghurst and Sydenhurst which indicate the woods of Frith, Cylla and Sutta respectively, and there are also other wood names in the parish; the earliest known forms of these names are post-Domesday, but, whether they be older or not, they indicate the former existence of wood.[2] Or again, the parish of Hampton in Warwickshire contains the name of Kinwalsey which, when traced back to its earliest form (twelfth century) is seen to be nothing other than Cyneweald's 'haeg' or clearing.[3]

In the Scandinavian parts of England strange-sounding terminations, such as *lundr*, *skogr* and *viothr*, are incorporated in the place-names; all three mean 'wood'. Along the Pennine valley of the Ure lies Wensleydale. The name Wensley itself is English and means 'Waendel's clearing'. Higher up the valley there is sequence of parish and other names that also imply wood; some are English and some are Scandinavian. West Witton is 'the farm in the wood' (*wudu ton*); Ellerlands is 'alder wood' (*elri lundr*); Aysgarth is 'the open space by the oaks' (*eik scarth*); Lunds is 'wood' (*lundr*); Brindley is 'the clearing caused by fire' (*brende leah*); Litherskew is 'the slope with wood' (*lithr skogr*).[4] Similar names are encountered elsewhere in the Pennine valleys and around the North York Moors.

In the Celtic parts of England there is another group of un-English names that indicate the former presence of wood. The place-names Clesketts and Culgaith in Cumberland[5] and Culcheth and Penketh in Lancashire[6] incorporate the Celtic element 'ceto' which is cognate with the modern Welsh 'coed' meaning wood. The exposed plateau surfaces

---

[1] E. Ekwall, *The place-names of Lancashire* (Chetham Society, Manchester, 1922), 113–14, 104 and 131.

[2] J. E. B. Gover *et al.*, *The place-names of Surrey* (Cambridge, 1934), 186–94.

[3] J. E. B. Gover *et al.*, *The place-names of Warwickshire* (Cambridge, 1936), 61.

[4] A. H. Smith, *The place-names of the North Riding of Yorkshire* (Cambridge, 1928), 257–69.

[5] A. H. Armstrong *et al.*, *The place-names of Cumberland*, pt 1 (Cambridge, 1950), 84 and 184.

[6] E. Ekwall (1922), 97 and 106.

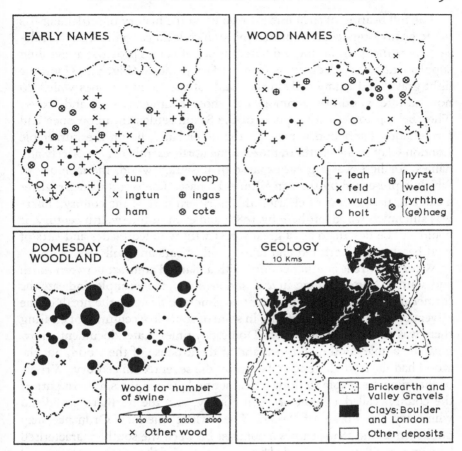

Fig. 7 The woodland of Middlesex
  Based on: (1) J. E. B. Gover *et al.*, *The place-names of Middlesex* (Cambridge,
  1942); (2) H. C. Darby and E. M. J. Campbell (eds.), *The Domesday geography
  of south-east England* (Cambridge, 1971), 100 and 124.

of Cornwall are not very favourable for the growth of trees, but there was
once a fair number in its valleys. The Old Cornish version of 'coed' was
'coit', and it appears in such forms as *cut*, *quite*, *coose* and *coys*. Thus the
name Trequite in the parishes of St Germans, St Mabyn and St Kew is the
same as Tregoose in the more western parishes of Probus, St Erth,
Sithney and the like. Penquite and Pencoose, meaning the 'end of the
wood', are other common Cornish names.[1]

[1] C. Henderson, *Essays in Cornish history* (Oxford, 1935), 135–51.

The full bearing of such information upon the history of settlement has yet to be investigated. A clear example of its relevance comes from Middlesex where the distribution of different types of names provides a revealing supplement to any deductions from surface geology (Fig. 7).[1] Upon the light gravel and loamy soils in the south of the county, names which do not indicate wood are common, e.g. those ending in 'ton' and 'cote'. They belong to the earlier phase of the Saxon settlement upon open and fertile land. The wood names, on the other hand, lie on the intractable London Clay to the north. In the extreme north-east there are scarcely any names, for here was a great expanse of unsettled wood, the memory of which is preserved by the name Enfield Chase. The wood names register a stage in the process of clearing this northern part of the county. There was still much wood left here by 1086; and even in the twelfth century, it could still be described as 'a great forest with wooded glades and lairs of wild beasts, deer both red and fallow, wild boars and bulls'.[2]

Warwickshire is another county with a marked contrast between north and south.[3] Names ending in *leah* and *ge(haeg)*, for example, indicate the former presence of wood, and they lie almost entirely to the north of the River Avon (Fig. 8). Here again, in spite of some five centuries of clearing there was still much wood left in Domesday times. Later documents show how the arable continued to expand at the expense of the wood, but the wood had far from disappeared even by the seventeenth century. Writers of that time were able to draw a distinction between the southern part of the county, called Feldon or open county, and the northern woodland where lay the Forest of Arden. The earliest one-inch Ordnance map published in the 1830s shows a contrast between the south, characterised by compact villages surrounded by open fields, and the north, characterised by dispersed houses, the dispersion resulting from scattered settlement proceeding piecemeal as the wood was cleared. These features of later times carry us back to a much earlier age.

One of the most well-known stretches of wood was that of the Weald which means 'wood'. The Anglo-Saxon Chronicle described it, in an entry for 893, as 'the great wood which we call Andred', or Andredesweald. It was a region with a distinctive economy. When something can be learned of its condition, from the later Anglo-Saxon charters, it

[1] J. E. B. Gover et al., *The place-names of Middlesex* (Cambridge, 1942), xv.
[2] F. M. Stenton et al., *Norman London* (Hist. Assoc. London, 1934), 27.
[3] J. E. B. Gover et al., *The place-names of Warwickshire* (Cambridge, 1936), xiii–xv, 315, 316.

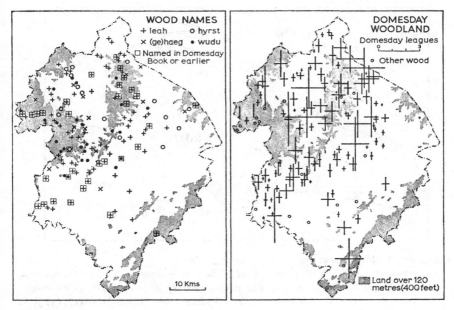

Fig. 8 The woodland of Warwickshire
     Based on: (1) J. E. B. Gover *et al.*, *The place-names of Warwickshire* (Cambridge, 1936); (2) H. C. Darby and I. B. Terrett (eds.), *The Domesday geography of Midland England* (2nd ed., Cambridge, 1971), 296.

appears as a land of swine pastures or 'denns' appurtenant to the villages around. The distribution of these denn names poses its own problems.[1] Fig. 9 shows very many denn names in the eastern part of the Weald of Kent; the area of frequent occurrence stops abruptly at the Kent–Sussex border, although some denn names are widely scattered throughout Sussex and there are a few in Surrey. Why should this be so? Does the restricted distribution mean that swine pastures were frequent in the Kentish Weald but not in the Wealds of Sussex and Surrey? This can hardly have been so. Wealden swine pastures were not restricted mainly to Kent, and pre-Conquest charters show that the names of those in Sussex and Surrey frequently did not end in 'den'. The distribution of the element 'denn' thus reflects not only geographical circumstances but also local peculiarities of dialect, and it may imply the settlement of the south-east peninsula by different groups of people. Another termination that almost

[1] H. C. Darby, 'Place-names and the geography of the past' in A. Brown and P. Foote (eds.), *Early English and Norse studies* (London, 1963), 14–18.

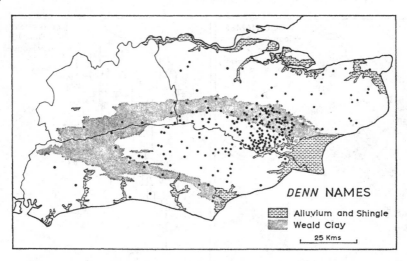

Fig. 9 The Wealden area: place-names ending in 'denn'
Based on: (1) J. E. B. Gover *et al.*, *The place-names of Surrey* (Cambridge, 1934); (2) A. Mawer *et al.*, *The place-names of Sussex*, 2 pts. (Cambridge, 1929–30); (3) J. K. Wallenberg, *Kentish place-names* (Uppsala, 1931); (4) J. K. Wallenberg, *The place-names of Kent* (Uppsala, 1934).

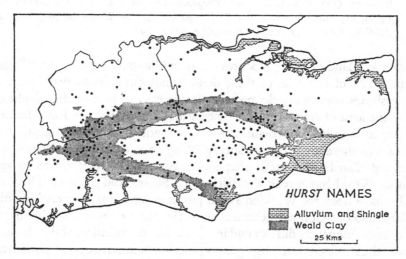

Fig. 10 The Wealden area: place-names ending in 'hurst'
Sources as for Fig. 9.

THE ANGLO-SCANDINAVIAN FOUNDATIONS

always implies the presence of wood is 'hurst'. Place-names which em-
body it are much more widely distributed, but even they tend to show
local concentrations (Fig. 10). Then there are other elements (such as
leah and *feld*) which fill out the picture of the wooded Weald. We can,
however, only guess at the variety within this woodland, a variety ranging
from the dense oakwoods of the heavy clay to the more open woods of the
lighter soils.

Taken together, this mass of place-name evidence shows how the
attack on the wooded countryside was prosecuted with vigour between
the fifth and the eleventh century. New farms and new hamlets sprang up
and grew, until by 1086, a large proportion of the villages we know today
had come into being. There was, it is true, still much wood left, but a great
part of the clayland had become arable and was tilled by the plough-teams
that feature so prominently in the entries of Domesday Book.

Another important element in the Anglo-Saxon landscape was marsh,
but references to its extent are very few. The biographer of St Guthlac of
Crowland, writing shortly after 700, told of a 'fen of immense size' that
stretched from Cambridge northwards to the sea. It was overhung by fog
and was characterised by wooded islands and tortuous streams.[1] But it
was not without inhabitants or profit; Bede explained how the island of
Ely received its name from the great quantity of eels to be found around,
and pre-Domesday Fenland charters refer to fishermen and nets and
boats.[2] Another expanse of marsh was that in Somerset, and King Alfred's
biographer, writing in the ninth century, described the island of Athelney
as surrounded by impassable marshes; it could be approached only by
boat until, after much effort, a causeway was built.[3] Apart from these two
accounts and a few scattered references in charters (e.g. to Romney Marsh)
we can only infer the existence of many other marshy areas from the
distribution of peat and alluvium on the geological map – of such marshes
as those of the Humber lowlands, of the Ancholme valley, of the moss-
lands of Lancashire and the peaty areas of north Shropshire, to say nothing
of the marshy tracts along many a river valley and many a stretch of
coast. What is clear is that the conquest of the marsh, unlike that of the
wood, did not begin until many centuries after the Anglo-Scandinavian
period.

[1] B. Colgrave (ed.), *Felix's Life of St Guthlac* (Cambridge, 1956), 87.
[2] H. C. Darby, *The medieval Fenland* (Cambridge, 1940), 24, 29.
[3] W. H. Stevenson (ed.), *Asser's Life of King Alfred* (Oxford, 1904), 79–80.

URBAN LIFE

Allied with the question of the survival of the Romano-Britons is that of the survival of Roman towns and cities. Gildas in the sixth century spoke of the destruction of the twenty-eight cities of Britain, and the evidence, so far as it goes, points to decline and disappearance. A few centres may have survived to carry on what little trade there was, but the trade must have been very small in amount and the towns very rudimentary. The fate of London, for example, has often been discussed. Some have believed that its organised life as a city continued throughout all the stress of the time, and it is possible that the site of the Roman city was far from derelict in the sixth and seventh centuries, but its life must have been at a very low ebb.

Before the end of the eighth century, the glimmerings of some revival can be discerned. There are indications of trans-Channel traffic in such things as glass, pottery, swords, metalware, woollen garments, wine, slaves, and from this time come references to trade between Offa and Charlemagne. In the eighth century, too, Bede described London as 'the market place of many peoples coming by land and sea', and we hear also of Frisian traders in York and of a market in Southampton. These are but stray fragments of information.[1] We know practically nothing of the early Anglo-Saxon towns themselves.

With the coming of the Scandinavians in the ninth century, urban life received a new stimulus. The Danish supremacy in the eastern Midlands rested upon a confederate group of five fortified centres or *burhs* (Fig. 6) – Derby, Leicester, Lincoln, Nottingham and Stamford – and these were not the only Danish boroughs. Here, today, survives an 'exceedingly neat and artificial scheme of political geography'. Each shire has taken its name from a fortified *burh*, which became its administrative centre.[2]

The example of the Danes was not lost upon the English who, in turn, proceeded to establish a series of strongholds. Our information about these comes mainly from two sources which are complementary in that they refer to different areas.[3] One is the curious document known as the 'Burghal Hidage' which seems to date from the years 911–919, and which

[1] G. C. Dunning, 'Trade relations between England and the Continent in the late Anglo-Saxon period', in D. B. Harden (eds.), *Dark Age Britain* (London, 1956), 218–33.

[2] F. W. Maitland, 187.

[3] J. F. Benton, *Town origins: the evidence from medieval England* (Boston, Mass., 1968), 48–53.

names some thirty or so burhs which Alfred had begun to build before his death in 899; all of them, except Buckingham, were to the south of the Thames or along it, and they had formed a defensive system for Wessex against Danish attack. The other source is the Anglo-Saxon Chronicle which mentions over a score of burhs established in the reign of Edward the Elder (899–924) of Wessex, some built by himself and his sister Aethelflaed of Mercia during the years 907–916 to secure English territory in the Midlands.[1] Many burhs were on sites where Roman towns had once flourished, e.g. Bath and Exeter; the place-name element ceaster or caster (from the Latin *castra*) is often embodied in the names of such Roman centres, e.g. Chichester and Towcester, but this must not be taken as necessarily implying continuity of occupation.

The fact that the term 'burh' or 'borough' later came to mean town does not justify the conclusion that all burh-building meant the creation of urban settlements. The Anglo-Danish burhs were forts, but very often the reasons that made them suitable defensive and administrative centres also made them commercial centres. There has been much debate about whether defence or trade provided the basic impetus to early urban development in England, but clearly both were important. The security of fortified centres encouraged trading activity, while, in turn, trading centres needed to be fortified for protection. The potentialities of the burhs varied. Not all survived to become the boroughs of a later age; the names of some cannot even be identified. Conversely, later boroughs developed on sites which had known no burhs. Some sites were especially favourable for trade, and here burhs, or maybe villages, developed into flourishing towns – at bridgeheads on rivers, at the confluences of tributaries, at the crossing-places of routeways, or near convenient gaps in hill country. Long before the Norman Conquest a force had begun to operate which was ultimately to give to the English borough its most lasting characteristic – i.e. that of a trading centre. We hear, for example, of merchants frequenting York about the year 1000, and they were said to have been mainly Danes. An account of the customs of the port of London about the year 1000 shows trade across the Channel and beyond to Lorraine; cargoes included fish, pepper and wine. Anglo-Saxon merchants also were to be found crossing the Alpine passes on their way to Rome.[2] There were Irish traders in Cambridge about 975.[3]

[1] See pp. 17–19 above.
[2] For a summary see J. Tait, *The medieval English borough* (Manchester, 1936), 118–19.
[3] *Ibid.*, 10 n.

The association of the early boroughs with markets and mints, difficult as it may be to interpret, indicates a commercial element. The evidence of the coins themselves is striking. The number of moneyers in a borough may be taken to indicate, in a very general way, its importance as a commercial centre. Between 1042 and 1066 London had as many as twenty moneyers working at the same time. York had more than ten; Lincoln and Winchester had at least nine each; Chester had at least eight; Canterbury and Oxford had at least seven each; Thetford, Gloucester and Worcester had at least six each.[1] This kind of evidence depends upon the chance discovery of coins, and may well do less than justice to many cities; but at any rate it prepares us for the indications of urban life to be found in Domesday England.

[1] F. M. Stenton (1947), 529. See also (1) G. C. Brooke, *Catalogue of English coins in the British Museum: the Norman Kings* 1 (London, 1916), clx–clxxxviii; (2) R. H. M. Dolley (ed.), *Anglo-Saxon Coins* (London, 1960).

Chapter 2

DOMESDAY ENGLAND

H. C. DARBY

The Norman Conquest in 1066, unlike the Anglo-Saxon and Scandinavian invasions, was not a mass movement of people but the work of a small power group. Twenty years after their coming, the Normans instituted the enquiry that resulted in the Domesday Book. With hindsight we can say that it came at a fortunate moment for us because it enables us to inspect the economic and social foundations of the geography of England after the Anglo-Saxons and Scandinavians had firmly established themselves in the land to which they had come.

We speak of the Domesday Book, but it is really in two volumes. The smaller volume describes Essex, Norfolk and Suffolk; the larger volume, in somewhat less detail, describes the rest of England with the exception of the four northern counties. There are also a number of subsidiary documents that must have been composed from the original returns and that sometimes help in the elucidation of the main survey. Among these is the so-called Exeter Domesday Book dealing with the counties of Cornwall, Somerset, most of Devonshire, parts of Dorset and a solitary Wiltshire manor. It was from this Exeter Domesday Book that the relevant portions of the main Domesday Book were made. Obviously an account that is nearer to the original returns than the main Domesday Book must be of especial interest. The word 'Domesday' does not occur in the Book itself, and there are varying opinions about how it got this name. The Treasurer of England, writing in 1179, said: 'This book is called by the natives Domesday – that is, metaphorically speaking, the day of judgement.' So thorough was the enquiry that its result may well have seemed comparable to the Book by which one day all will be judged (*Revelations*, 20:12).

Domesday Book is far from being a straightforward document. The exact method of its compilation is the subject of controversy, and many of its entries bristle with difficulties. Even so, it is probably the most remarkable statistical document in the history of Europe. Its information is arranged under the heading of each county. Within each county, it is set out under the names of the principal landowners, beginning with the king himself and continuing with the great ecclesiastical lords, then with the lay lords. The holding of each landowner is described village by village. If, say, three lords held land in a village, the three sets of information must be brought together to obtain a picture of the village as a whole.[1] The account of Buckden in Huntingdonshire is fairly representative. It was held entirely by the bishop of Lincoln, and so is described in a single entry on folio 203b:

In Buckden the bishop of Lincoln had 20 hides that paid geld. Land for 20 plough-teams. There, now on the demesne 5 plough-teams, and 37 villeins and 25 bordars having 14 plough-teams. There, a church and a priest and one mill yielding 30s [a year] and 84 acres of meadow. Wood for pannage one league long and one broad. In the time of King Edward [i.e. in 1066] it was worth £20 [a year], and now £16. 10s.

The variety of detail in such entries as this falls into two categories. In the first place there are those items that recur in almost every entry: hides (or carucates in the Danish districts) which were units of taxation, ploughlands, plough-teams, various categories of population, and annual values usually for 1066 and 1086, but sometimes also for an intermediate date. The second group comprises such items as the mill, meadow and wood entered for Buckden, and also, where relevant, pasture, salt-pans, fisheries, waste and vineyards. It is from these two groups of information that the geography of England in 1086, in all its regional diversity, can be reconstructed.

COUNTIES, HUNDREDS, WAPENTAKES AND VILLS

The land that William the Conqueror took over already possessed a highly developed territorial organisation. England south of the Tees was completely divided into shires, and these were divided into hundreds or wapentakes, and these, in turn, comprised vills. The Inquest itself, and the Book that resulted from it, was primarily organised on the basis of shires or, to use the Norman word, of counties. Lancashire, it is true, did not appear under that name until towards the end of the twelfth century; the area south

[1] For a general account of the making and contents of the Domesday Book, see R. Welldon Finn, *An introduction to Domesday Book* (London, 1963).

of the Ribble was described in a kind of appendix to the Domesday account of Cheshire, and the northern part of the county was included with Yorkshire. Rutland likewise did not appear as a county until the thirteenth century. Its eastern part was in Domesday Northamptonshire, and its western part formed an anomalous unit called 'Roteland' which was described at the end of the Nottinghamshire section of the Book. There have also been various other less important changes in the inter-county boundaries. Beyond Domesday England, the four northern counties were in the nature of border provinces, and 'it is probable that responsibility for their internal order, as for their defence, rested with the great lords of the country'.[1]

The Old English word 'scir' meant division, and it came to be applied to administrative divisions. Today, the word has survived not only in the names of many counties but also in those of some smaller districts that were also once administrative units, e.g. Hallamshire in the West Riding and Richmondshire in the North Riding. The shire system first emerged in Wessex, apparently by the early years of the ninth century.[2] Some shires corresponded to the areas occupied by groups of people organised in dependence upon a local capital – Dorset upon Dorchester, Somerset upon Somerton, Hampshire upon Southampton, Wiltshire upon Wilton. Former independent kingdoms also appeared as shires. In the west were Devon and Cornwall, the descendants of the old Celtic kingdom of Dumnonia or West Wales. In the east, Sussex and Kent also corresponded to old kingdoms. Other shires to emerge were Surrey and Berkshire. Thus it was that the supremacy of Wessex in the ninth century had made possible 'the establishment of a uniform scheme of local administration throughout southern England'.[3]

The shires to the north of the Thames did not appear until a later date. The expansion of Wessex at the expense of the Danes in the tenth century resulted in the emergence as shires of the three ancient units of the Danish kingdom of East Anglia – Norfolk, Suffolk and Essex. The Danish kingdom of York likewise became a shire. The remaining shires of England are largely artificial in the sense that they bear no relation to the earlier territorial divisions of the Anglo-Saxon period.[4]

[1] F. M. Stenton, *Anglo-Saxon England* (Oxford, 1947), 496.

[2] *Ibid.*, 289–90 and 332–4.

[3] *Ibid.*, 290.

[4] C. S. Taylor, 'The origin of the Mercian shires', *Trans. Bristol and Gloucs. Archaeol. Soc.*, XXI (Bristol, 1898), 32–57. Reprinted in a modified form in H. P. R. Finberg (ed.), *Gloucestershire studies* (Leicester, 1957), 17–51.

Although their names as shires did not appear until the early years of the eleventh century, their origins were older. Those of the west Midlands seem to have resulted from a reorganisation of existing arrangements by the Wessex kings, Edward the Elder (899–924) and Athelstan (924–39), who had destroyed Mercian independence. Thus Shropshire, centred on Shrewsbury, was an artificial union of lands belonging to two groups of people – the Wreocensætan in the north and the Magesætan in the south. Other districts such as Warwick, Oxford, Gloucester were likewise created and named after central fortresses. The shires of the east Midlands were equally artificial and represented districts occupied by the various Danish armies, each district being centred on a town – Derby, Nottingham, Cambridge and the like. What Maitland described as an 'exceedingly neat and artificial scheme of political geography' was then incorporated into the Wessex scheme of things.[1]

Shires were subdivided into smaller units called 'hundreds'. These appeared as units of local government in the tenth century, and they provided a basis for the administration of justice and of public finance. In theory they were districts assessed for the purposes of taxation at 100 hides but this correspondence was rare. The assessments of the West Saxon hundreds ranged from about 20 to 150 or so hides, and the complexities that lie behind these variations are lost to us. In the Midlands, hundred assessments more often approximated to 100 hides, which may suggest that the division was imposed as part of the organisation into shires in the first half of the tenth century. It is impossible to say to what extent it replaced a more primitive arrangement.[2] The place of the hundred was taken by the wapentake in the highly Scandinavianised counties of Derby, Leicester, Lincoln, Nottingham and in the North and West Ridings (but not in the East Riding). The word is of Scandinavian origin and was first used to denote a territorial division in 962. This use seems to have been an innovation, 'for no divisions so named are known from Scandinavia'.[3] Functionally, the wapentake was the same as the hundred, and the terms are occasionally used interchangeably, e.g. twice in the Domesday folios for Northamptonshire and once in those for the East

---

[1] F. W. Maitland, *Domesday Book and beyond* (Cambridge, 1897), 187.

[2] O. S. Anderson, 'On the origin of the hundredal division', being pp. 209–17 of *The English hundred names: the south-eastern counties* (Lund, 1939). See also F. M. Stenton, 296–7.

[3] O. S. Anderson, *The English hundred-names* (Lund, 1934), xxi.

Riding.[1] The four northern counties were without hundreds or wapen-takes, but in their place the unit of the 'ward' appeared in the thirteenth century.

Other territorial units are also named in the Domesday Book. Some were units intermediate between those of the shire and the hundred or wapentake. Thus Yorkshire was divided into three ridings (the Scandinavian word 'riding' implies a third part). There were also ridings in Lindsey, itself a third part of Lincolnshire; the other parts were those of Kesteven and Holland. Kent had 'lathes' and Sussex had 'rapes', which may have represented older divisions of these ancient kingdoms.

Hundreds were composed of villages or vills, but alongside the physical reality of the village there was the institutional reality of the manor which features so prominently in the Domesday folios. Sometimes, a vill coincided with a manor. Sometimes it contained two or three manors belonging to different lords. Sometimes vills themselves were components of a large manorial complex which their lord treated as one unit. Clearly, manors varied greatly in size. There were enormous manors, such as that of Rothley in Leicestershire with holdings or members (*membra*) in twenty-two separate villages. Other manors were minute, such as Fernhill in Devonshire with only one recorded man and a solitary ox. Manors also varied in character; much has been written about the place of the manor in the feudal system and about its varying nature in different parts of the country. The normal Domesday entry distinguishes between the plough-teams of the manorial demesne, on which the local peasantry were obliged to work, and the plough-teams of the peasantry themselves. But there are many variations, and any attempt to assess the varying degree of manoriali-sation over the country as a whole is fraught with difficulties.[2] Important as the manor was in the social and legal organisation of the realm, the village itself was the unit prominent in the landscape of Domesday England

About 13,000 separate vills are named in the Domesday Book, but this

---

[1] Throughout the chapter no references are given to Domesday folios. These may be found in the relevant volumes of the Domesday geography of England: H. C. Darby, *The Domesday geography of eastern England* (3rd ed. 1971); H. C. Darby and I. B. Terrett (eds.), *The Domesday geography of midland England* (2nd ed. 1971); H. C. Darby and E. M. J. Campbell (eds.), *The Domesday geography of south-east England* (1962); H. C. Darby and I. S. Maxwell (eds.), *The Domesday geography of northern England* (1962); H. C. Darby and R. Welldon Finn (eds.), *The Domesday geography of south-west England* (1967); H. C. Darby and G. R. Versey, *Domesday Gazetteer* (1975); H. C. Darby, *Domesday England* (1977).

[2] R. Lennard, *Rural England, 1086–1135* (Oxford, 1959), 213–36.

cannot be the total number in the area it surveys. Some Domesday names covered more than one settlement. A number of these came to be represented in later times by groups of two or more adjoining places with distinguishing appellations such as Great or Little and East or West; and some of these subdivisions were already in existence by 1086. Or again, the constituent members of some large manors were not named; they were described as *berewichae, appendicii* or *membra,* or they were not even indicated at all. Thus the large manor of Sonning in Berkshire seems to have included a number of places unnamed in Domesday Book;[1] so did the large manor of Farnham in Surrey.[2]

Where we can be reasonably sure that one name stood for one settlement, it is clear that there was much variation in size even in the same locality. The recorded populations of the dozen vills in the Cambridgeshire hundred of Wetherley ranged from 12 to 82. Whatever factor we use (whether 5 or some other) to obtain the actual populations, the fact of variety remains. Some places were very small indeed. Radworthy in Devon had only four recorded people with 1½ plough-teams, and Lank Combe had only one without a team. Other places were completely uninhabited in 1086 and some were said to be waste. It would appear that village churches constituted a familiar feature of the countryside and that most were built of wood. They were, however, only irregularly entered in the Domesday Book. They appear for 352 out of 634 recorded places in Suffolk, but for only 17 out of 440 places in Essex, and for only two out of 334 places in Staffordshire.

Vills varied greatly not only in size but in form. Maitland, following Meitzen, drew attention to the fact that 'at least two types of vill must be in our eyes when we are reading Domesday Book' – nucleated and dispersed.[3] The number of place-names mentioned for Devonshire amounts to 983, but in this land of dispersed settlement, the number of hamlets and isolated farmsteads in 1086 has been placed at many times this number.[4] We can only argue retrospectively from the evidence of later ages about the varying types of settlement in Domesday England. The same is true of the size and shape of the parishes or tracts of territory attached to these names, which may preserve in a fossilised form vestiges of older arrangements.

[1] A. Mawer and F. M. Stenton, *Introduction to the study of English place-names* (Cambridge, 1929), 39; *V.C.H. Berkshire*, 1 (1906), 301.
[2] F. W. Maitland, 13–14.
[3] F. W. Maitland, 16.
[4] W. G. Hoskins, *Provincial England* (London, 1963), 21.

In spite of these limitations, what the details recorded for the Domesday vills can do is provide information about their respective resources.

POPULATION

The details of population in successive Domesday entries cover a whole series of categories, ranging from freemen to slaves. Between these extremes come villeins, bordars, cottars and many other groups. The total number of people so entered amounts to about 275,000. This recorded population is usually taken to refer to heads of households, and, in order to obtain the actual population, we must multiply it by some factor according to our ideas of the size of a medieval family. Various multipliers have been suggested, but the most likely seems to be 5.[1] The total that results, however, is subject to many doubts. The information about towns, for example, is often fragmentary. Then there are occasional entries which seem to be incomplete; we hear, for example, of plough-teams and other resources but not of people. There are also places about which we are told practically nothing beyond their names. There are, moreover, a number of entries which refer to unspecified numbers of men, and these cannot be included in a total; and there are also a few blanks where figures were never inserted. Such omissions and imperfections seem to indicate the presence of unrecorded households. Taking these doubts into consideration, it may be fair to say that the total population of England, excluding the four northern counties, amounted to about 1½ million in 1086; the northern counties must have been very sparsely peopled. Merely to be able to make such an estimate for such a remote date is in itself an indication of the value of the Domesday Book as a source of information.

On Fig. 11 East Anglia stands out as the most densely populated area, much of it with over 15 recorded people per square mile, and sometimes with over 20. The concentrations along the fertile coastlands of Sussex and Kent are also notable. On the other hand, there are a number of agriculturally unrewarding districts with very low densities in southern and eastern England – the Weald, the Bagshot Sands area of Surrey and north-eastern Hampshire, the Burnham gravel area of the southern

[1] Maitland (p. 437) suggested 5 'for the sake of argument'. J. C. Russell more recently suggested 3.5 – *British medieval population* (University of New Mexico, Albuquerque, U.S.A., 1948), 38 and 52. But for evidence in support of the traditional multiplier of 5 or something near it, see J. Krause, 'The medieval household: large or small', *Econ. Hist. Rev.*, 2nd ser., IX (1957), 420–32.

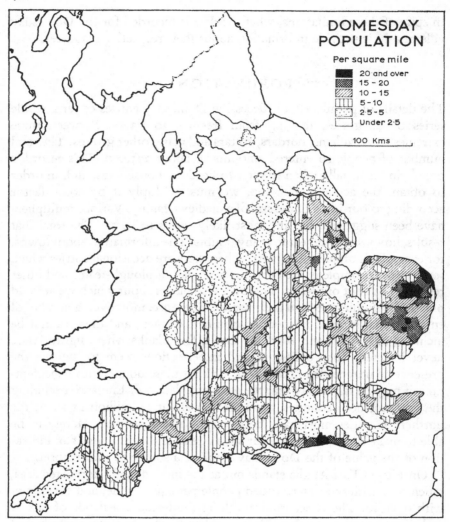

Fig. 11 Population in 1086
        Based on H. C. Darby (ed.), *The Domesday geography of England,* 7 vols
        (1952–77). See p. 43 n above.

Chilterns, the Fenland, the Breckland, the New Forest, the Dorset heathlands, Dartmoor and Exmoor. Apart from these sparsely occupied areas,
the density of population, in general, decreases towards the west and the
north. Nowhere in the Midlands is a density of over 15 reached, and north-

ward the figure falls to 5 and in places even to below 2.5. The variations in density of population reflected to a large extent the fertility of the land. Upland areas were inherently infertile, so were the light soils of the low-lands before the days of improved farming. But natural conditions do not explain all of the low density areas. Some of them, towards the northern parts of the realm, owed their poverty to the fact that they had been deliberately devastated, particularly by the armies of the king.[1]

THE COUNTRYSIDE

Arable land

From numerous hints in late Saxon documents, we may suppose that the arable, or much of it at any rate, was arranged in open-field strips. There is, however, only a solitary Domesday entry (that for Garsington in Oxfordshire) that seems to refer to scattered strips, and we have to rely upon later evidence for details about variations in field arrangements from district to district. Although Domesday Book itself gives us no information about these matters, it does enable us to indicate the relative distribution of the arable over the face of the country.

Entry after entry states: (*a*) the amount of land for which there were teams (*terra n carucis*); (*b*) the number of teams (*carucae*) actually at work, both on the demesne and on the land of the peasants. The first statement sounds very straightforward, especially as the Exeter version expands it to indicate the number of teams which could plough the land of an estate (*hanc terram possunt arare n carucae*). But, in fact, the implications of the Domesday teamland are far from clear and have provoked considerable discussion. The number of teams sometimes exceeds that of teamlands; the formula varies in detail; it is absent for some counties; and for other counties (e.g. Lincoln, Northampton and Nottingham) the figures are often related to those for the assessments in such a way as to suggest that they are artificial. Some scholars have thought that teamlands refer to the teams of 1066; others that they indicate the arable (actual and potential) of 1086; and yet others that, for some counties certainly, they refer to an earlier and obsolete assessment. In view of these difficulties and the present state of our knowledge, it would be unwise to regard teamlands as reflecting geographical realities generally over the face of the country in 1086.

The second statement also sounds straightforward, and may well be so. It is reasonable to suppose that we are being given an idea of the land

---

[1] See pp. 58–62 below.

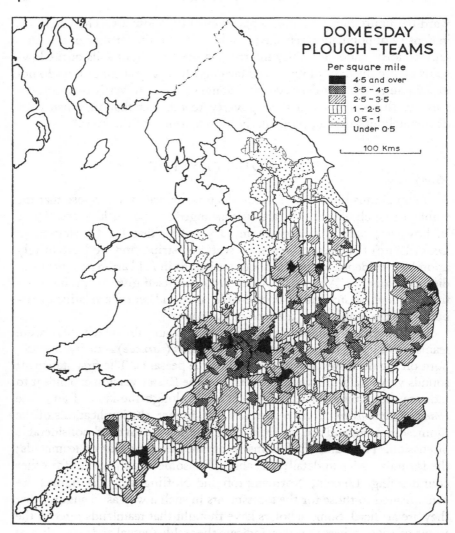

Fig. 12 Plough-teams in 1086
Sources as for Fig. 11.

actually cultivated. Although the teams at work in the field may have
varied in size on different soils, a comparison of parallel Domesday
entries show that, for the purposes of computation, eight oxen made
a Domesday team just as, wrote Maitland, 'twelve pence make a shilling'.[1]

[1] F. W. Maitland, 417.

It is true that this has been challenged, particularly for the south-western counties for which a variable Domesday team has been postulated, but the balance of evidence seems to be in favour of a standard eight-ox team.[1]

On Fig. 12 variations in the distribution of plough-teams reflect, to a great extent, the fertility of the land. The areas with the most arable, those with, say, over 3½ teams per square mile included the coastal plain of Sussex, the eastern part of East Anglia, and districts in the Midlands and other places where, in the context of the time, soils were favourable – districts such as the 'red lands' of north Oxfordshire, the Vale of Evesham, south-east Herefordshire, south Warwickshire and the lower Exe Basin. Conversely, districts with less than a team per square mile included the Weald, the New Forest, the Dorset and Surrey heathland, the Fenland and the Breckland. To these must be added much of the northern counties of Domesday England – Yorkshire, Cheshire and the northern parts of Staffordshire and Derbyshire. Natural conditions do not explain all these low density areas; those of the north, in particular, owed their lack of teams very largely to the fact that they had been deliberately devastated.

This statistical picture can be supplemented by very little other evidence about the agriculture of eleventh-century England. Dating from a generation or so before the Conquest, there is the remarkable *Rectitudines* which deals with the management of an estate and with the duties of the peasantry. Supplementing this is the *Gerefa* which sets out the duties of a reeve throughout the year. Both documents note that conditions of husbandry varied from place to place, but without giving any details.[2] Crops and people were very much at the mercy of bad weather. The year of the Inquest, according to the Anglo-Saxon Chronicle, was 'very disastrous and sorrowful'; grain and fruits ripened late and there was a murrain amongst cattle. The following year was also a bad one in which 'many hundreds died of hunger'. Then again, 1089 was a backward year in which crops were sometimes not reaped until Martinmas (11 November).

[1] R. Lennard, 'Domesday plough-teams: the south-western evidence', *Eng. Hist. Rev.*, LX (1945), 217–33; H. P. R. Finberg, 'The Domesday plough-team', *Eng. Hist. Rev.*, LXVI (1951), 67–71; R. Lennard, 'The composition of the Domesday caruca', *Eng. Hist. Rev.*, LXXXI (1966), 770–5.

[2] H. R. Loyn, *Anglo-Saxon England and the Norman Conquest* (London, 1962), 189–94.

Grassland

The distinction between meadow and pasture was clear. Meadow denoted land bordering a stream, liable to flood, and producing hay; pasture denoted land available all the year round for feeding cattle and sheep. The two varieties of grassland merged into one another, but they were usually entered separately. The amount of meadow recorded for the average Domesday vill was not great, but it was important; thirteenth-century evidence shows that an acre of meadow was frequently two, three or more times as valuable as an acre of arable. Meadow, like wood, was measured in a variety of ways. For the majority of counties it was measured in terms of acres, but we do not know what area was implied by such an 'acre'; for the counties of Bedford, Buckingham, Cambridge, Hertford and Middlesex, it was entered in terms of the teams of oxen which its hay could support; occasionally, it was expressed in terms of linear dimensions or of money payments or in some other way. As in the case of wood, it is impossible to reduce these various types of measurement to a common denominator. Very little meadow was entered for Cornwall which may reflect geographical conditions, and none at all for Shropshire which may indicate some local idiosyncrasy in the record.

Fig. 13 shows the varying incidence of meadow over a great part of England. Much lay along the alluvial valleys of the larger river courses – in the south, along the Thames itself and along the Lea and the Kennet; in the north, along the Nene and the Trent together with its tributaries the Soar and the Dove. It was also widely distributed along the many streams of the clay belt that extends from north Berkshire through Oxfordshire, Buckinghamshire, Bedfordshire and Huntingdonshire. In northern Berkshire, in the western part of the Vale of White Horse, renders of cheese were entered for three places, at two of which dairies are also mentioned; here may be the hint of a local dairying economy. Elsewhere, dairies and renders of cheese were entered for only a few scattered places.

Pasture was much more irregularly entered. It never, or only rarely, appears in the folios for the counties of the north and the Midlands. For the three counties of Cambridge, Hertford and Middlesex, its presence is indicated by the formula 'pasture for the cattle of the village'. For Oxfordshire and the south-western counties, it was measured in terms either of acres or of linear dimensions. Other variants, including money renders, occasionally appear for the south-eastern counties. This irregular record makes it impossible to reconstruct the distribution of pasture over the

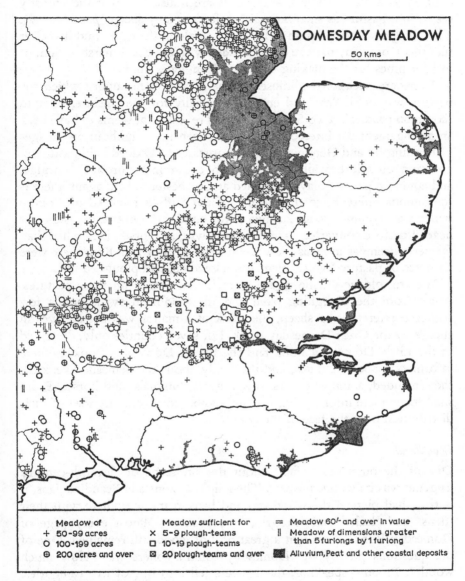

Fig. 13 Meadow in 1086
  Sources as for Fig. 11.
  There was some meadow in almost every village. Only the larger quantities
  are plotted in order to emphasise the areas with considerable amounts. It is
  impossible to equate the four methods of enumeration.

3-2

country as a whole. The Essex entries are unusual in that they mostly record not pasture as such but pasture for so many sheep (*pastura oves*). The villages with such pasture lay in a belt along the coast, and it is clear that this Domesday pasture corresponded with the Essex marshes, famous in later times for the making of cheese from ewes' milk.

There are occasional glimpses of the arrangements under which the pasture was used. Scattered entries, especially for Devonshire, refer to 'common pasture'. Two Somerset entries reveal what seems to have been an arrangement for intercommoning between the neighbouring villages of Hardington and Hemington. The burgesses of Oxford held pasture in common 'outside the wall'; we also hear of common pasture at Cambridge, and common pasture may be implied at Colchester. Then, again, there is the famous reference to the pasture of the Suffolk hundred of Colneis which was 'common to all the men of the hundred'. On two occasions we hear of pasture converted to arable; at Swyre, in Dorset, land which had formerly been pasture was 'now sown', and at Bourne in Kent there were six acres of pasture 'which men from elsewhere had ploughed up'.

We are told very little of the animals that grazed on these pastures. Apart from the ploughing oxen and the woodland swine and a few scattered references to sheep, the stock of a manor was passed over in silence by the Great Domesday Book. For seven counties only, described in the Little Domesday Book and the Exeter Domesday Book, is there information about animals, and then only about the demesne livestock; they included, amongst others, sheep, goats, 'animals' and horses. From such incomplete information, it is impossible to obtain any idea of their distribution over the face of the country.

Woodland

One of the outstanding facts about the landscape of eleventh-century England was its wooded aspect. The Anglo-Saxons and Scandinavians, it is true, had pierced the woodland and broken it everywhere with their 'dens' and 'leahs' and 'skogrs', but, even so, almost every page of Domesday Book shows that a great deal of wood still remained. One of the questions put by the Domesday commissioners was 'How much wood?'. Broadly speaking, the answers fell into one of five categories. Sometimes, they said that there was enough wood to support a given number of swine, for the swine fed upon acorns and beechmast. A variant of this was a statement not of total swine but of annual renders of swine in return for pannage. A third type of answer gave the length and breadth of wood in terms of leagues, furlongs and perches, but whether the information referred to mean diameters or to extreme diameters or conveyed some other notion, we cannot say. A fourth type stated the size of the wood in terms of acres, but we do not know what area was implied by such an eleventh-century 'acre'. The fifth category of answer was a miscellaneous one that included a number of variants and idiosyncrasies occasionally encountered in the text, e.g. wood for fuel, for the repair of houses or for the making of fences. Normally, each county was characterised by one main type of entry but also included a few other entries of a different character.

The difficulty presented by this array of information can be simply stated. It is impossible satisfactorily to equate swine, acres and linear dimensions, and so reduce them to a common denominator. Any map of Domesday woodland covering a number of counties must therefore suffer from this restriction, and we cannot be sure that the visual impression as between one set of symbols and another is correct. There are also other difficulties such as the fact that some woods may not have been recorded. Despite these limitations, much can be learned from Fig. 14. With all its problems, it leaves us in no doubt about the wooded aspect of large tracts of England in 1086. One surprising feature is the absence of wood from the Weald. This arises from the fact that much of the wood of the Weald was entered under the names of surrounding villages. It must therefore be 'spread out', so to speak, by eye. But we may well suspect the existence of unrecorded wood here, and also in other districts such as northern Berkshire.

Domesday Book tells us little or nothing about the process of clearing. The fuller information of the Little Domesday Book for Essex, Norfolk and Suffolk enables us to know that wood had been cut down between 1066–86 in at least 109 out of the 1,807 villages in those three counties. The circumstantial evidence about Wealden 'denes' or swine pastures in the folios for Kent also indicate clearing; so does some circumstantial evidence for Sussex; so also do the references to cartloads of wood that fed the Droitwich salt industry in Worcestershire. Ironworks must have consumed more wood, but we are told nothing of this. It is certain that clearings for cultivation were already known as 'assarts', a word derived from the French *essarter*, meaning to grub up or clear land of bushes and trees. Hereford is the only county for which the Domesday entries mention them, but there is every reason to believe that what was happening in Herefordshire was also happening in other counties.

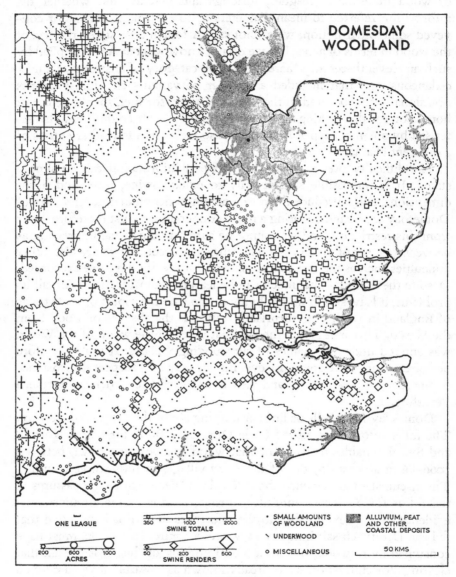

Fig. 14 Woodland in 1086
Sources as for Fig. 11.

Forests

Pre-Domesday kings had hunted and possessed game preserves, but with the Norman Conquest these royal activities greatly increased. The forest law and forest courts of Normandy were introduced into England and there was a rapid and violent extension of forest land.[1] The Norman kings had a passionate love of the chase and, as the Anglo-Saxon chronicler wrote under the year 1087, King William 'made large forests for deer'. The word 'forest' is neither a botanical nor a geographical term, but a legal one. It implied land outside (*foris*) the common law and subject to a special law that safeguarded the king's hunting. Forest and woodland were thus not synonymous terms, for the forested areas included land that was neither wood nor waste, and they sometimes covered whole counties. Even so, a forested area usually contained some wood and often large tracts of wood. Within the forests, no animals could be taken without express permission and the right to cut wood and to make assarts was severely restricted. In this way, much forest land was kept free from the plough and maintained in its primitive condition.

As forests were not liable for renders or geld, they were rarely mentioned and specifically described in Domesday Book. But we frequently hear that part of a manor had been placed 'in the king's forest', or 'in the forest', or 'in the king's wood' or 'in the king's enclosure' (*in defensione regis*). Sometimes it was the wood itself that had been taken out of a manor and afforested (*Silva hujus maneriae foris est missa ad silvam regis*). Or again at *Haswic* in Staffordshire there was land for eight teams, but the entry adds, 'it is now waste on account of the king's forest'; and at Ellington in Huntingdon one out of ten hides was waste on account of the king's wood (*per silvam regis*). Such references occur in the entries for many counties. There is also mention of 'hays' or enclosures (*haiae*) for regulating the chase, particularly in entries relating to Cheshire, Herefordshire, Shropshire and Worcestershire. Finally, parks were recorded for 33 places; and sometimes these were specifically for wild beasts (*parcus bestiarum silvaticarum*). Some belonged to the king; others to various lay or ecclesiastical lords.

There is one forest, and only one, about which the Domesday Book gives a wealth of detail. It is the New Forest, and the account of it occupies a special section of the Domesday description of Hampshire. The amount

[1] C. Petit-Dutaillis, *Studies and notes supplementary to Stubbs' Constitutional History*, 2 vols. (Manchester, 1915), II, 166–70.

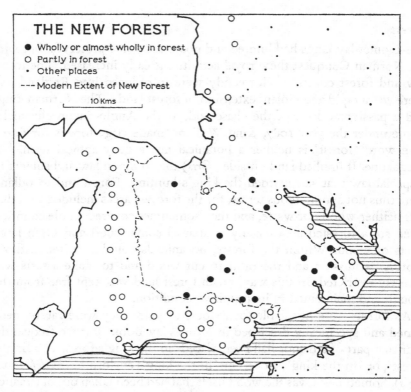

Fig. 15 The making of the New Forest
Based on H. C. Darby and E. M. J. Campbell (eds.), *The Domesday geography
of south-east England* (Cambridge, 1971), 325.

of destruction involved in the making of the Forest was once a matter of
controversy. The chroniclers of the twelfth century declared that William
reduced a flourishing district to a waste by the wholesale destruction of
villages and churches. But this is not entirely borne out either by the poor
soils of the district or by the evidence of the Domesday Book itself. It
would appear that the area was but thinly occupied in 1066; there were
some villages of moderate size but many were very small. About 30 to 40
of these villages were placed wholly or almost wholly under forest law,
leaving no record of their former population. Only about one-half of these
appear on Fig. 15 because the rest are unidentified. They lay almost entirely in the centre of the forest or to the south-east. Varying portions of
about 40 other villages on the borders of the main part of the forest were

also included. The limits of this eleventh-century forest have shrunk but it still covers some 92,000 acres and remains a unique memorial of Norman England.

Miscellaneous resources

Fish, salted or otherwise, must have played an important part in the life of an eleventh-century community, and the presence of fisheries is noted in many Domesday entries. They are sometimes recorded in association with mills, being derived from mill-ponds, and sometimes separately. It is very unlikely, however, that the Domesday record gives a complete picture of the fishing activity of the time. Maybe some fisheries were unrecorded because they returned no profit to the lord, or perhaps because of the absence of any contrivance such as a weir (*gurgites*) for catching fish.[1] The species normally mentioned is that of eels, and these often appear as eel renders from mills. Salmon were recorded for places along the Severn, the Dee and the Dart.

The fisheries were located along the main rivers and their tributaries – the Thames, the Severn, the Nene, the Trent, the Great Ouse, the Dee, the Medway, the Avon and others. But the chief assemblage was in and around the Fenland. Here, some of the north Cambridgeshire vills returned great numbers of eels to their lords – as many as 33,260 from Wisbech; 27,150 from Doddington; 24,000 from Stuntney; and 17,000 from Littleport. Clearly, here was a district with distinctive regional economy. Few of the entries tell us anything about the operations involved. There is, it is true, a very interesting description of Whittlesey Mere, on the western edge of the Fenland, with its fisheries, its fishermen and its fishing boats owned by the abbots of Ramsey, Thorney and Peterborough. We hear also of nets, fish-traps and weirs in the Fenland and elsewhere. Of the marsh itself, there is only occasional indication in entries relating to the Fenland, the Isle of Axholme, the Somerset Levels and Romney Marsh; but the evidence of these entries provides scarcely more than a hint of the economies of these regions.

Sea-fisheries were hardly ever mentioned. Herring renders were recorded for a number of villages along or near the Suffolk coast; there is also reference to a sea weir (*heiemaris*) at Southwold and to 24 fishermen at Yarmouth. Herring renders were also occasionally entered for London and Southwark and for places along the coasts of Kent and Sussex, and there is mention even of porpoises for Southease in Sussex. We can only

[1] R. Lennard (1959), 248–9.

conclude that these are but stray indications of what must have been a considerable activity around the coasts of the country.

Among the other miscellaneous resources of a manor were vineyards. They are recorded for 55 places, the most northerly being Ely. J. H. Round argued that the culture of the vine was reintroduced (since Roman times) by the Normans, and he based his view upon the fact that Domesday vineyards were normally on holdings in the direct hands of Norman tenants-in-chief; that they were usually measured by the foreign unit of the arpent; and that some had but lately (*noviter* or *nuperrime*) been planted.[1] But this view cannot be maintained, because we hear of vineyards in England in the eighth, ninth and tenth centuries.[2] It looks as if the Normans had not so much reintroduced the vine as extended, possibly greatly extended, its culture. We may see in the Domesday references to vineyards, as in the references to forests, indications of the new order that William brought to the land he conquered.

Devastated land

Devastated land had formed an important element in the geography of England during the period of Anglo-Danish warfare; and it did not cease to be important after 1042 when a member of the royal house of Wessex, Edward the Confessor, once more ruled over all England. During his reign, the complicated politics of the great earldoms of the realm resulted in much conflict. Earl Godwin and his sons, banished overseas, returned in 1052 to raid along the south coast. There were also raids from Ireland, from Wales and from Norway that resulted in the devastation of various parts of the realm. The sources are obscure, and we can only repeat what the Anglo-Saxon Chronicle says under the year 1058: 'it is tedious to relate how it all happened'. An echo of some of the complications comes in the Domesday Book which tells us that Edward the Confessor had reduced the geld liability of Fareham in Hampshire from 30 hides to 20 'on account of the Vikings, because it is by the sea'. From the north, Earl Morcar of Northumbria raided south, in 1065, to Northamptonshire, and killed and plundered so that, says the Anglo-Saxon Chronicle, that shire and the other shires around 'were for many years the worse'. One indication of this is seen in the low 1066 values for many Northamptonshire estates, and in the Northamptonshire Geld Roll

[1] J. H. Round in H. R. Doubleday and W. Page (eds.), *V.C.H. Essex*, I (1903), 282–3.

[2] G. Ordish, *Wine growing in England* (London, 1953), 20–1.

(not later than 1075) which records about a third of the county as waste;[1] by 1086, however, recovery was general. Then again, the Domesday folios for Cheshire, Shropshire and Herefordshire show that many villages lay waste in 1066, the result of Welsh raiding. Along the northern border the raids of the Scots also left much devastation.

The Norman Conquest brought further devastation. William landed at Pevensey and, after the battle of Hastings, marched to Dover, then via Canterbury towards London. He did not cross the Thames into the city itself but, after some of his forces had burnt Southwark, he began an encircling movement westward and then northward across the Thames at Wallingford. From here he continued in a general north-east direction, and then turned to approach London from the north. The Anglo-Saxon Chronicle summarises these events by saying that 'he harried all that part which he over-ran until he came to *Beorh-hamstede*'. This is generally identified with Berkhamsted but a case has also been made for Little Berkhamsted to the east. An army such as this lived largely on the countryside it passed through, and inevitably left a trail of destruction. The differences in the values of many manors just before and then after the Conquest have been held to indicate the 'footprints' of the Conqueror's march.[2] The evidence does not lend itself to any rigid interpretation for there was not one army but several forces, and also there may have been foraging bands. A number of villages around Hastings and Dover still lay entirely waste in 1070 or so, and there were many more around London generally that had been reduced in value. But by 1086 most had recovered their former values, wholly or in part.

The south-eastern counties might have borne the first burden of the Conquest, but they did not bear the heaviest brunt. To a large extent, wasting by armies and raiders was the reason for the low densities of plough-teams and population in the west and particularly in the north. The term waste (*wasta*) found in many Domesday entries implies not the natural waste of mountain, heath and marsh, but land that had gone out of cultivation mainly as a result of deliberate devastation, but also perhaps because of some local vicissitude that is lost to us. Some of the wasted

[1] J. H. Round, *Feudal England* (1895), 147–56. See also D. C. Douglas and G. W. Greenway, *English historical documents, 1042–1189* (London, 1953), 483–6.

[2] F. H. Baring, 'The Conqueror's footprints in Domesday', *Eng. Hist. Rev.*, XIII (1898), 17–25. Reprinted with 'some additions and alterations' in F. H. Baring, *Domesday Tables* (London, 1909), 207–16. For a detailed account of the changes in values, see R. Welldon Finn, *The Norman Conquest and its effects on the economy: 1066–86* (London, 1971).

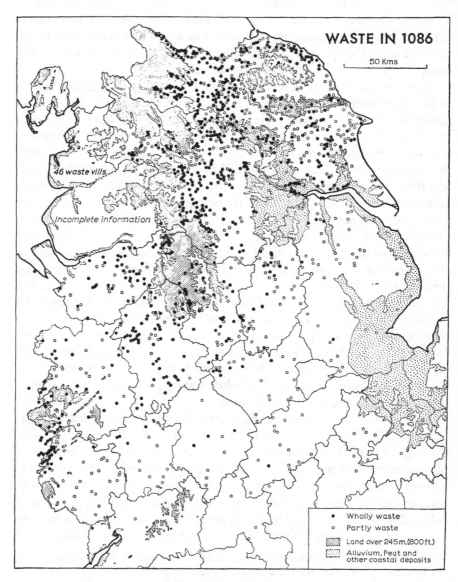

Fig. 16 Waste in 1086
Sources as for Fig. 11.

villages of Herefordshire, Shropshire and Cheshire were, we may suppose, the result of Welsh raiders, but many were the result of the crushing of rebellion by the king's armies during the years 1068–70. Much of the waste of those years was still evident in 1086. The account of Cheshire is unusual in that the Domesday folios give details of waste not only for 1066 (i.e. before the Conquest) and for 1086 (the year of the enquiry) but also for the date when the existing owner received an estate, i.e. about 1070. Out of a total of 264 there were 52 villages wholly or partly waste in 1066, presumably as a result of Welsh raiding. By 1070, the figure had increased to 162, the result of William's campaign of 1069–70. By 1086, there were still 58 villages wholly or partly waste. What we hear of in 1086 is a countryside not yet recovered but on its way to recovery.

It was in the north that William took the most terrible revenge, and he left the countryside in a condition in which it could give him no trouble again. The entry under the year 1069 in the Anglo-Saxon Chronicle is brief but eloquent; it merely says of Yorkshire that the king 'laid waste all the shire'. Yorkshire suffered most, but we are told in one early account that the harrying also extended over Derbyshire, Cheshire, Shropshire and Staffordshire, nor did Nottinghamshire escape. The general statements of the chronicles are borne out in a vivid way when the Domesday entries are plotted on a map (Fig. 16). Seventeen years or so had not been enough to obliterate the effects of the harrying. Entry after entry for these northern villages reads: *Wasta est*. As well as villages specifically described as waste, there were others without population, and most of these, too, we must suppose, were the result of William's campaigns.

There was also other post-Conquest devastation. Thus, in an entry for 1067, the Anglo-Saxon Chronicle tells how one of the sons of King Harold, from whom William had won the kingdom, came with a fleet from Ireland to the mouth of the Avon, and plundered the countryside around. That there were also other raiders across the western seas we know from a marginal note in the Exeter Domesday Book which refers to nine manors in the extreme south of Devon 'devastated by Irishmen'; seven of the manors had far from recovered by 1086, and three of these were each worth only a quarter of their 1066 values. Whatever its cause Welsh raiding or Norman strategy, the sum total of the evidence set out in Fig. 16 leaves us in no doubt about the importance of devastation as an element in the economic geography of eleventh-century England. Were it possible to reconstruct a similar map for, say, 1070, it is certain that the wasted villages would be more numerous and more widespread. The picture we see on Fig. 16 is that of a countryside already on its way to recovery.

INDUSTRY

Industry was not important in the eleventh century, but, even so, we might expect to hear more of it than we do in the Domesday Book. Although iron was worked in Roman times, there is remarkably little evidence of Anglo-Saxon working, and only a few scattered Domesday references. An ironworks (*ferraria*) is entered for a holding in East Grinstead hundred in Sussex, and there had been others at Corby and Gretton in Northamptonshire. Renders of iron in Gloucestershire may have reflected the presence of iron-working in the Forest of Dean; the renders at Gloucester itself included 'iron rods for making nails for the king's ships'. The *ferrariae* and *fabricae ferri* at Stow and at Castle and Little Bytham in Lincolnshire were not at places on iron-bearing outcrops. Likewise there were *ferrariae* at Chertsey (Surrey), Fifield Bavant (Wilts), Stratfield (Hants), Wilnecote (Warwick), and iron-workers (*ferrarii*) at Hessle (W. Riding) and North Molton (Devon); these may imply the presence of forges. Smiths (*fabri*) are also mentioned for a few places, but the number is not large and, presumably, many smiths and iron-workers were entered as villeins or borders. At Rhuddlan, described in the Cheshire folios, we hear specifically of mines (*mineriae ferri*).

Lead works (*plumbariae*) are entered for six places in the Peak District of Derbyshire; all lie on or near the outcrop of Carboniferous Limestone which contains veins of metalliferous ores from which the lead must have been derived, but we are told nothing of the methods by which the finished lead was produced. There is also mention of a *fabrica plumbi* in the entry for Northwick and Tibberton, possibly a leadworks for making the vats used in the manufacture of salt at Droitwich nearby. That there was trade in lead throughout the country we may infer from a casual but interesting reference in the *Inquisitio Eliensis* which is near-contemporary with the Domesday Book; this compares the weight of a 'fodder' of Peak lead (*carreta plumbi del pec*) with that of a London 'fodder', and it suggests that Ely Cathedral may have had its supply of lead for roofing from the Peak District.[1] Of other lead-mining areas such as Shropshire, the Pennine Dales and the Mendips, known to have been worked in Roman

[1] N.E.S.A. Hamilton, *Inquisitio Comitatus Cantabrigiensis...subjicitur Inquisitio Eliensis* (London, 1876), 191.

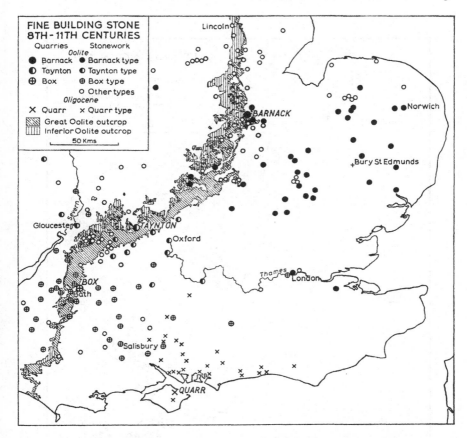

Fig. 17 Southern England: fine building stone, eighth to eleventh centuries
Based on E. M. Jope, 'The Saxon building-stone industry in southern and Midland England', *Medieval Archaeology*, VIII (1964), 92; and on additional information provided by Professor Jope. Navigable portions of the Thames and Severn are shown.

times, the Domesday Book says nothing. Nor does it mention Cornish tin.

Quarries are occasionally mentioned. That at Taynton in Oxfordshire was well known for its Great Oolite freestone which is to be found in the surviving pre-Domesday masonry in villages up to thirty miles and more from the outcrop (Fig. 17). The quarries at Whatton in Nottinghamshire and Bignor in Sussex produced millstones; but these, and a few other quarries mentioned for other places, can only have been a fraction of the

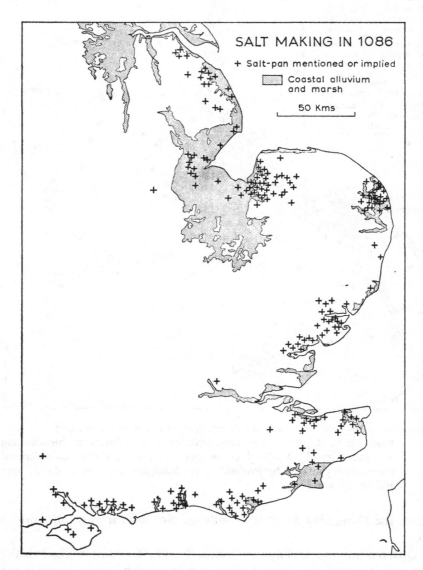

Fig. 18 Southern and eastern England: coastal salt-making in 1086
Sources as for Fig. 11.

total worked throughout the country. So too, the potteries and potters entered for Bladon (Oxon), Haresfield (Gloucs) and Westbury (Wilts) must stand for a large number in the country as a whole.

There is one industry, however, about which information is relatively abundant. Salt was an indispensable item in medieval economy, especially for the preservation of meat and fish. Rock salt was not worked in Britain until 1670, so that the commodity was obtained by evaporation either from sea water or from inland brine springs. The chief areas for the production of maritime salt were along the marshes and estuaries of the east and south coasts. The Domesday record usually states the number of salt-pans on a holding, and sometimes also includes their render either in money or in fish or in loads of salt. Caister in eastern Norfolk, for example, had as many as 45 salt-pans. Some villages for which salt-pans appear are situated inland, from which we must conclude that their lords held the pans along the coast some distance away (Fig. 18).

The inland centres of production were in Worcestershire and Cheshire, and the brine springs were derived from the Keuper Marl beds of the Triassic system. The centre of the Worcestershire industry was at Droit-wich for which brine-pits (*putei*) and salt-pans (*salinae*) are recorded. There is no information about the processes of manufacture, but we do hear of leaden pans or vats (*plumbi*) and of furnaces (*furni*); we also hear occasionally of fuel for the latter in the form of 'wood for the salt-pans'. Some of the Droitwich salt-pans belonged not only to nearby Worcester-shire villages, but to villages of other counties (Fig. 19). We even hear of a salt-worker of Droitwich rendering loads of salt as far away as Princes Risborough in Buckinghamshire. All this suggests some interesting reflections about the movement of commodities in Domesday England. The salt was presumably transported by pack-horse; names such as Saltway, Salter's Corner and Salford are preserved on the One-Inch Ordnance maps of today, and it is possible to reconstruct a system of ways leading from Droitwich.[1] The Cheshire industry was centred at Northwich, Middlewich and Nantwich which had only partly recovered from the disturbances of 1070. We hear of salt-pans and of 'boilings of salt' (*bulliones salis*). We are also told of a few nearby manors with salt-pans at one of the Wiches, but that the trade in salt extended into other counties is apparent from the details of the tolls levied on those who transported the salt away. The tolls increased with the distance from which a purchaser

[1] A. Mawer *et al.*, *The place-names of Worcestershire* (Cambridge, 1927), 4–9; A. H. Smith, *The place-names of Gloucestershire*, pt 1 (Cambridge, 1964), 19–20.

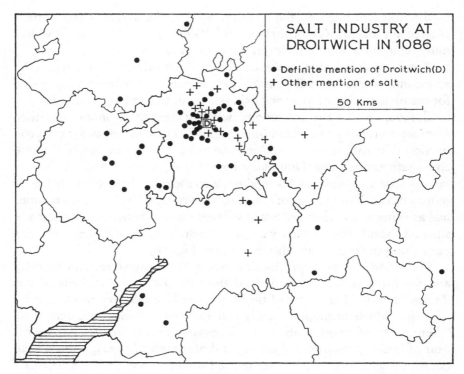

Fig. 19 The Droitwich salt industry in 1086
    Based on H. C. Darby and I. B. Terrett, *The Domesday geography of Midland
    England* (2nd ed., Cambridge, 1971), 258.

and his wagons (*carri*) came, whether from the same hundred as that of the
Wich, or from another hundred, or from another shire. There is also
reference to the toll paid by those who carried salt about the county to
sell. These glimpses, like those for Droitwich, imply an organised trade
in salt that extended far beyond the vicinity of the Wiches themselves.

TOWNS AND COMMERCE

Whatever the difficulties of interpretation, the Domesday information
for rural England is systematically presented and is remarkable for the evidence it provides. When we turn from the countryside to the towns, all is
different. The information is so incomplete and so unsystematic that it is
often impossible to form any clear idea of the size of a town or of the

economic and other activities that sustained it. Altogether 112 places (including Rhuddlan in North Wales) seem to have been boroughs in 1086 (Fig. 20). The Anglo-Saxon word 'burh' signified a fortified centre, and the test of burghal status was neither size nor general prosperity. There has been debate about whether defence or trade provided the impetus to urban development, but surely both were important, and it is clear that by Domesday times a commercial element had become an important feature of many boroughs. It is also clear that burghal status was not constant. Not all the boroughs of the tenth century survived to appear as such in the Domesday Book, and not all Domesday boroughs retained their standing as such in the twelfth century. Conversely, in each century, new boroughs emerged.

By far the largest borough must have been London. It is therefore particularly unfortunate that Domesday Book contains no account of it. The 126th folio, where this should have come, is blank, and there are only a few incidental references to it elsewhere in the Book. It had been a Roman city, and, after the confusion of the Anglo-Saxon invasions was over, its advantages of site and location re-asserted themselves. By the eighth century it had become, in the words of Bede, 'the market place of many peoples coming by land and sea'; and, for the eleventh century, there is evidence of its wide trading connections with the Continent. It is true that the idea of a capital city had not yet become current in western Europe, and that the centre of government moved about with the court of the king. But London already had a distinct place among the boroughs of England, a place emphasised by its role as a centre of resistance against Danish invasion in the early years of the eleventh century.[1] A guess might place the number of its inhabitants at over 10,000, but any attempt to estimate how much over becomes even more hazardous. Fire was always a danger in London, as in other cities. The Anglo-Saxon Chronicle, under the year 1077, records an extensive fire in London, greater than any 'since the town was built'; and ten years later there was another fire which destroyed 'the greatest and fairest part of the whole city'.

There are likewise only incidental Domesday references to Winchester, the city to which the results of the Inquest were brought and in which the Domesday Book was at first kept. There are, however, two early surveys of the city, dating from 1103–5 and 1148, which tell us a little about it. We hear, among other things, of a merchants' guildhall, of a market, of shops and stalls, of mints, of forges and of a prison. Its maximum

[1] F. M. Stenton, 531.

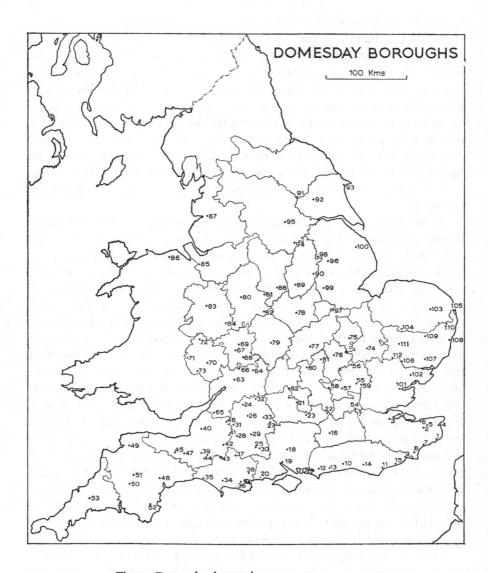

Fig. 20 Domesday boroughs
Sources as for Fig. 11.
Names of boroughs are given on pp. 69–70.

Key to Fig. 20

The counties and the boroughs are set out in the order in which they appear in the Domesday Book, and then in the Little Domesday Book

Kent
   1 Dover
   2 Canterbury
   3 Rochester
   4 Sandwich
   5 Fordwich
   6 Seasalter
   7 Hythe
   8 Romney

Sussex
   9 Rye
  10 Steyning
  11 Pevensey
  12 Chichester
  13 Arundel
  14 Lewes
  15 Hastings

Surrey
  16 Guildford
  17 Southwark

Hampshire
  18 Winchester
  19 Southampton
  20 Twynham

Berkshire
  21 Wallingford
  22 Windsor
  23 Reading

Wiltshire
  24 Malmesbury
  25 Wilton
  26 Cricklade
  27 Bedwyn
  28 Warminster
  29 Tilshead
  30 Salisbury
  31 Bradford on Avon
  32 Calne
  33 Marlborough

Dorset
  34 Dorchester
  35 Bridport
  36 Wareham
  37 Shaftesbury
  38 Wimborne Minster

Somerset
  39 Langport
  40 Axbridge
  41 Frome
  42 Bruton
  43 Milborne Port
  44 Ilchester
  45 Milverton
  46 Bath
  47 Taunton

Devonshire
  48 Exeter
  49 Barnstaple
  50 Lydford
  51 Okehampton
  52 Totnes

Cornwall
  53 Bodmin

Middlesex
  54 London

Hertfordshire
  55 Hertford
  56 Ashwell
  57 St Albans
  58 Berkhamsted
  59 Stanstead Abbots

Buckinghamshire
  60 Buckingham
  61 Newport Pagnell

Oxfordshire
  62 Oxford

[*continued overleaf*

Gloucestershire
63   Gloucester
64   Winchcomb
65   Bristol
66   Tewkesbury

Worcestershire
67   Worcester
68   Pershore
69   Droitwich

Herefordshire
70   Hereford
71   Clifford
72   Wigmore
73   Ewias Harold

Cambridgeshire
74   Cambridge

Huntingdonshire
75   Huntingdon

Bedfordshire
76   Bedford

Northamptonshire
77   Northampton

Leicestershire
78   Leicester

Warwickshire
79   Warwick

Staffordshire
80   Stafford
81   Tutbury
82   Tamworth

Shropshire
83   Shrewsbury
84   Quatford

Cheshire
85   Chester
86   Rhuddlan

(Lancashire)
87   Penwortham

Derbyshire
88   Derby

Nottinghamshire
89   Nottingham
90   Newark

(Rutland)
Nil

Yorkshire
91   York
92   Pocklington
93   Bridlington
94   Dadsley
95   Tanshelf

Lincolnshire
96   Lincoln
97   Stamford
98   Torksey
99   Grantham
100 Louth

Essex
101 Maldon
102 Colchester

Norfolk
103 Norwich
104 Thetford
105 Yarmouth

Suffolk
106 Sudbury
107 Ipswich
108 Dunwich
109 Eye
110 Beccles
111 Bury St Edmunds
112 Clare

population at this time has been estimated at between 6,000 and 8,000, but this seems large.[1] Of the other boroughs, we can conjecture from the unsatisfactory Domesday evidence that the following had at least 4,000 and maybe over 5,000 inhabitants in 1086: York, Lincoln, Norwich and possibly Thetford. Below them came a group with at least 2,000 and maybe over 3,000 each. This group included:

Bury St Edmunds	Huntingdon
Canterbury	Lewes
Colchester	Oxford
Dunwich	Stamford
Exeter	Wallingford

How many others should be counted in the group, we cannot say. Of towns such as Bedford, Bristol, Gloucester and Salisbury, the Domesday evidence is too fragmentary to allow us to make a guess. In the case of Gloucester, the evidence of a subsidiary document (the Evesham Abbey Survey) shows that the city had ten churches and a population of certainly over 3,000. Below the 2,000 mark came a variety of boroughs; some were very small and poor, such as Bruton and Langport in Somerset, both of which ceased to be boroughs in the twelfth century. We must also remember that most, if not all, boroughs, small or large, had an agricultural flavour. Arable, meadow and pasture are sometimes entered for them, and also such categories of population as villeins and bordars as well as burgesses. Thus Cambridge, with a population of at least 1,600, was a substantial settlement in the context of the time, yet its burgesses had been accustomed to lend their plough-teams to the sheriff at least three times a year, and in 1086 he was demanding that this be increased to nine times.

The immediate effect of the Norman Conquest on some boroughs was the destruction of houses for the building of castles. At York, one of its seven wards had been so wasted; at Lincoln, 166 houses had been destroyed; at Norwich, 98 houses; at Shrewsbury, 51 houses; and there had been similar destruction for castles on a smaller scale at Cambridge, Canterbury, Gloucester, Huntingdon, Stamford, Wallingford and Warwick. We hear also of houses destroyed by fire at Exeter, Norwich and Lincoln. At these and other boroughs there were also wasted or unoccupied houses for which no reason was given. Waste, due to one cause or another, was entered for as many as 33 of the 112 boroughs. Oxford, for example, had

---

[1] G. W. Kitchin, *Winchester* (London, 1890), 79.

suffered badly. Out of a total of about 1,000 properties in 1066, about one half were so wasted or destroyed that they rendered nothing; it does not follow that all their inhabitants had disappeared, but, whatever view we take, the Oxford of 1086 was clearly in reduced circumstances. The cause is not stated; some of the waste may have been the result of clearance to provide a site for the castle which had been built in 1070 but about which the Domesday Book says nothing.

Alongside these setbacks there were long-term tendencies of a different kind at work. Some people have believed that the Anglo-Saxon borough had little or no really urban character and that the origin of commercial towns (as distinct from fortified boroughs) was essentially a development of the years after 1066 and owed much to the Normans.[1] Others have challenged this view which, they have said, underestimates the amount of trade and urban growth before 1066; they point to the development of the Anglo-Saxon coinage and to various hints of internal and foreign trade, and they have concluded that there was no new urban concentration in the years immediately after the Conquest.[2] While this may be true in a general sense, the fact remains that there are many indications of commercial activity within the Domesday boroughs, some of it certainly new. The urban trend of the later Anglo-Saxon period continued to develop and perhaps even to accelerate.[3]

One indication is the establishment of new boroughs alongside the old at Norwich, Northampton and Nottingham; that at Norwich included 125 French burgesses, and at Nottingham we hear of 48 merchants' houses. A new borough (*novus burgus*) had also been established at Rye in the manor of *Rameslie* in Sussex; and, incidentally, another at Rhuddlan in North Wales. Groups of Frenchmen or French burgesses had also settled at Dunwich, Hereford, Shrewsbury, Southampton, Stanstead Abbots, Wallingford and York. In the borough around the castle (*In burgo circa castellum*) at Tutbury, in Staffordshire, there were 42 men who devoted themselves wholly to trade and who rendered £4. 10s. At Eye in Suffolk there were 25 burgesses who lived in or around the market place (*in*

---

[1] C. Stephenson, *Borough and town: a study of urban origins in England* (Cambridge, Mass., 1933), 70–1.

[2] J. Tait, *The medieval English borough* (Manchester, 1936), 132–8.

[3] For a convenient summary of the problem, see J. F. Benton, *Town origins: the evidence from medieval England* (Boston, Mass., 1968). See also R. P. Beckinsale, 'Urbanization in England to A.D. 1420', being ch. 1 (pp. 1–46) of R. P. Beckinsale and J. M. Houston (eds.), *Urbanization and its problems* (Oxford, 1968).

mercato manent xxv burgenses). Then again, there were 25 houses (rendering 100s. a year) in or around the market place of Worcester (*in foro Wirecestre*). At York, we hear of stalls in the provision market (*banci in macello*). Furthermore, there are hints of external trade at other boroughs. The account of Chester refers to the coming and going of ships, and to a trade in marten pelts which we know, from other sources, came from Ireland.[1] In the south, we are told of dues from ships at Southwark, of foreign merchants (*extranei mercatorum*) at Canterbury, and of a guildhall (*gihalla burgensium*) at Dover and of its harbour and ships. At Frostenden, in Suffolk, there was a seaport (*i portus maris*), and hithes or harbours are mentioned at four places in Kent along the shores of the Thames estuary – at Dartford, Gravesend, Milton and Swanscombe.

Of the 58 markets specifically mentioned in the Domesday Book, only fourteen were entered for boroughs, an example of the incompleteness of the Domesday urban record. Thus of the four boroughs in Gloucestershire (Gloucester, Bristol, Tewkesbury and Winchcomb) a market was recorded only for Tewkesbury, and it was said to have been established recently by Queen Matilda; but there were also three other places with markets – Thornbury, Cirencester with 'a new market', and Berkeley where there was a *forum* in which dwelt 17 men who paid a rent, but we are not told how much. We hear, moreover, not of a market but of 10 traders at Abingdon dwelling in front of the door of the church, i.e. Abingdon Abbey; there were also 10 traders at Cheshunt. There are only two Domesday references to fairs – one at Aspall in Suffolk, and an 'annual fair' at Methleigh in Cornwall. There must have been many others. There is, for example, non-Domesday evidence of a fair and of a Thursday market at or near St Michael's Mount in Cornwall.[2]

The outstanding Domesday example of new commercial growth comes from what was recorded merely as the *villa* 'where rests enshrined Saint Edmund, King and Martyr of glorious memory'. Accretion around the monastery at Bury St Edmunds had been considerable between 1066 and 1086:

Now the town [*villa*] is contained in a greater circle, including [*de*] land which then used to be ploughed and sown; whereon are 30 priests, deacons, and clerks together [*inter*]; 28 nuns and poor persons who daily utter prayers for the king

[1] J. H. Round (1895), pp. 465–7.
[2] H. P. R. Finberg, 'The castle of Cornwall', *Devon and Cornwall Notes and Queries* XXIII (1949), 123.

and for all Christian people; 75 bakers, ale-brewers, tailors, washer-women, shoemakers, robe-makers [*parmentarii*], cooks, porters, stewards together. And all these daily wait upon the Saint, the abbot, and the brethren. Besides whom there are 13 reeves over the land, who have their houses in the said town, and under them 5 bordars. Now, 34 knights, French and English together, and under them 22 bordars. Now altogether [there are] 342 houses in the demesne on the land of St Edmund which was under the plough in the time of King Edward.

This is about a place which was not technically a borough in 1086, but which has been included in the total of 112. Obviously the town as an economic reality was capable of growing up in places other than legal boroughs. Soon the time was to come when the creation or confirmation of markets was an essential element in the granting of borough charters.

This catalogue of the miscellaneous hints of the commercial life that appears in the Domesday Book points to the new age which the Norman Conquest had inaugurated. While the Conquest can hardly have meant a revolution in commerce, it must have given a great impulse to that trade and urban development that had already begun to grow in Anglo-Saxon times.

Chapter 3

CHANGES IN THE EARLY MIDDLE AGES

If the compilers of Domesday Book had been able to retrace their steps in the early 1300s they would have been greatly impressed by the changes that had taken place in the English countryside and in English life generally in the course of two centuries or so. Everywhere there was evidence of growth – in population and in the area of improved land, in industry and trade, and in the number and size of towns. The power of the central government also had increased and, while churchmen still played an important part in affairs of state, lay officials were becoming ever more numerous and influential.[1] This partly reflected changes in the system of education: many grammar schools[2] and two universities had come into being, taking the place of the monastic schools. Finally, wider external contacts – through increased commerce, the spread of the new monastic Orders, and participation in the 'twelfth-century renaissance'[3] – strengthened those ties with continental Europe that had been established by the Norman Conquest.

POPULATION

The population of England more than doubled between 1086 and the beginning of the fourteenth century, from approximately 1½ million to 4 or even 4½ million. There were, however, wide variations in rates of increase between one area and another. This is apparent in comparing the

[1] J. R. Strayer, 'Laicization of French and English society in the thirteenth century', *Speculum*, XV (1940), 76–86; H. G. Richardson and G. O. Sayles, *The governance of medieval England* (Edinburgh, 1963), 167, 283, 319.

[2] L. Thorndike, 'Elementary and secondary education in the Middle Ages', *Speculum*, XV (1940), 400–8; A. B. Emden, 'Learning and education' in A. L. Poole (ed.), *Medieval England*, II (Oxford, 1958), 515–40.

[3] R. W. Southern, 'The place of England in the twelfth-century renaissance', *History*, n.s., XLV (1960), 201–16; C. H. Haskins, 'England and Sicily in the twelfth century', *Eng. Hist. Rev.*, XXVI (1911), 433–47; A. C. Crombie, *Robert Grosseteste and the origins of experimental science, 1100–1700* (Oxford, 1953), especially 16–43.

evidence of Domesday Book with that of surveys of estates dating from
the twelfth and thirteenth centuries.[1] Many long-established places
probably grew at far less than the national rate; the villages of south
Warwickshire, for example, appear to have been much the same size in
1279 as in 1086.[2] On the other hand, the Forest of Arden, to the north of
the river Avon, was attracting population at this time.[3] In the Fenland,
where reclamation and population growth were clearly interdependent,
some phenomenal increases occurred. Between 1086 and the late thirteenth
or early fourteenth century, the number of tenants in Spalding township
multiplied six and a half times, in Pinchbeck township over eleven times,
and in Fleet sixty-one times.[4] A particularly rapid upward trend in developing areas was very likely the result, not merely of immigration, but of
a lowering of the average age of marriage and thus an increase in fertility.

Some of the highest rural population densities were found in association
with the custom of partible inheritance.[5] The latter was a notable feature
of Norfolk and Kent, and was also of some importance in parts of Cambridgeshire, Leicestershire, Nottinghamshire, Lincolnshire, Suffolk, Essex
and Middlesex. Partition could not go on indefinitely, but it went furthest
where there were extensive common rights and/or new land that could be
reclaimed. Where a man inherited, or was certain to inherit, a viable holding, even though the nucleus of it was very small, he tended to marry and

---

[1] J. C. Russell, *British medieval population* (Albuquerque, 1948), 70–6, 246–9. See
also R. R. West (ed.), *The eleventh and twelfth-century sections of Cott. MS. Galba E.
ii: The register of the abbey of St Benet of Holme*, Norfolk Record Soc., 2 (1932), 244,
246; J. A. Raftis, *The estates of Ramsey Abbey* (Toronto, 1957), 66; F. Baring,
'Domesday Book and the Burton Cartulary', *Eng. Hist. Rev.*, XI (1896), 98–100.

[2] J. B. Harley, 'Population trends and agricultural developments from the Warwickshire Hundred Rolls of 1279', *Econ. Hist. Rev.*, 2nd ser., XI (1958), 13.

[3] J. B. Harley, 'The settlement geography of early medieval Warwickshire', *Trans.
and Papers, Inst. Brit. Geog.*, XXXIV (1964), 115–30; B. K. Roberts, 'A study of medieval colonization in the Forest of Arden, Warwickshire', *Agric. Hist. Rev.*, XVI (1968),
101–13.

[4] H. E. Hallam, 'Some thirteenth-century censuses', *Econ. Hist. Rev.*, 2nd ser., X
(1958), 340; 'Population density in the medieval Fenland', *Econ. Hist. Rev.*, 2nd ser.,
XIV (1961), 79; and *Settlement and society: a study of the early agrarian history of south
Lincolnshire* (Cambridge, 1965), 198–200.

[5] G. C. Homans, 'Partible inheritance of villagers' holdings', *Econ. Hist. Rev.*,
VIII (1937–8), 48–56; D. S. Pitkin, 'Partible inheritance and the open fields', *Agric.
Hist.*, XXV (1961), 65–9; R. J. Faith, 'Peasant families and inheritance customs in
medieval England', *Agric. Hist. Rev.*, XIV (1966), 77–95; B. Dodwell, 'Holdings and
inheritance in medieval East Anglia', *Econ. Hist. Rev.*, 2nd ser., XX (1967), 53–66.

to settle down. The custom of primogeniture and of Borough English (succession through the youngest son) stabilised the population earlier and then encouraged migration. At the same time, an undivided holding might support more than the recognised tenant and his immediate family.[1] By the addition of a cottage or simply a few extra rooms (the usual solution in towns), aged parents, unmarried brothers and sisters, and even undersettlers and co-parceners were sometimes accommodated.[2]

English towns probably grew at about the same rate as the population as a whole in the 250 years after the Conquest. But, as the urban population may not even have been reproducing itself, any substantial increase in numbers depended upon immigration. 'The villages were the primary seedbeds of population'[3] which generally moved first towards local or regional centres and thence from smaller to larger places. Population movement in the twelfth and thirteenth centuries is difficult to quantify, but the place elements in personal names and the many references to *chevagium*, a tax on those living away from their home manor, suggest that it was considerable around towns and within areas of agricultural opportunity.[4] On the estates of Ramsey abbey, the emigration of serfs 'was a regular feature of manorial life from the time of the earliest extant court rolls'.[5]

Also of considerable social significance was the rapid increase in the clerical population.[6] There were possibly twenty times as many regular clergy in 1300 as in 1066. The rate of increase in the number of parish clergy, although less than this, was still comparatively high. The Church drew heavily upon the sons of the lesser nobility and, more important, it was one of the very few avenues of advancement open to the peasant. At one time in the village of Weston in Lincolnshire, 12 out of 68 known adult sons became clergymen.[7]

[1] J. Krause, 'The medieval household: large or small?', *Econ. Hist. Rev.*, 2nd ser., IX (1957), 420–32; H. E. Hallam (1961), 71.

[2] J. Amphlett *et al.* (eds.), *Court rolls of the manor of Hales*, Worcester Hist. Soc., 28 (1910), xliv, 167; M. Morgan, *The English lands of the abbey of Bec* (Oxford, 1946), 92.

[3] J. C. Russell, 'Late medieval population patterns', *Speculum*, XX (1945), 164.

[4] See pp. 127, 135 below.

[5] J. A. Raftis, *Tenure and mobility: studies in the social history of the medieval English village* (Toronto, 1964), 139.

[6] J. C. Russell, 'The clerical population of medieval England', *Traditio*, II (1944), 177–212.

[7] H. E. Hallam (1958), 356.

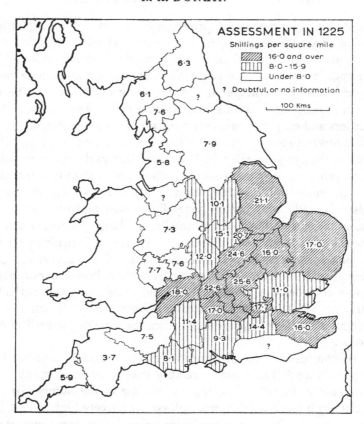

Fig. 21 Assessment in 1225
        Based on F. A. Cazel, 'The fifteenth of 1225', *Bull. Inst. Hist. Research*,
        XXXIV (1961), 66–81.

The famines of 1315–17 and 1321[1] are generally regarded as marking
the end of the 'economic thirteenth century' and as a significant turning
point in the population history of England. The incidence of famine is
also relevant to the question of short-term fluctuations in growth and to
the problem of 'overpopulation'. The numerous 'pestilences' before the
Black Death were mainly the result of famine,[2] and affected, first and

[1] H. S. Lucas, 'The great European famine of 1315, 1316 and 1317', *Speculum*, V
(1930), 341–77; J. C. Russell, 'Effects of pestilence and plague, 1315–85', *Comparative
Studies in History and Society*, VIII (1966), 464–73; S. L. Thrupp, 'Plague effects in
medieval Europe', *ibid.*, 474–83.
[2] C. Creighton, *A history of epidemics in Britain*, I (Cambridge, 1891), 8.

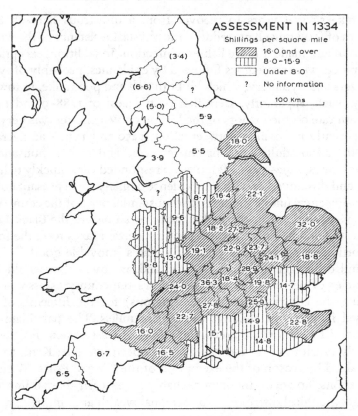

Fig. 22 Assessment in 1334
Based on J. F. Willard, 'The taxes upon moveables in the reign of Edward III',
Eng. Hist. Rev., XXX (1915), 73.
For the purpose of comparison with Fig. 21 the assessment at a tenth has been
converted to, and added to, that at a fifteenth. The figures for Cumberland,
Westmorland and Northumberland are for 1336; see p. 137 below. London
has been excluded in order to make the map comparable with Fig. 21; the
inclusion of London would bring the figure for Middlesex up to 77s. For a
more detailed map of the assessed wealth itself, see Fig. 35.

foremost, the poor. Famine, in turn, has been partly correlated with ad-
verse weather conditions,[1] especially heavy rainfall in late autumn and
early winter, leading to low yields or even to no yields at all. An inadequate return meant skimping on seed corn the following year, so it

[1] J. Z. Titow, 'Evidence of weather in the account rolls of the bishopric of Winchester, 1209–1350', *Econ. Hist. Rev.*, 2nd ser., XII (1960), 360–407.

might take several years to recover from a disastrous harvest. Severe shortages of food must have reduced the physical resistance of the working population and made it more liable to infection. In addition, exceptionally dry summers were sometimes followed by epidemics, probably of typhus-type fevers and dysentery, which hit both rich and poor alike, in town and country; such, apparently, were the 'pestilences' of 1288–9 and 1328–9. The death rate on the manors of the bishop of Winchester was very high, and apparently increasing, between about 1250 and 1350 – an average of 40 per thousand adults over the entire period and 52 per thousand over the last 55 years, 1292–1347.[1] Population responded very quickly to harvest failure; and this, together with the evidence of many small peasant holdings and of rising entry fines,[2] suggests that substantial parts of the country were indeed overpopulated in the half-century or so before the Black Death.

From the middle of the twelfth century, certain taxes took the form of a proportion of the assessed value of a man's 'movable goods' – chiefly grain and the larger domestic animals, and, in towns, household effects and articles of personal use. Fig. 21 shows the county assessments (i.e. excluding the contributions of the religious) for the fifteenth of 1225, expressed in terms of shillings per square mile.[3] The palatinates were exempt and London made no contribution. The assessment for Sussex is obviously much too low and has therefore been ignored; Kent, too, was undervalued by reason of the exemption of the Cinque Ports. With these qualifications, however, the map probably gives a fair overall impression of the geographical distribution of personal wealth and, indirectly, of the density of population in the early thirteenth century. It does not, of course, show the average wealth of individual taxpayers. The most notable feature is a band of medium and high values stretching from Gloucestershire, through the south Midlands to Lincolnshire and East Anglia, with very high values in a central block of country comprising Northamptonshire,

[1] M. M. Postan and J. Titow, 'Heriots and prices on Winchester manors', *Econ. Hist. Rev.*, 2nd ser., XI (1959), 399. See also S. L. Thrupp, 'The problem of replacement rates in late medieval English population', *Econ. Hist. Rev.*, 2nd ser., XVIII (1965), 107.

[2] M. M. Postan, 'Medieval agrarian society in its prime: England', in M. M. Postan (ed.), *Cambridge Economic History of Europe*, I (2nd ed., Cambridge, 1966), 553; J. Z. Titow, *English rural society, 1200–1350* (London, 1969), 73–8. See also B. F. Harvey, 'The population trend in England between 1300 and 1348', *Trans. Roy. Hist. Soc.*, 5th ser., XVI (1966), 23–42.

[3] F. A. Cazel, 'The fifteenth of 1225', *Bull. Inst. Hist. Research*, XXXIV (1961), 66–81; S. K. Mitchell, *Studies in taxation under John and Henry III* (New Haven, 1914), 159–69; and *Taxation in medieval England* (New Haven, 1951), 20ff.

Buckinghamshire and Bedfordshire (the last two were, however, grouped together, like some other counties).

In 1334, rural communities were taxed at the rate of one-fifteenth and boroughs and ancient demesnes at the rate of one-tenth.[1] For the purpose of Fig. 22, the tenth has been converted to a fifteenth and added to the rural assessment. Cheshire and Durham were exempt; Northumberland, Cumberland and Westmorland were excused as they had recently been devastated by the Scots; Kent and Sussex again were undervalued; and London has been excluded. In 1334, the axis of greater 'wealth' still lay diagonally across the country, from the Severn to the Wash. More significantly, the zone of higher values had expanded since 1225, notably towards the north-west and the south-west; only the counties of the north of England and of the extreme south-west stood apart, chiefly, perhaps, on account of the comparatively low density of population in these areas.

THE COUNTRYSIDE

Field arrangements

At the beginning of the fourteenth century, field arrangements fell into two broad categories: (i) those in which the arable was grouped into two or three, or occasionally more, large 'fields', and (ii) those in which such 'fields' were either absent or were organised on very special lines. The area covered by the latter included Kent[2] and East Anglia,[3] both settled very early and both characterised by large numbers of freemen and a comparatively fluid social structure. Here a man's holding, although probably divided, usually lay in one part of the total arable area. This arrangement perhaps developed from family tenements that were originally compact, subsequently breaking up under the influence of partible inheritance, and later being regrouped within the same general area. In Kent, where hamlets were characteristic, most holdings appear to have comprised small, enclosed pieces of land that were worked in severalty. Furthermore, in both

---

[1] J. F. Willard, 'The taxes upon moveables in the reign of Edward III', *Eng. Hist. Rev.*, XXX (1915), 69–74, and *Parliamentary taxes on personal property 1290 to 1334* (Cambridge, Mass., 1934); R. S. Schofield, 'The geographical distribution of wealth in England', *Econ. Hist. Rev.*, 2nd ser., XVIII (1965), 483–510.

[2] A. R. H. Baker, 'The field system of an East Kent parish', *Archaeologia Cantiana*, LXXVIII (1963), 96–117.

[3] D. C. Douglas, *The social structure of medieval East Anglia* (Oxford, 1927), 17–67; M. R. Postgate, 'The field systems of the Breckland', *Agric. Hist. Rev.*, X (1962), 80–101.

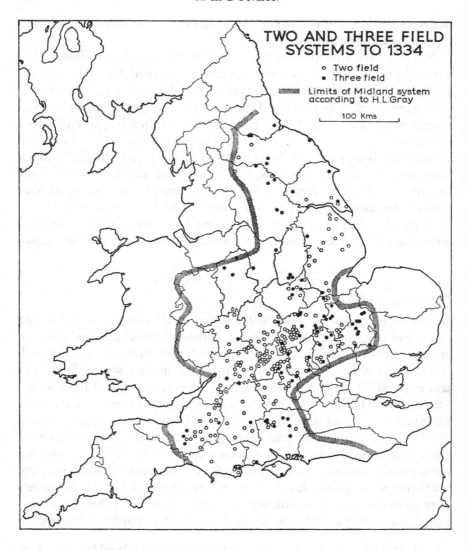

Fig. 23 Two- and three-field systems to 1334
         Based on H. L. Gray, *English field systems* (Cambridge, Mass., 1915), 450–509.

Kent and East Anglia, as well as in some northern and western districts,
varieties of infield-outfield cultivation existed. Under this system, an intensively cultivated central area was surrounded by periodically cultivated
'intakes' or 'breaks'.

The main area of the two- and three-field system was the midland belt of England (Fig. 23). The primary unit of plough-land was the 'land' or selion (*selio*), of varying width and length, but usually less than half an acre in area;[1] alone, or grouped with others, it formed a 'strip' – a tenurial term – where property was intermixed. Selions were often ridged, which had the effect of throwing off water, and Walter of Henley referred, about 1250, to the ridging of wet land.[2]

The next division up the scale was the furlong or *cultura*, a bundle of 'lands', uniformly cultivated. And, finally, over a large part of medieval England, furlongs were grouped into two or three open and common fields. How and when this came about are questions that are still unresolved.[3] The level of co-operation implied by the existence of two or three great 'fields' may be no older than the eleventh or twelfth century. Only then or later, it can be argued, did various *ad hoc* agreements between tenants with intermixed holdings give way, under the influence of population growth, to community-regulated cropping and the joint grazing of fallow. But in any event, where most of the village arable was arranged in 'fields', a man's strips were fairly equally distributed between them. At first, strips may also have lain in the same relation to one another within furlongs, but if so, this was all but obscured by partition and exchange by the late thirteenth century.[4]

In the twelfth century there were usually only two 'fields', half the cultivated area being fallow at any one time, but towards 1200 there is evidence of village territories organized into three 'fields'.[5] H. L. Gray was probably right in thinking that the one system normally developed from the other, and that the change first assumed importance during the thirteenth

[1] H. M. Clark, 'Selion size and soil type', *Agric. Hist. Rev.*, VIII (1960), 91–8.

[2] E. Lamond (ed.), *Walter of Henley's Husbandry* (London, 1890), 17.

[3] J. Thirsk, 'The common fields', *Past and Present*, XXIX (1964), 3–25; J. Z. Titow, 'Medieval England and the open field system', *Past and Present*, XXXII (1965), 86–102; J. Thirsk, 'The origin of the common fields', *Past and Present*, XXXIII (1966), 142–7; G. C. Homans, 'The explanation of English regional differences', *Past and Present*, XLII (1969), 32–4.

[4] S. Goransson, 'Regular open-field pattern in England and Scandinavian *Solskifte*' in *Morphogenesis of the agrarian cultural landscape, Geografiska Annaler*, XLIII (1961), 80–104; G. C. Homans, *English villagers in the thirteenth century* (Cambridge, Mass., 1942), 97.

[5] K. Major (ed.), *The registrum antiquissimum of the cathedral church of Lincoln: VII*, Lincoln Record Soc., 46 (1953), 58; D. M. Williamson, 'Kesteven villages in the Middle Ages', *Lincolnshire Historian*, II (1955), 11; D. Roden, 'Demesne farming in the Chiltern Hills', *Agric. Hist. Rev.*, XVII (1969), 17.

century.[1] Certainly by 1334 the three-field system had been widely adopted, as can be seen from the sample of evidence plotted on Fig. 23.

Three 'fields', implying a three-course husbandry, had several advantages over two.[2] Assuming that the fallow field was ploughed twice in both cases, the three-field system increased the area a peasant could cultivate by one-eighth and his productivity by one-third, or even by a half if he could apply his 'surplus' ploughing capacity to newly cleared land; it distributed field work more evenly over the year; and by leading to a more nearly balanced pattern of sowing in autumn and spring, it lessened the risk of famine. On the other hand, the three-field system, by reducing the period of fallow, might lead in time to impoverished soils and falling yields.

It is necessary to distinguish clearly between a three-course system of husbandry on the one hand, and arable arranged in three 'fields' on the other. Obviously the latter lent itself to the former, but a three-course rotation was also found in East Anglia[3] and Kent,[4] and indeed in places where there were more than three 'fields'. From the late thirteenth century, the term 'field' was often used very loosely. References to from five to forty 'fields' are fairly common, especially, but not exclusively, in areas of late colonisation. These 'fields' do not necessarily imply a more elaborate field course. A four-course rotation including leguminous crops, which were becoming more important,[5] was sometimes practised, but a three-course system was normal – winter wheat/rye, spring barley/oats/ legumes, and fallow; legumes were occasionally planted in place of fallow, a procedure known as *inhoking*. This three-course rotation could be imposed on any number of 'fields', from two upwards, provided that they were not regarded as fixed cropping or rotational units. During the thirteenth century, the furlong emerges as functionally the most important division of the arable. Furlongs might be grouped differently in different years to vary the proportions of spring and winter corn, or

[1] H. L. Gray, *English field systems* (Cambridge, Mass., 1915), 80.

[2] E. Lamond (ed.) (1890), 8; L. White, *Medieval technology and social change* (Oxford, 1962), 72.

[3] F. G. Davenport, *The economic development of a Norfolk manor, 1086–1565* (Cambridge, 1906), 27.

[4] T. A. M. Bishop, 'The rotation of crops at Westerham, 1297–1350', *Econ. Hist. Rev.*, IX (1938), 41.

[5] F. B. Stitt (ed.), *Lenton priory estate accounts, 1296–98*, Thoroton Soc., Record Series, 19 (1959), xl; W. O. Ault, 'Open field husbandry and the village community', *Trans. American Philos. Soc.*, LV (1965), 19.

a poor furlong might be put down to permanent pasture. Indeed where spring-sown crops predominated, as was commonly the case, a fixed pattern of 'fields' could not possibly have been maintained.

If, by exchange or purchase, a peasant managed to acquire an entire furlong, he might cultivate the land independently. To enclose it was, however, another matter, for this would interfere with grazing rights, arranged either between neighbours or involving the community as a whole. There was undoubtedly a great deal of regrouping of strips for convenience sake, but for a peasant to be able to enclose part of an open field was probably very rare. On the other hand, demesne that had been sufficiently consolidated might be enclosed and withdrawn from common field arrangements by the lord of the manor.[1]

Open and common fields also existed (and were possibly typical in the early fourteenth century) on the margins of the Celtic west – towards the south-west and along the Welsh border – and in parts of Lancashire and Northumberland. Furthermore, where lordship was strong in areas settled or resettled, even in post-Conquest times (for example, parts of the Vale of York and east Cheshire), the demesne was carved out and cast with peasant holdings into two or three 'fields'. The multiplication of 'fields' beyond this was significant only in that it implied greater flexibility of management; and such flexibility, clearly perceptible in the late thirteenth century, tended to reduce the contrast between the south-east and the rest of the country.

Not only arable but pasture was essential to the economy of a village. Animals were turned on to the harvest stubble; and beyond the open fields rights of rough grazing and common over the 'waste' were usually linked with arable holdings. Where, however, there was an abundance of pasture, as in the Fenland and parts of the south-west and the north, such arrangements were unnecessary and intercommoning between villages was also practised.[2]

In the absence of rotation grasses and root crops, the hay from meadows, bordering streams and enriched by floods, was a valuable commodity. Meadow was often worth four or five times as much as arable. Thus on the manor of Laughton (Sussex) in 1325, arable was valued at between 3d. and

---

[1] See, for example, S. R. Scargill-Bird (ed.), *Custumals of Battle abbey in the reigns of Edward I and Edward II*, Camden Soc., n.s. 41 (1887); A. M. Woodcock (ed.), *Cartulary of the priory of St Gregory, Canterbury*, Camden Soc., 3rd ser., 88 (1956), xvi, xvii, 83, 106.

[2] N. Neilson (ed.), *A terrier of Fleet, Lincolnshire* (British Academy, London, 1920); and 'Early English woodland and waste', *Jour. Econ. Hist.*, II (1942), 57–8.

6*d.* an acre, pasture at between 6*d.* and 18*d.*, and meadow at between 24*d.* and 30*d.*[1] Meadow was usually redistributed annually and held in severalty until the hay harvest when it reverted to common use.

Demesne and peasant holdings

An important distinction for the majority of people in the early Middle Ages was that between the strips held by the lord, the demesne, and the strips held directly by the villeins who also provided the demesne with the greater part of its labour. Freeholders paid rent for their holdings and rendered little or no services on the demesne. E. A. Kosminsky estimated from the Hundred Rolls of 1279 (covering parts of seven counties, mainly in the south Midlands) that demesne occupied 32% of the total arable, villein land 40%, and freeholdings 28%.[2] The proportions, however, varied greatly between manors. Moreover some manors included several vills and others amounted to only a fraction of a vill. To add to the complexity, there were sub-manors, formed by persons other than the *capitalis dominus* dividing a substantial holding into 'demesne' and allotments for sub-tenants. Finally, in East Anglia, the Danelaw and Northumbria, where there were large numbers of freemen and sokemen, the village rather than the manor was 'the essential form of rural organization'.[3]

By 1300 there were also wide differences in the size of holdings of both freemen and villeins. This was the combined result of: (i) sub-letting and the buying and selling of freehold land, even by villeins;[4] (ii) inheritance

[1] A. E. Wilson (ed.), *Custumals of the manors of Laughton, Willingdon and Goring*, Sussex Record Soc., 60 (1961), 79.

[2] E. A. Kosminsky, *Studies in the agrarian history of England in the thirteenth century* (Oxford, 1956), 89. See also E. A. Kosminsky, 'The Hundred Rolls of 1279–80 as a source for English agrarian history', *Econ. Hist. Rev.*, III (1931–2), 16–44; and 'Services and money rents in the thirteenth century', *Econ. Hist. Rev.*, V (1935), 24–45; M. M. Postan, 'The manor in the Hundred Rolls: essays in bibliography and criticism, XV', *Econ. Hist. Rev.*, 2nd ser., III (1950–1), 119–25.

[3] F. M. Stenton (ed.), *Documents illustrative of the social and economic history of the Danelaw* (British Academy, London, 1920), lxi.

[4] C. N. L. Brooke and M. M. Postan, *Carte nativorum: a Peterborough abbey cartulary of the fourteenth century*, Northamptonshire Record Soc., 20 (1946), xxixff; D. C. Douglas (ed.), *Feudal documents from the abbey of Bury St Edmunds* (British Academy, London, 1932), cliv; E. Toms (ed.), *Chertsey abbey court rolls abstract*, Surrey Record Soc., 21 (1937), xvi; D. G. Watts, 'A model for the early fourteenth century', *Econ. Hist. Rev.*, 2nd ser., XX (1967), 543–7; E. Miller, *The abbey and bishopric of Ely* (Cambridge, 1951), 144; P. R. Hyams, 'The origins of the peasant land market in England', *Econ. Hist. Rev.*, 2nd ser., XXIII (1970), 18–31.

customs, marriage settlements and piecemeal gifts of land; (iii) assarting and reclamation, which, however, did not generally keep pace with the growth of population. While a few peasants managed to acquire substantial holdings of up to 100 acres, the majority of holdings were getting smaller or at least were supporting an increasing number of folk.[1] Many were very small indeed, especially in south-eastern England. Thus at Wykes in Bardwell, Suffolk, in the latter half of the thirteenth century, there were 76 freeholders of whom 7 held 116 acres between them; 6 had messuages without land, and the remaining 63 possessed little more than $2\frac{1}{2}$ acres apiece.[2] Again at Chippenham, Cambridgeshire, in 1279, 59 of the 143 tenants (more than four times the number recorded in Domesday Book) had less than 2 acres apiece and must have depended on wage labour.[3] Forty-six per cent of the 13,500 peasant holdings examined by Kosminsky in the Hundred Rolls amounted to a quarter-virgate (a *ferling*, about 8 acres) or less.[4] In almost any part of the country, 8 acres of arable was probably near or below the minimum required for subsistence.[5]

The whole or part of the demesne of a manor was either worked directly, under a bailiff, by the immediate feudal lord, or it was leased or 'farmed' (*ad firmam*) for an agreed period at a fixed rent in money or kind or both. The former alternative was generally adopted when agricultural prices were rising, the latter when they were declining and a stable income was desirable. Demesnes were mostly being leased and/or were contracting in area in the late eleventh and the twelfth centuries.[6] Over this period there

[1] J. Z. Titow, 'Some evidence of thirteenth century population increase', *Econ. Hist. Rev.*, 2nd ser., XIV (1961), 222–3; and 'Some differences between manors and their effects on the condition of the peasants in the thirteenth century', *Agric. Hist. Rev.*, X (1962), 4.

[2] W. Hudson, 'Three manorial extents of the thirteenth century', *Norfolk Archaeology*, XIV (1901), 17.

[3] M. Spufford, *A Cambridgeshire community: Chippenham from settlement to enclosure* (Leicester, 1965), 29–30.

[4] E. A. Kosminsky (1956), 228; M. M. Postan (1966), 619, 625.

[5] J. Z. Titow (1969), 79.

[6] M. M. Postan, 'The chronology of labour services', *Trans. Roy. Hist. Soc.*, 4th ser., XX (1937), 169–93; 'Glastonbury estates in the twelfth century', *Econ. Hist. Rev.*, 2nd ser., V (1953), 359; and 'Glastonbury estates in the twelfth century: a reply', *Econ. Hist. Rev.*, 2nd ser., IX (1956–7), 106–18; R. Lennard, 'The demesnes of Glastonbury abbey in the eleventh and twelfth centuries', *Econ. Hist. Rev.*, 2nd ser., VIII (1955–6), 355–63; W. H. Hale (ed.), *The Domesday of St Paul's of 1222*, Camden Soc., 69 (1858), xxii; B. A. Lees (ed.), *Records of the Templars in England in the twelfth century* (British Academy, London, 1935), cix, cxx; B. Lyon, 'Encore le problème de la chronologie des corvées', *Le Moyen Age*, LXIX (1963), 615–30; J. A. Raftis (1957), 58, 86, 89.

Table 3.1.   *The Great Pipe Roll of 2 Henry II*

Shires	Danegeld due	In waste	Proportion of waste to total
	£ s. d.	£ s. d.	
Warwick	128 12 6	80 11 0	nearly 2/3
Notts and Derby	112 1 11	58 11 6	over 1/2
Leicester	99 19 11	51 8 2	over 1/2
Oxford	249 6 5	96 2 10	about 2/5
Bucks and Beds	316 6 8	107 14 3	over 1/3
Berks	205 11 4	77 16 7	over 1/3
Cambridge	114 14 9	34 3 0	about 1/3
Gloucester	184 1 6	59 3 6	nearly 1/3
Northampton	119 10 9	38 12 1	nearly 1/3
Hereford	93 15 6	19 3 6	over 1/4
Worcester	102 5 9	27 14 3	over 1/4
Wilts	389 13 0	99 16 9	about 1/4
Essex	236 8 0	61 4 0	about 1/4
Herts	110 1 3	29 17 4	nearly 3/11
Huntingdon	70 5 0	14 0 6	about 1/5
Stafford	44 1 0	8 8 0	nearly 1/5
Somerset	277 10 4	54 5 0	nearly 1/5
Surrey	184 16 0	30 12 9	nearly 1/6
Middlesex	85 0 6	10 0 0	nearly 1/8
Sussex	216 10 6	9 2 0	nearly 1/23
Kent	105 16 10	0 8 0	nearly 1/270

was much that was 'specially unfavourable to the direct management of the demesne'.[1] The years of civil war in Stephen's reign (1137–54) wrought havoc in the countryside. Areas were devastated at different times and the rate at which they recovered also varied, but the figures in table 3.1, taken from the Pipe Roll of 1156, leave no doubt about the wasted character of much of the country around the middle of the century.[2]

Towards the end of the twelfth century, the prices of grain and livestock began to rise[3] (Fig. 24). Bad weather and low yields, as in 1201 and

[1] M. M. Postan (1966), 585.

[2] H. W. C. Davis, 'The anarchy of Stephen's reign', *Eng. Hist. Rev.*, XVIII (1903), 630.

[3] D. L. Farmer, 'Some grain price movements in thirteenth-century England', *Econ. Hist. Rev.*, 2nd ser., X (1957), 214; 'Some livestock price movements in the thirteenth century', *Econ. Hist. Rev.*, 2nd ser., XXII (1969), 1–16; and 'Some price

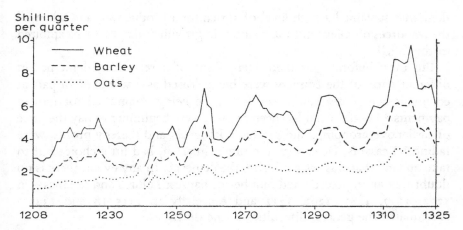

Fig. 24 Movement of grain prices, 1208–1325
Based on D. L. Farmer, 'Some grain price movements in thirteenth-century England', *Econ. Hist. Rev.*, 2nd ser., X (1957), 214.
Seven-year moving averages for sales of major grains.

1205,[1] led to sudden rises; then after a good year or two, prices dropped, but not usually to their previous level. The broad inflationary tendency has been attributed sometimes to the increasing amount of money in circulation, sometimes to a greater increase in population than in production, and sometimes to both.[2] But whatever the cause, as prices rose and real wages fell, demesne cultivation became more profitable. Lords began to recover lost portions of their demesnes and to expand them at the expense of any available villein holdings. The thirteenth century generally was an age of prosperous demesne farming when labour services were re-enforced and when the value of demesne arable as much as doubled. 'It was, above all, an age when landlords managed their estates as speculative enterprises geared to expanding markets, when "buoyant"

fluctuations in Angevin England', *Econ. Hist. Rev.*, 2nd ser., IX (1956), 34–9; A. L. Poole, 'Livestock prices in the twelfth century', *Eng. Hist. Rev.*, LV (1940), 285.

[1] D. L. Farmer (1956), 38; C. Easton, *Les hivers dans l'Europe occidentale* (Leyden, 1928), 57.

[2] M. M. Postan, 'The rise of a money economy', *Econ. Hist. Rev.*, XIV (1944), 123–34; J. Schreiner, 'Wages and prices in England in the later Middle Ages', *Scandinavian Econ. Hist. Rev.*, II (1954), 61–73; W. C. Robinson, 'Money, population and economic change in late medieval Europe', *Econ. Hist. Rev.*, 2nd ser., XII (1959), 63, 75; C. M. Cipola, 'Currency depreciation in medieval Europe', *Econ. Hist. Rev.*, 2nd ser., XV (1963), 413–22.

demesnes sustained a high level of agricultural production, and when all the resources of villein and wage labour in growing villages were exploited to the full.'[1]

But even before 1300 there were signs of a recession; demesnes in different parts of the country were being leased as a whole or in part at economic rents, and labour services were being commuted for money payments.[2] Landlords, it has been argued, were beginning to pay the price of earlier overcropping, and productivity levels, and thereby profits, were falling.[3] Peasants, faced with the same situation, had little choice but to take up land that was released from the demesne. In any event, there is no doubt that an unprecedented number of harvest failures and famines – in 1272, 1277, 1283, 1292, 1311 and especially in 1315–18 and 1321 – accentuated the general difficulties of the time.

Arable farming

Perhaps 20% of English peasants in the last quarter of the thirteenth century had holdings of approximately one virgate or 30 acres, i.e. at best 20 acres in cultivation at any one time. Such a holding probably produced at least 70 bushels of grain (over and above seed required for the following year), of which approximately 30 bushels would have been available for sale and to meet any fixed payments in kind.[4] During the thirteenth and early fourteenth centuries, demesne land, as we have seen, was producing for the market,[5] and sales of well over 50% of a harvest were common.

[1] E. Miller, 'England in the twelfth and thirteenth centuries: An economic contrast?', *Econ. Hist. Rev.*, 2nd ser., XXIV (1971), 2. See also N. Denholme Young, 'The Yorkshire estates of Isabella de Fortibus', *Yorks. Archaeol. Jour.*, XXXI (1934), 398; F. M. Page (ed.), *Wellingborough manorial accounts, 1258–1323*, Northants. Record Soc., 8 (1936), xxx; R. H. Britnell, 'Production for the market on a small fourteenth-century estate', *Econ. Hist. Rev.*, 2nd ser., XIX (1966), 380–7; J. A. Raftis (1957), 97, 103, 110–12.

[2] See, for example, T. A. M. Bishop, 'The distribution of manorial demesne in the vale of Yorkshire', *Eng. Hist. Rev.*, XLIX (1934), 395; R. H. Hilton, *The economic development of some Leicestershire estates in the fourteenth and fifteenth centuries* (Oxford, 1947), 40, 88; F. B. Stitt (ed.), *Lenton priory estate accounts, 1296–98*, Thoroton Soc., Record Series, 19 (1959), xviii.

[3] M. M. Postan (1966), 556–9, 588; J. Z. Titow (1969), 52–4.

[4] A. Ballard, 'Woodstock manor in the thirteenth century', *Vierteljahrschrift für Social- und Wirtschaftsgeschichte*, VI (1908), 439; A. T. Gaydon (ed.), *The taxation of 1297 for Barford, Biggleswade and Flitt hundreds*, Beds. Hist. Record Soc., 39 (1959), xx.

[5] R. H. Hilton, *A medieval society: the west Midlands at the end of the thirteenth century* (London, 1966), 78–9; J. A. Raftis (1957), 114.

Yet even lords of manors closely connected with the market, such as the bishop of Winchester, were drawing more from rents than from the sale of produce. 'This shows that the market was supplied in the first place by the peasant holdings.'[1] Money rents could be met only by selling surplus products, of which grain was by far the most likely.

Wheat and barley were the preferred bread grains, but, as they were the principal cash crops too, it is difficult to say how far they – and particularly wheat – were consumed in peasant households. Oats also was widely grown, but its use as human food probably varied more than that of the other two. It was the leading crop in Lancashire, and perhaps in other parts of the north and west, in the early fourteenth century. Rye was generally confined to marginal land or was mixed with other grain. Peas, beans, vetch, and occasionally flax,[2] made up the common range of field crops. In house gardens, cabbages, onions, leeks, lettuce, spinach, parsley and a few herbs were grown.[3]

Open-field husbandry and relatively inflexible rotations must often have made it difficult to take full advantage of local variations in soil and terrain.[4] Nevertheless, the value of the arable of a particular township might vary from a few pence per acre to several shillings, depending on whether the land was held in severalty or in common, and, presumably, on its accessibility and upon the nature of the soil. Work that has been done on the question of yields (based on manorial accounts of demesne) suggests (i) that average yields were very low by present standards, and (ii) that there was probably a downward trend from about the middle of the thirteenth century.[5] Lord Beveridge, using the account rolls of nine manors (in seven counties) covering the period 1200 to 1450, found that the average yield in bushels per acre was 9.36 for wheat, 14.32 for barley,

[1] E. A. Kosminsky, 'The evolution of feudal rent in England from the eleventh to the fifteenth centuries', *Past and Present*, VII (1955), 20; and E. A. Kosminsky (1956), 324.

[2] C. W. Foster and K. Major (eds.), *The registrum antiquissimum of the cathedral church of Lincoln: IV*, Lincoln Record Soc., 32 (1937), 36; B. Dodwell (ed.), *Feet of fines, Suffolk, 1199–1214*, Pipe Roll Soc., n.s., 32 (London, 1958), 259; L. C. Loyd and D. M. Stenton (eds.), *Sir Christopher Hatton's Book of Seals* (Oxford, 1950), 98.

[3] J. Clapham, *A concise economic history of Britain* (Cambridge, 1949), 85; J. J. Hunt, 'Two medieval gardens', *Proc. Somerset Archaeol. and Nat. Hist. Soc.*, CIV (1960), 91–101.

[4] See, however, A. T. Gaydon (ed.), xxxi; and A. Smith, 'Regional differences in crop production in medieval Kent', *Archaeologia Cantiana*, LXXVIII (1963), 147–60.

[5] M. M. Postan (1966), 557; J. Z. Titow (1969), 52–3. J. Z. Titow, *Winchester yields: a study in medieval agricultural productivity* (Cambridge, 1972), 12–33.

and 10.56 for oats, the ratios of seed to yield being 1:3.9, 1:3.8, and 1:2.4 respectively.[1] Allowing for a margin of error of 10 to 15%,[2] the general ratio may be put at about 1:3.5, and there is no reason to suppose that it differed much over the country as a whole. The average yield, then, was only one-quarter to one-third of what might reasonably be expected today, and the inevitable fluctuations from year to year were enough to make all the difference between profit or loss on the demesne and between bare subsistence or hunger for the small peasant. It is against a background of low and fluctuating yields from overcropped land that much of the work of medieval reclamation must be viewed.

Here and there, efforts were made to maintain and even to increase yields on demesne arable. Forms of centralised accounting were adopted,[3] and officials might refer to several treatises on estate management.[4] Targets were sometimes set; these varied from manor to manor, and occasionally from year to year for the same manor, and officials were 'fined' when they were not met.[5] The amount of seed sown per acre was related to the importance of the crop and to the quality of the land,[6] and records, even of yield ratios, were kept. Marling was frequently mentioned (circa 1095 is the earliest known date[7]) and chalk, sea sand and seaweed were applied when appropriate and available. We read of animal dung being collected and sold[8] and also being specially reserved when pasture was granted or

[1] W. Beveridge, 'The yield and price of corn in the Middle Ages' in E. M. Carus-Wilson (ed.), *Essays in economic history*, I (London, 1954), 16.

[2] R. Lennard, 'Statistics of corn yields in medieval-England', *Econ. Hist.*, III (1934–7), 173–92, 325–49; M. K. Bennett, 'British wheat yield per acre for seven centuries', *Econ. Hist.*, III (1934–7), 19.

[3] E. Stone, 'Profit-and-loss accountancy at Norwich cathedral', *Trans. Roy. Hist. Soc.*, 5th ser., XII (1962), 25–48; R. A. L. Smith, 'The central financial system of Christ Church, Canterbury, 1186–1512', *Eng. Hist. Rev.*, LV (1940), 253–69; and 'The *Regimen Scaccarii* in English monasteries', *Trans. Roy. Hist. Soc.*, 4th ser., XXIV (1942), 73–94.

[4] D. Oschinsky, 'Medieval treatises on estate management', *Econ. Hist. Rev.*, 2nd ser., VIII (1955–6), 296–309.

[5] J. S. Drew, 'Manorial accounts of St Swithun's priory, Winchester', *Eng. Hist. Rev.*, LXII (1947), 29ff.

[6] A. E. Wilson (ed.) (1961), xxviii.

[7] L. F. Salzman (ed.), *The chartulary of the priory of St Pancras of Lewes*, Sussex Record Soc., 38 (1932), 21.

[8] C. G. O. Bridgeman, 'The Burton abbey twelfth-century surveys', *William Salt Collections for Staffordshire* (1916), 273; H. E. Butler (ed.), *The chronicle of Jocelin of Brakelond* (London, 1949), 103; F. W. Weaver (ed.), *A cartulary of Buckland priory*, Somerset Record Soc., 25 (1909), 149; R. H. C. Davis (ed.), *The Kalendar of Abbot*

leased.[1] Grain straw was either ploughed in or burned in the fields. Composting was practised; and on assarts it was apparently a sign of permanent ownership. Arable might even be irrigated in particularly dry years.[2] The significance of all this must not, however, be exaggerated. For most of the time and in most places, not enough mineral and organic matter was being returned to the soil to produce any rise in output. Even the efforts of the most progressive landlords of the period of 'high farming' in the thirteenth century, efforts that have attracted much attention,[3] do not appear to have involved any really considerable capital investment in stock, buildings and equipment.[4]

Lay lords may be said to have eaten their way round their estates. A great lord would not, however, regularly visit all his manors in demesne, and produce was sometimes sent up to chosen centres. This was very characteristic of monastic estates. A religious house would always find it necessary to purchase some commodities, but the bulk came from its estates. A manor or group of manors would be responsible for a week or two's supplies.[5] The canons of St Paul's, about 1300, received each year 5,760 bushels of wheat, the same quantity of oats, and 1,080 bushels of barley.[6] Apart from such renders (*firmae*), grain was also moved between manors to make up deficiencies of one sort or other – the consequence, perhaps, of specialisation within the framework of a great estate. In some cases, it was more convenient or profitable to sell the entire surplus of one

Samson of Bury St Edmunds and related documents, Camden Soc., 3rd ser., 84 (1954), 76, 137–8; E. L. Sabine, 'City cleaning in medieval London', *Speculum*, XII (1937), 24–5.

[1] W. T. Lancaster (ed.), *Abstract of charters of Bridlington priory* (London, 1912), 94.

[2] K. Ugawa, 'The economic development of some Devon manors in the thirteenth century', *Trans. Devon Assoc.*, XCIV (1962), 644.

[3] D. Knowles, *The religious orders in England*, I (Cambridge, 1956), 45–54, 314; R. A. L. Smith, 'The Benedictine contribution to medieval English agriculture' in *Collected Papers* (London, 1947), 103–16, and *Canterbury cathedral priory* (Cambridge, 1943), 133–8.

[4] R. H. Hilton, 'Rent and capital formation in feudal society', *Deuxième Conférence Internationale d'Histoire Economique, Aix-en-Provence, 1962* (Paris, 1965), 33–68; M. M. Postan, 'Investment in medieval agriculture', *Jour. Econ. Hist.*, XXVII (1967), 576–87.

[5] R. H. C. Davis (ed.), *The Kalendar of Abbot Samson of Bury St Edmunds and related documents*, Camden Soc., 3rd ser., 84 (1954), 1.

[6] H. H. Hale (ed.), *The Domesday of St Paul's of 1222*, Camden Soc., 69 (London, 1858), xlviii.

manor and to bring in seed grain from outside which, in any case, accorded with what the agricultural writers advised.

Grain milling using water power was made more efficient by the introduction of the overshot wheel, probably in the late twelfth century.[1] The first mention of a windmill occurs in 1191 on the estates of the abbot of Bury St Edmunds.[2] It became common in the thirteenth century, particularly in eastern England. Hand mills were used by the peasants although they were usually expected to take their grain to the lord's mill.[3]

Pasture farming

Peasant livestock was of domestic rather than commercial importance; and peasant holdings, like much demesne, were usually understocked. M. M. Postan, using detailed tax assessments for parts of southern Wiltshire (1225) and the double Hundred of Blackbourne in Suffolk (1283), came to the following conclusions: (i) A bias towards sheep farming among some vills in both areas (one traditionally pastoral, the other arable) was more apparent than any broad contrast between the two. (ii) A disproportionate number of sheep were in the hands of a few men. (iii) In Wiltshire, a much higher proportion of downland was being used for grain, and a smaller proportion for sheep, than in the eighteenth century and perhaps before the thirteenth century. (iv) In both areas, cows were very evenly distributed and numbers were not necessarily smaller where there were comparatively large flocks of sheep. (v) The average number of all animals per taxpayer was low in relation to the needs of thirteenth-century husbandry – 15.6 sheep and 2.8 cows or calves in Wiltshire; 10.5 and 3.2 respectively in Suffolk.[4] Understocking appears to have been even more marked in parts of Bedfordshire,[5] and almost all enquiries show that very few peasants could put out a ploughing team of four oxen or horses, and

[1] M. Bloch, 'Avenement et conquêtes du moulin à eau', *Annales d'Histoire Economique et Sociale*, VII (1935), 539–40, 557–8; B. Gille, 'Les développements technologiques en Europe, de 1150 à 1400', *Cahiers d'Histoire Mondiale*, III (1956), 63–108.

[2] H. E. Butler (ed.), *The chronicle of Jocelin of Brakelond* (London, 1949), 59; J. Salmon, 'The windmill in English medieval art', *Jour. British Archaeol. Assoc.*, 3rd ser., VI (1941), 88–9; L. Deslisle, 'On the origin of windmills in Normandy and England', *Jour. British Archaeol. Assoc.*, VI (1851), 403–6.

[3] T. Stapleton (ed.), *Chronicon Petroburgense*, Camden Soc., 47 (1849), 66–7.

[4] M. M. Postan, 'Village livestock in the thirteenth century', *Econ. Hist. Rev.*, 2nd ser., XV (1962), 219–49.

[5] A. T. Gaydon (ed.), *The taxation of 1297 for Barford, Biggleswade and Flitt hundreds*, Beds. Hist. Record Soc., 39 (1959), 107–8.

that they were forced to combine with others or make do with fewer beasts.[1] The chief problem, over much of eastern, central and southern England, was shortage of pasture, and this was felt increasingly as village 'waste' and 'pasture' retreated before the plough.[2]

Demesne livestock comprised beasts of burden and traction, and animals kept primarily for their milk, hides, wool and flesh. The ox was the traditional field animal and it remained ubiquitous.[3] The nailed horseshoe had appeared in western Europe about the beginning of the tenth century, and the solid horse collar and modern tandem harness just a little later. Together they made the horse an economic as well as a military asset,[4] and in England the horse appears to have graduated from the harrow to the plough about the close of the twelfth century.[5] Thereafter, we hear of horses used alone, but more commonly in mixed teams, as Walter of Henley recommended.[6]

As demesnes contracted during the twelfth century, the numbers of livestock they carried also decreased. But this is not to say that the total number of animals, within the country or even within the limits of a particular estate, declined. In fact, the grand total was probably rising, while being differently distributed between lords, tenants and 'farmers', and, to some extent, between different kinds of lords. The houses of the new religious Orders, for example, were building up their flocks and herds at this time. During the thirteenth century, the numbers of demesne livestock increased with the trend towards 'high farming'.[7] Reproduction rates were low by modern standards and disease took a periodic toll, but there is little evidence of heavy autumn killings.[8] With what hay was available, supplemented occasionally by legumes and evergreens,

[1] H. G. Richardson, 'The medieval plough team', *History*, n.s., XXVI (1941–2), 287–96.

[2] See, for example, D. M. Stenton (ed.), *The earliest Lincolnshire assize rolls*, Lincoln Record Soc., 22 (1926), 10; W. Farrer (ed.), *Cartulary of Cockersand abbey*, Chetham Soc., n.s., 38 (1898), 85.

[3] J. H. Moore, 'The ox in the Middle Ages', *Agric. Hist.*, XXXV (1961), 90–3.

[4] G. Duby, 'La révolution agricole médiévale', *Revue de Géographie de Lyon*, XXIX (1954), 362; L. des Noettes, *L'Attelage et le cheval de selle à travers les âges* (Paris, 1931), 122, 237.

[5] B. A. Lees (ed.), *Records of the Templars in England in the twelfth century* (British Academy, London, 1935), lxxxii.

[6] E. Lamond (ed.) (1890), 11.

[7] See, for example, J. A. Raftis (1957), 117.

[8] F. M. Page, 'Bidentes hoylandie', *Econ. Hist.*, I (1929), 609.

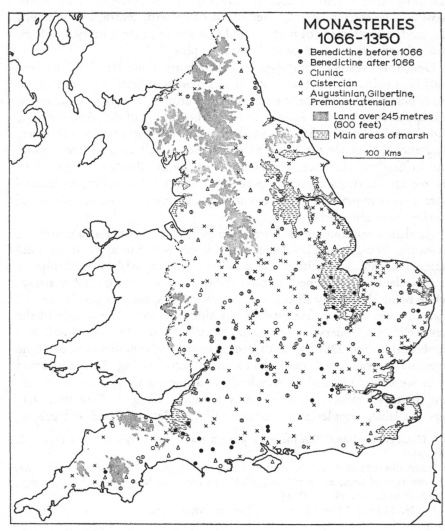

Fig. 25 Monasteries, 1066–1350
Based on D. Knowles and R. N. Hadcock, *Medieval religious houses: England and Wales* (London, 1953).

the great majority of animals were brought safely through the winter.[1]

Sheep were kept chiefly for their wool, and some of the larger land-owners grazed well over 10,000 in the late thirteenth and early fourteenth centuries. The bishop of Winchester had 15,000 or more as early as 1208–9, and nearly double this number by 1259.[2] Only Fountains among the Cistercian houses ever possessed anything like 15,000 sheep. The Cistercians (Fig. 25) were probably more deeply committed to wool growing, and certainly to supplying the overseas trade, than any of the other monastic Orders, but they by no means overshadowed all contemporary producers. At best, they supplied only 3 to 4% of all the wool exported at the close of the thirteenth century. Much of the overseas demand, not to speak of the continuing home market, must have been met by the small and middle-order producers, ranging from the prosperous peasant with a hundred or so sheep to the lord of several manors with two or three thousand. Descriptions of grants of pasture strongly suggest that a high proportion of these animals were fed on arable stubble, village commons, and enclosed blocks of lowland demesne pasture. The Cistercians grazed the high Pennines, but their arable granges also played an important part in sheep rearing.

The attention accorded sheep in connection with the overseas wool trade has tended to mask the importance of cattle as milk and hide producers. Cattle thrived on the rich sward of freshwater marsh; and in areas like Romney Marsh, the Isle of Thanet and the Fens, the *vaccaria* or cattle farm was an important feature of demesne farming. Cattle were also prominent in certain areas of woodland, scrub and rough pasture. In private chases and the royal forests they were considered less harmful than either sheep or goats. An unusually interesting development was the siting of vaccaries within and around the Pennines.[3] So far as we know, this commenced late in the twelfth century when there were 15 cattle stations in Wyresdale, and several in Nidderdale (belonging to Fountains) and Wensleydale (belonging to Jervalux). About the same time, another

[1] W. Farrer (ed.), *Early Yorkshire charters*, III (Edinburgh, 1916), 205; J. Radley, 'Holly as winter feed', *Agric. Hist. Rev.*, IX (1961), 89–92; J. Tait (ed.), *The chartulary or register of the abbey of St Werburgh, Chester*, Chetham Soc., n.s., 82 (1923), 422; W. Farrer (ed.), *Lancashire inquests, extents and feudal aids, 1205–1307*, Record Soc. Lancs. and Cheshire, 48 (1903), 285.

[2] E. Power, *The wool trade in English medieval history* (London, 1941), 34.

[3] R. A. Donkin, 'Cattle on the estates of medieval Cistercian monasteries in England and Wales', *Econ. Hist. Rev.*, 2nd ser., XV (1962), 31–53.

Cistercian house, Stanlaw, began to use the south-eastern corner of the forest of Rossendale for cattle rearing, and by 1295–6 the valleys of Blackburnshire, which included Rossendale, were dotted with 28 vaccaries handling nearly 2,500 head of cattle belonging to the earl of Lancaster alone.[1] In the early fourteenth century, the Pennine chases had a characteristically 'pioneer' economy, consisting of lead and iron working, stone quarrying, and cattle and horse rearing.

Among the smaller animals, the hare was native to England, and there is a Domesday reference to 'warren for hares' at Gelston in Lincolnshire. The rabbit, on the other hand, was a newcomer. It seems to have been introduced from France or possibly Spain in the twelfth century.[2] The consequences of its introduction for stretches of the English countryside were important. Rabbits were valued for their meat and skins, and in the thirteenth century we begin to hear of grants of 'free warren' from the king. That they were also destructive is seen, for example, in the complaint that 100 acres of arable at Ovingdean, in Sussex, were 'lying annihilated' as a result of their activities, and again that in West Wittering wheat had been devoured 'year after year by the rabbits of the bishop of Chichester'.[3]

The expansion of improved land

Clearing the wood and waste. Around the arable nucleus of the medieval township there normally lay clumps of high woodland and blocks of unimproved pasture and scrub (known collectively as 'waste'). These provided such natural products as firewood and turbary and common grazing for cattle, sheep and pigs. But throughout much of the twelfth and thirteenth centuries, and in virtually every part of the country, woodland and 'waste' were also being cut or reclaimed to provide, in particular, more arable. It has been estimated, for example, that about 1,000 acres were cleared in the manor of Witney (Oxfordshire) and the same amount in the manor of Wargrave (Berkshire) in the first half of the thirteenth century, and a further 660 acres and 680 acres respectively between 1256 and 1306 'when for all practical purposes reclamation may be said to have ended on both manors'.[4] In Laughton (Sussex), 975 acres were added to

---

[1] P. A. Lyons (ed.), *Two compoti of the Lancashire and Cheshire manors of Henry de Lacy, Earl of Lincoln, 24 and 33 Edward I*, Chetham Soc., 112 (1884), 129–42; R. B. Smith, *Blackburnshire* (Leicester, 1961), 8.

[2] E. M. Veale, 'The rabbit in England', *Agric. Hist. Rev.*, V (1957), 85–90.

[3] W. H. Blaauw, 'Remarks on the *Nonae* of 1340, as relating to Sussex', *Sussex Archaeol. Coll.*, I (1848), 62.

[4] J. Z. Titow (1962), 8.

the cultivated area between 1216 and 1325.[1] As the margins of possible cultivation were reached, in these and other places, township boundaries and common rights became more closely defined.

Not all inter-village waste was freely available to the peasantry. Lords of manors had rights there too, and perhaps the clearest witness to the pressure on unimproved land, and to the land shortage generally, was the Statute of Merton (1236) which permitted lords to enclose waste on condition that adequate amounts were left to their freeholders. Monasteries, especially those of the new Orders, were also given extensive grazing rights, even in places where they possessed no arable. And, finally, covering large parts of the country, there were the royal forests and private parks within which grazing and assarting were controlled in the interests of the chase. The royal forests were probably most extensive at the end of Henry II's reign (1189). Only Norfolk, Suffolk and Kent – areas of generally high population density and considerable numbers of freemen – were then without afforested land. During the thirteenth century, the larger tracts tended to disintegrate through piecemeal disafforestations and the granting of special franchises, and the overall area of forest was reduced by approximately one-third between 1250 and 1325.[2]

Even apart from the village lands that lay within their bounds, forests and chases (which included open moor and marsh as well as woodland) were by no means left unexploited.[3] The monasteries greatly benefited by grazing privileges, and by grants of land, timber, underwood and stone, or simply 'necessities'.[4] Eight of the ten royal foundations of the Cistercian Order were sited within or very close to forest bounds.[5] Permission

[1] J. S. Moore, *Laughton: a study in the evolution of the Wealden Landscape* (Leicester, 1965), 41–2. See also J. A. Raftis (1957), 71–4.

[2] The following contain useful maps: M. L. Bazeley, 'The extent of the English forest in the thirteenth century', *Trans. Roy. Hist. Soc.*, 4th ser., IV (1921), 140–72; N. Neilson, 'The forests' in J. F. Willard and W. A. Morris (eds.), *The English government at work, 1327–36*, I (Cambridge, Mass., 1940), 394–448; J. C. Holt, *The Northerners* (Oxford, 1961).

[3] J. R. Birrell, 'The forest economy of the honour of Tutbury in the fourteenth and fifteenth centuries', *Univ. Birmingham Hist. Jour.*, viii (1962), 114–34, and 'Peasant craftsmen in the medieval forest', *Agric. Hist. Rev.*, XVII (1969), 91–107. The most elaborate study of a particular forest is C. E. Hart, *Royal forest: a history of Dean's woods as producers of timber* (Oxford, 1966).

[4] H. A. Cronne, 'The royal forest in the reign of Henry I' in H. A. Cronne (ed.), *Essays in British and Irish history in honour of J. E. Todd* (London, 1949), 10.

[5] R. A. Donkin, 'The Cistercian settlement and the English royal forests', *Cîteaux: Commentarii Cistercienses*, XI (1960), 39–55, 117–32.

to 'assart and cultivate' was given to these and other houses, and also to many individuals; but, judging by the fines imposed for assarting, the amount of land actually cleared far outstripped what was permitted in advance. Where forest rolls of the late twelfth and early thirteenth centuries survive, 'they present an impressive picture of agrarian expansion at the expense of the king's rights'.[1] Yet it is probably true to say that the Crown, in its usual impecunious state, was not unwilling that assarting should continue, or other privileges be obtained, provided that they were defined and supervised and, of course, duly paid for.

Open fields were often enlarged during the twelfth and thirteenth centuries. Shares in assarts and assarts held 'in common' are occasionally mentioned. More frequent are references to assarts 'in the fields' or to furlongs bearing 'clearing' names. The addition of new furlongs, in which not necessarily the whole village had an interest, appears to have been the usual way in which fields were extended at this time.[2] This meant taking in some 10 to 30 acres at a time to form extra cropping units. The spread of three-field arrangements during the thirteenth century, permitting more land to be cultivated for the same amount of labour as before, and the more flexible grouping of furlongs at this and later times, may have encouraged joint assarting.

Areas that were once heavily wooded (the Weald, Wychwood, Feckenham, Arden and the like) or where there was much exploitable 'waste' (for example, parts of Cannock and of the Pennine and Cheviot foothills) often contain an unusually large number of hamlets and villages with name elements suggesting secondary settlement or 'clearing' (-stoc, -rydding, -leah, -feld, amongst others), and a fair proportion of these probably date from the twelfth or early thirteenth century.[3] In some parts of the country, for example in Suffolk, Essex and north Warwickshire, the presence of 'moated settlements' also appears to be associated with the colonisation of woodland.[4]

There is not much evidence of lords deliberately introducing settlers,

[1] J. C. Holt, 160; G. Wrottesley, *The pleas of the forest, Staffordshire*, William Salt Collections for Staffordshire, 5 (1884), 137ff.

[2] T. A. M. Bishop, 'Assarting and the growth of open fields', *Econ. Hist. Rev.*, VI (1935), 17.

[3] B. C. Redwood and A. E. Wildon (eds.), *Custumals of the Sussex manors of the archbishop of Canterbury*, Sussex Record Soc., 57 (1958), xxxi–xxxii; B. F. Brandon, Medieval clearances in the east Sussex Weald', *Trans. and Papers, Inst. Brit. Geog.*, XLVIII (1969), 136; J. B. Harley (1964), 122–5; B. K. Roberts, 101–13.

[4] F. V. Emery, 'Moated settlements in England', *Geography*, XLVII (1962), 385.

apart from monastic communities, to develop their estates, but some hamlets that appear to have been entirely informal growths may well have been encouraged, just as assarting by individuals was sometimes encouraged, by low initial rents.[1] About 1212–13, Robert Arsic settled a group of freemen on his demesne at Cogges in Oxfordshire, and each man was required to erect a house on his new holding.[2] The monks of La Charité (Caen) had *hospites* on their lands in Northamptonshire in 1107.[3] Temple Bruer, Lincolnshire, seems to have been promoted by the Templars about 1165.[4] In a late twelfth-century confirmation of land near Kniveton, Derbyshire, to one Sewall de Mungei we read, 'if Sewall wishes to settle (*erburgare*) this land, the men who dwell there shall have free common in wood and fields'.[5] Early in the following century (*circa* 1215), the earl of Chester permitted his barons to 'settle strangers' on their lands.[6] Some Cistercian houses also endeavoured to foster existing settlement by introducing new men.[7]

Lay and ecclesiatical lords cleared fresh land partly in order to expand or to consolidate their demesnes, and most of the large blocks of land that appear to have been cleared as a whole and kept in severalty turn out to be additions to the demesne. Lords organised assarting in several ways: (i) directly, by using permanent estate workers, hired labour or, in the case of the monasteries, lay brothers; (ii) by leasing cultivable land to peasants for a single lifetime;[8] (iii) by permitting peasants, villeins as well as freemen, to assart on condition of receiving back a proportion, usually one-third, of the improved land.[9] Land granted for the purpose of clearing

---

[1] P. A. Lyons (ed.), *Two compoti of the Lancashire and Cheshire manors of Henry de Lacy, Earl of Lincoln, 24 and 33 Edward I*, Chetham Soc., 112 (1884), xxi, 150. See also N. Neilson (1942), 61.   [2] L. C. Loyd and D. M. Stenton (eds.) (1950), 76–8.

[3] C. Johnson and H. A. Cronne (eds.), *Regesta Henrici Primi* (Oxford, 1956), 70.

[4] B. A. Lees (ed.), *Records of the Templars in England in the twelfth century* (British Academy, London, 1935), cxxxviii, clxxxii.

[5] F. M. Stenton, 'The free peasantry of the northern Danelaw', *Bulletin de la Société Royale des Lettres de Lund*, XXVI (1926), 164.

[6] J. Tait (ed.), *The chartulary or register of the abbey of St Werburgh, Chester*, Chetham Soc., n.s., 79 (1920), 103. See also J. E. A. Joliffe, 'Northumbrian institutions', *Eng. Hist. Rev.*, XLI (1926), 13.

[7] R. A. Donkin, 'Settlement and depopulation on Cistercian estates during the twelfth and thirteenth centuries', *Bull. Inst. Hist. Research*, XXXIII (1960), 141–65.

[8] R. Holmes (ed.), *The chartulary of St John of Pontefract*, Yorks. Archaeol. Soc., Record Series, 30 (1902), 409.

[9] N. Neilson (ed.) (1920), lxxxvi; R. P. Littledale (ed.), *Pudsay deeds*, Yorks. Archaeol. Soc., Record Series, 56 (1916), 168.

often had to be cultivated within a specified time and/or be made available for common grazing after cropping.[1]

An important example of clearing by direct methods is provided by the work of several Cistercian houses in the Vale of York about 1140. Here, devastated village lands, in some cases 1,000 acres or more in extent, were reclaimed and converted into large arable granges.[2] Much piecemeal assarting was also undertaken to consolidate holdings;[3] and scores of charters, in referring to 'old' and 'new' assarts, to assarts 'next to each other' or 'next to the moor', testify to the progress of reclamation. Furthermore, a good deal of the land granted to the Cistercian monks had only recently been cleared by peasant families.

Assarting and improvement chiefly benefited the older Benedictine foundations through increases in rent, whether or not the initiative had been theirs or the peasants. But a new house had to assemble an adequate demesne. The Benedictine community at Battle in Sussex, founded in 1066, vigorously developed its *banlieu*, to the extent of about 550 acres of fresh arable within a generation, of which just under half may have been recovered from peasants who had cleared and first cultivated the land.[4] The Church generally encouraged the extension of cultivation for, apart from what it gained directly, the amount of tithe was thereby increased. In the closing years of the twelfth century, Abbot Samson of the Benedictine house of Bury St Edmunds, 'cleared many lands and brought them into cultivation', so the abbey chronicle tells us.[5]

Peasant assarts tended to be small, a few acres at most. They could however be accumulated, and this helps to explain the occasional references to large peasant holdings.[6] In areas of late colonisation, assarts, hedged and dyked and worked in severalty, were frequently more im-

[1] A. Saltman (ed.), *The chartulary of Tutbury priory*, Hist. MSS. Commission (London, 1962), 133, 194; E. Toms (ed.), *Chertsey abbey court rolls abstract*, Surrey Record Soc., 21 (1954), 72.

[2] R. A. Donkin, 'The Cistercian grange in England in the twelfth and thirteenth centuries, with special reference to Yorkshire', *Studia Monastica*, VI (1964), 95–144; and 'The Cistercian Order and the settlement of the north of England', *Geog. Rev.*, LIX (1969), 403–16.

[3] R. A. Donkin, 'The English Cistercians and assarting, c. 1128–1350', *Analecta Sacri Ordinis Cisterciensis*, XX (1964), 49–75.

[4] E. Searle, 'Hides, virgates and tenant settlement at Battle abbey', *Econ. Hist. Rev.*, 2nd ser., XVI (1963), 290–300.

[5] H. E. Butler (ed.), *The chronicle of Jocelin of Brakelond* (London, 1949), 27.

[6] See, for example, J. F. Nichols, 'The extent of Lawling, A.D. 1310', *Trans. Essex Archaeol. Soc.*, XX (1931), 184, 187; B. K. Roberts, 112.

portant than intermixed strips, and elsewhere they formed a normal appurtenance to open-field holdings. Around the margins of many older settlements, and where peasant communities were still spreading into areas of woodland and waste, the landscape can hardly have appeared open. There is perhaps a general tendency to underestimate the amount of enclosed land in the England of 1300.

Most references to assarts do not mention the kind of land cleared. When they do, it is sometimes 'heath' or 'waste' or 'pasture' or what had earlier been abandoned – in other words, land that was probably 'marginal not only in location but also in quality'.[1] Nevertheless, the most common description is 'woodland'. A good stand of timber would generally indicate superior soils, although not all assarts were cultivated, and there was also the wood itself to be used or burned. But the most important reason why woodland was more often mentioned is to be found in the fact that regulations about its clearing, and above all about the clearing of oak,[2] were comparatively strict. Before 1300, in areas thinly wooded or closely settled, some concern was being expressed for supplies of constructional timber. There are one or two references to planned cutting, as at Charlbury in Oxfordshire where the woods were divided into seven parts, one to be cleared each year.[3] In places, even hedgerows, a ready source of firewood had to be protected.[4] Fortunately, in this respect, the tide of medieval colonisation was about to turn.[5] 'A progressive fall in the return for capital and labour expended' on clearing and reclamation may have been widely experienced.[6] Lord and peasant alike had sometimes been over-ambitious (or too hard pressed), and soon after 1300, parcels of poorer land, of the demesne at least, appear to have dropped out of cultivation altogether.[7]

---

[1] M. M. Postan (1966), 551.   [2] L. C. Loyd and D. M. Stenton (eds.) (1950), 322.

[3] H. E. Salter (ed.), *Eynsham Cartulary*, Oxford Hist. Soc., 51 (1908), xxxvii. See also P. L. Hall (ed.), *The cartulary of St Michael's Mount, Cornwall*, Devon and Cornwall Record Soc., 5 (1962), xxii; E. W. Crawley-Boevey (ed.), *Chartulary of Flaxley abbey* (Exeter, 1887), 32; C. W. Foster (ed.), *Final concords of the county of Lincoln*, II, Lincoln Record Soc., 17 (1920), 46; *Calendar Close Rolls, 1234–37* (London, 1908), 416.

[4] H. H. Hale (ed.), cxxvi.

[5] A. R. Lewis, 'The closing of the medieval frontier, 1250–1350', *Speculum*, XXXIII (1958), 475–83.

[6] E. Miller, 'The English economy in the thirteenth century: implications of recent research', *Past and Present*, XXVIII (1964), 31.

[7] A. R. H. Baker, 'Evidence in the *Nonarum Inquisitiones* of contracting arable lands in England during the early fourteenth century', *Econ. Hist. Rev.*, 2nd ser., XIX (1966), 518–32; E. Miller (1951), 100; M. M. Postan (1966), 558–9; B. F. Harvey, 23–42.

Draining the marsh. About 1135, the monks of Tutbury priory were allowed to make drainage ditches in the moor of Uttoxeter 'and to take branches off the willows and osiers overhanging the water for the improving of marshy fields'.[1] Work of this kind was going on all over the country in the twelfth and thirteenth centuries, sometimes more than doubling the value of small pieces of land. But the main scenes of reclamation were the great stretches of marsh in eastern and southern England: Holderness and the Humber Levels, the Fenland, the Somerset Levels, and certain coastal marshes, in particular Romney Marsh and Pevensey Levels. Altogether much land was improved and protected at great expense, and the chief driving force was the buoyant market for agricultural produce during the thirteenth century. Reclaimed marsh was either rented out on a contractual basis or kept in demesne. Common-field arrangements were very rarely introduced, but a special form of co-operation between tenants and landowners, combined with specific responsibility for the upkeep of drains and walls, was everywhere essential.

The improvement of Pevensey Levels proceeded piecemeal from the late twelfth century. But after 150 years the grazing of sheep on salt marsh and the production of salt in coastal pans were still typical features of the area.[2] New Romney Marsh, to the south-west of the Rhee Wall, and adjacent parts of Walland Marsh, were reclaimed in a series of 'innings' in the late twelfth and thirteenth centuries.[3] To protect the *terra conquesta*, walls of packed clay, stiffened with timber and hurdles and faced with stone and turf, were constructed. Already in the twelfth century, a body of marsh custom, the *consuetudo marisci*, concerned with responsibilities for water control and general maintenance, was in being.[4] In the late thirteenth century special officers administered this code, which was confirmed by the Crown in 1252 and used as a model for other areas, such as Holderness.[5] Before royal interest manifested itself in the middle of the thirteenth century, the work of organising reclamation fell largely on a group of ecclesiatics headed by the archbishop of Canterbury, and

[1] A. Saltman (ed.) (1962), 71.

[2] L. F. Salzmann, 'The inning of Pevensey Levels', *Sussex Archaeol. Coll.*, LIII (1910), 32–60.

[3] N. Harvey, 'The inning and the winning of the Romney Marshes', *Agriculture*, LXII (1955), 334–8; R. A. L. Smith, 'Marsh embankment and sea defence in medieval Kent', *Econ. Hist. Rev.*, X (1940), 29–37.

[4] M. T. Derville, *The Level and Liberty of Romney Marsh* (London, 1936), 6ff.

[5] S. G. E. Lythe, 'The organization of drainage and embankment in medieval Holderness', *Yorkshire Archaeol. Jour.*, XXXIV (1939), 282.

including the abbot of St Augustine's, Canterbury, the abbots of Roberts-bridge and Lesnes, and the prior of Bilsington. In all about 23,000 acres of New Romney and Walland Marsh were recovered by the end of the thirteenth century.[1]

Churchmen were also prominent improvers elsewhere. Meaux abbey was responsible for a whole network of drains in Holderness, some to make cultivation possible, but mostly to improve pasture and meadow.[2] A large part of the Somerset Levels was controlled by the cathedral church of Wells and the abbeys of Glastonbury, Muchelney and Athelney, and here again effort was mainly directed towards improving pasture. Agreements giving exclusive rights within blocks of land to particular villages were often a preliminary to systematic improvement by burning and by cutting drains. At the same time, the natural resources of the fen might be increased by the planting of osiers, alder and willow.[3]

In the Fenland itself, where perhaps as much was achieved as in all the other areas put together, the work of winning new land proceeded from the closely settled siltlands and from points within and around the margins of the peat fen. Seaward reclamation in the Lincolnshire Fenland commenced before the Norman Conquest, but the earliest recorded intake belongs to the period 1090–1110.[4] Work continued throughout the twelfth and thirteenth centuries – notwithstanding frequent extensive flooding – new land being divided into strips and held in severalty. Sea banks were co-operative ventures, though not necessarily the work of entire villages. The successive banks built out into the fen were, however, sometimes the work of several villages. H. E. Hallam in his study of the wapentake of Elloe has stressed the role of the peasantry. The local monasteries – Spalding, Crowland, Swineshead – were 'not alone in taking the initiative. They were regarded, at most, as very senior partners in the adventure.'[5] New land went largely into rent-paying tenancies

[1] N. Neilson (ed.), *The cartulary and terrier of the priory of Bilsington, Kent* (British Academy, London, 1928), 41.

[2] R. A. Donkin, 'The marshland holdings of the English Cistercians before c. 1350', *Cîteaux in de Nederlanden*, IX (1958), 262–75.

[3] P. J. Helm, 'The Somerset Levels in the Middle Ages', *Jour. British Archaeol. Assoc.*, 3rd ser., XII (1959), 39, 49; M. Williams, *The draining of the Somerset Levels* (Cambridge, 1970), 25–81.

[4] H. E. Hallam, *The new lands of Elloe* (Leicester, 1954), 18. See also H. E. Hallam (1965).

[5] H. E. Hallam (1954), 34. See also B. Lyon, 'Medieval real estate development and freedom', *American Hist. Rev.*, LXIII (1957–8), 47–61.

rather than demesne. Where the opportunities for reclamation were greatest, there partible inheritance worked best, and population was encouraged to remain. In the fens of Elloe wapentake, approximately 50 square miles were won between 1170 and 1240, and more than double this amount in the Lincolnshire Fenland as a whole. Nevertheless, population increased at such a rate that in Elloe about 1300 there were only 1 to 1½ acres of enclosed arable, pasture and meadow per head, after making due allowance for tithe payments and land in demesne and fallow.[1] Clearly many peasants must have had to rely heavily on open grazing and the natural products of the fen – on fish and dairy products, turbary and reeds and salt.[2] The technical resources of the time were limited, and great stretches of peat lay unreclaimed in the southern Fenland, in the Somerset Levels and elsewhere.

INDUSTRY

England was clearly a more industrial country in the fourteenth century than it had been in the eleventh. Most villages and all towns included some craftsmen. The manufacture of pottery and the tanning of leather were widespread and growing activities; so too was metal working and contemporary surnames suggest considerable specialisation.[3] The production of pewter was an essentially English craft, for which detailed regulations were drawn up in 1348. Bell-founding developed locally but served distant markets, and it led to the casting of cannon in the fourteenth century.[4] Shipbuilding also was making progress. In about 1300 a sea-going vessel of 200 tons was exceptional, but by the second half of the fourteenth century the average size was 250 to 300 tons.[5] Other developing industries included glass-making, and we hear of substantial quantities being produced at Chiddingford and neighbouring villages in Surrey and Sussex in the thirteenth and fourteenth centuries.[6]

[1] H. E. Hallam (1961), 78.

[2] D. J. Siddle, 'The rural economy of medieval Holderness', Agric. Hist. Rev., xv (1967), 40–5.

[3] See G. Frannson, Middle English surnames of occupation, 1100–1350 (Lund, 1935).

[4] L. F. Salzman, English industries of the Middle Ages (Oxford, 1923), 151, 156.

[5] K. M. E. Murray, 'Shipping' in A. L. Poole (ed.), Medieval England, 1 (Oxford, 1958), 185.

[6] L. F. Salzman (1923), 183. See also Calendar Patent Rolls, 1307–13 (London, 1894), 129.

Salt, a vital commodity in every household, was obtained either from sea water or from brine springs. The chief producing areas of maritime salt seem to have been in Lincolnshire[1] and Norfolk, but salterns (*salinae*) were widely distributed along the south and east coasts.[2] Men from the Cinque Ports supplied the London market.[3] The inland centres were in Cheshire[4] and Worcestershire;[5] they were frequently mentioned in the twelfth century and almost certainly increased in relative importance later.

Among the variety of industrial activities, three deserve to be singled out for special attention – mining, building, and the making of cloth.

Mining

Iron mining and smelting were located mainly in the Forest of Dean until the second half of the thirteenth century (Fig. 26). From the forges of St Briavels, picks, shovels, axes and nails were dispatched to Ireland on the occasion of Henry II's expedition in 1172, and horseshoes and other ironware went with Richard I's crusade in 1189–91.[6] While Forest of Dean iron remained important long after this, other centres of production, notably the central and southern Pennines, Cleveland[7] and the Weald of Sussex,[8] gradually came to the fore. Nevertheless, the total production

[1] E. H. Rudkin and D. M. Owen, 'The medieval salt industry in the Lindsey marshland', *Rep. and Papers, Lincs. Archit. and Archaeol. Soc.*, n.s., VIII (1959–60), 76–84; H. E. Hallam, 'Salt-making in the Lincolnshire Fenland during the Middle Ages', *ibid.*, 85–112.

[2] See, for example, L. Fleming (ed.), *The chartulary of Boxgrove Priory*, Sussex Record Soc., 59 (1960), 57, 161; W. Farrer (ed.), *Early Yorkshire charters*, II (Edinburgh, 1915), 69, 96, 98.

[3] H. T. Riley (ed.), *Munimenta Gildhallae Londoniensis, I (Liber Albus)* (Rolls Series, London, 1859), lxxxv.

[4] J. Tait (ed.), *The chartulary or register of the abbey of St Werburgh, Chester*, Chetham Soc., n.s., 79 (1920), 215–16; H. J. Hewitt, *Medieval Cheshire* (Manchester, 1929), 108–21.

[5] R. R. Darlington (ed.), *The cartulary of Worcester Cathedral priory*, Pipe Roll Soc., n.s., 38 (1968), 1, 42, 47, 83, 90, 225, 308; E. K. Berry, 'The borough of Droitwich and its salt industry, 1215–1700', *Univ. Birmingham Hist. Jour.*, VI (1957–8), 39–61.

[6] A. L. Poole, *From Domesday Book to Magna Carta* (Oxford, 1951), 82.

[7] B. Waites, 'Medieval iron working in northeast Yorkshire', *Geography*, XLIX (1964), 33–43.

[8] G. S. Sweeting, 'Wealden iron ore and the history of its industry', *Proc. Geol. Assoc.*, LV (1944), 2.

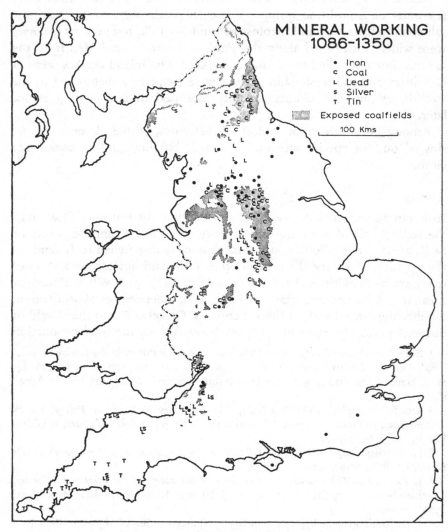

Fig. 26 Mineral working, 1086–1350
Based on miscellaneous sources.

fell well below the country's needs, and in the late thirteenth century England was still importing iron from Spain[1] and Sweden.

Coal had been worked in Roman times, but it was not until the end of the twelfth century that it seems to have been re-discovered.[2] It is mentioned among the exports to Bruges in 1200.[3] The thirteenth century saw a great expansion of coalmining, and coal was used increasingly in forging, in burning lime, in evaporating brine and in brewing and baking. Almost all the English coalfields had some mining activity by the end of the century. The coal of southern Northumberland was most often referred to on account of its export from the Tyne, but Fig. 26 shows that the Pennine fields also were important.

Lead was much used for roofing as well as for piping. There were four main producing areas: (i) Alston Moor on the borders of Cumberland and Northumberland; (ii) Derbyshire, from which lead was carried to Boston and Lynn for shipment; (iii) the Mendips; and (iv) the district around Bere Ferrers in South Devon where, between 1290 and 1340, there was a silver-lead mining boom to which miners came from other lead centres, notably Derbyshire.[4] Lead was also mined to some extent at other places – at Combe Martin in North Devonshire, in Shropshire and in Flintshire in North Wales.

Tin was produced in the south-west peninsula. It was worked in west Cornwall in pre-historic and Roman times. Documentary evidence begins in 1156 when tin was worked in Devon on the south-western flank of Dartmoor. 'From then until the end of the century the rich alluvial deposits of South-West Devon produced nearly all the tin of Europe.'[5] By the early thirteenth century, however, the main centres of production were in Cornwall, where they remained. Tin working was organised into 'stannaries', each with its own warden or supervisor, five in Cornwall and four in Devon.[6]

Quarrying and building

The Norman Conquest was followed by an enormous amount of new building, from humble and unrecorded extensions to most ambitious work

---

[1] H. F. Salzmann in W. Page (ed.), *V.C.H.*, *Sussex*, II (London, 1907), 242.
[2] W. Greenwell (ed.), *Boldon Buke*, Surtees Soc., 25 (1852), 5, 24.
[3] A. L. Poole, 81.
[4] W. G. Hoskins, *Devon* (London, 1954), 136.
[5] *Ibid.*, 131.
[6] G. R. Lewis, *The stannaries* (London, 1908).

on castles,[1] churches[2] and monasteries. All this represented an incalculably large investment of labour and wealth. The use of stone revolutionised military architecture, producing, first, the rectangular keep, and later, in the thirteenth century, the great curtain wall with half-round mural towers. English architecture generally was now fully exposed to powerful influences emanating from Normandy, the Paris Basin, and Burgundy. Much fine Anglo-Norman work still survives at Winchester, Norwich, Durham and Tewkesbury, for example, but even more was replaced during the thirteenth and fourteenth centuries by succeeding, and distinctively English, forms of Gothic. The records of the period abound in references to the right to take and to transport stone, often by water, and toll exemptions on building materials were freely given to religious communities. These included the Cluniacs, the Cistercians and the Canons Regular, who together were largely responsible for a more than tenfold increase in religious houses of men, excluding friaries, in the two and a half centuries after Domesday (Fig. 25). The building accounts of one Cistercian monastery for the period 1278–81 refer, in all, to more than 40,000 cartloads of stone.[3] Even in domestic building of a superior kind, stone was increasingly used for more than foundation work.[4]

The great majority of English parish churches were in existence by about 1200, but outlying chapels continued to be built long after this. Thus late in the thirteenth century, a certain Adam of Arden was allowed 'a private chapel in his manor house at Gayton [?-le-Marsh, Lincolnshire] since the road to his parish church was long and difficult to traverse, especially in winter'.[5] Churches and chapels, originally mainly of wood,

[1] R. A. Brown, 'Royal castle-building in England, 1154–1216', *Eng. Hist. Rev.* LXX (1955), 353–98; and 'A list of castles, 1154–1216', *Ibid.*, LXXIV (1959), 249–80; A. J. Taylor, 'Military architecture' in A. L. Poole (ed.), *Medieval England*, I (Oxford, 1958), 98–127; H. M. Colvin (ed.), *The history of the king's works*, 2 vols. (London, 1963).

[2] C. R. Cheney, 'Church building in the Middle Ages', *Bull. John Rylands Library*, XXXIV (1957); H. T. Johnson, 'Cathedral building and the medieval economy', *Explorations in Entrepreneurial History*, IV (1967), 191–210.

[3] J. Brownbill (ed.), *The ledger book of Vale Royal abbey*, Record Soc. Lancs. and Cheshire, 68 (1914), 198–203.

[4] M. E. Wood, *Thirteenth-century domestic architecture in England*, Supplement to *Archaeol. Jour.*, CV (1950); P. A. Faulkner, 'Domestic planning from the twelfth to the fourteenth century', *Archaeol. Jour.*, CXV (1958), 150–83.

[5] R. M. T. Hill (ed.), *The rolls and register of bishop Oliver Sutton, 1280–99*, Lincoln Record Soc., 48 (1954), 47. See also M. Gibbs (ed.), *Early charters of the cathedral church of St Paul, London*, Camden Soc., 3rd ser., 58 (1939), 148; F. Bradshaw,

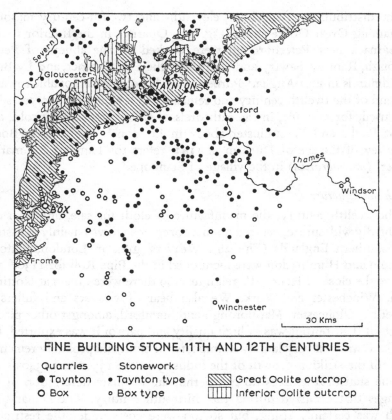

Fig. 27 Fine building stone from the Taynton–Box area in the eleventh and twelfth
centuries
Based on the map by E. M. Jope in A. F. Martin and R. W. Steel (eds.), *The
Oxford region* (Oxford, 1954), 114, and on additional information provided by
Professor Jope. Navigable portions of the Thames and Severn are shown.

were generally rebuilt in stone between the twelfth and the fourteenth
centuries, and in this work the influence of local schools of craftsmen can
sometimes be detected.[1]

The quarrying of building stone to meet distant demand can be seen

'The lay subsidy roll of 1296: Northumberland at the end of the thirteenth century',
Archaeologia Aeliana, 3rd ser., XIII (1916), 256.
[1] S. A. Jeavons, 'The pattern of ecclesiastical building in Staffordshire during the
Norman period', *Trans. Lichfield and South Staffs. Archaeol. and Hist. Soc.*, IV
(1962–3), 9–11.

in the distribution of surviving eleventh- and twelfth-century masonry containing Great Oolite stone (Fig. 27). Quarries in the Inferior Oolite at Barnack near Peterborough were worked by the abbeys of Peterborough, Ramsey, Sawtry and Crowland the Fen District, and by Bury St Edmunds in East Anglia.[1] Purbeck marble became fashionable towards the end of the twelfth century and remained so for the next 200 years.[2] It was used, for example, in the cathedrals of Chichester and Lincoln, and in St Paul's and Westminster abbey in London.[3] The Boldon Book, a survey of the see of Durham in 1183, refers to 'Lambert the marble cutter' (*marmorarius*) in the village of Stanhope.[4]

Cloth manufacture

In the twelfth century, the manufacture of cloth for sale, although undoubtedly widespread, centred upon a dozen or so towns mainly in eastern and southern England[5] (Fig. 28). Weavers' gilds in London, Oxford, Lincoln and Huntingdon were mentioned in the Pipe Roll for 1130,[6] and before the close of Henry II's reign in 1189 there were others at Nottingham, Winchester and York. We also hear of weavers and fullers at Beverley, Gloucester, Marlborough and Stamford, amongst other places. Much of their output was of high quality and part of it was exported, but England was still very largely an importer of cloth and probably remained so until the striking growth of the industry in the 1330s and 1340s.[7]

The statistical sources do not permit an adequate description of the changes that occurred during the thirteenth century. The demand for cloth was certainly rising, but so too were costs under the restrictive influence of the gilds; furthermore, the great Flemish centres of manu-

[1] L. F. Salzman (1923), 85; and *Building in England down to 1540* (Oxford, 1967), 119–39; C. Johnson and H. A. Cronne (eds.), *Regesta Regum Anglo-Normannorum*, II (Oxford, 1956), 189.

[2] D. Knoop and G. P. Jones, *The medieval mason* (Manchester, 1949), 11.

[3] L. F. Salzman (1923), 91–4.

[4] W. Greenwell (ed.), *Boldon Buke*, Surtees Soc., XXV (1852), 30.

[5] E. M. Carus-Wilson, 'The English cloth industry in the late twelfth and early thirteenth centuries', *Econ. Hist. Rev.*, XIV (1944–5), 32–50; E. Miller, 'The fortunes of the English textile industry in the thirteenth century', *Econ. Hist. Rev.*, 2nd ser., XVIII (1965), 64–82. For an early (1202) list of cloth towns, see D. M. Stenton (ed.), *Pipe Roll 4 John*, Pipe Roll Soc., n.s., XV (London, 1937), xx.

[6] J. Hunter (ed.), *Pipe Roll 31 Henry I* (Record Commission, London, 1929), 2, 48, 109, 194.

[7] E. M. Carus-Wilson, 'Trends in the export of English woollens in the fourteenth century', *Econ. Hist. Rev.*, 2nd ser., III (1950–1), 162–79.

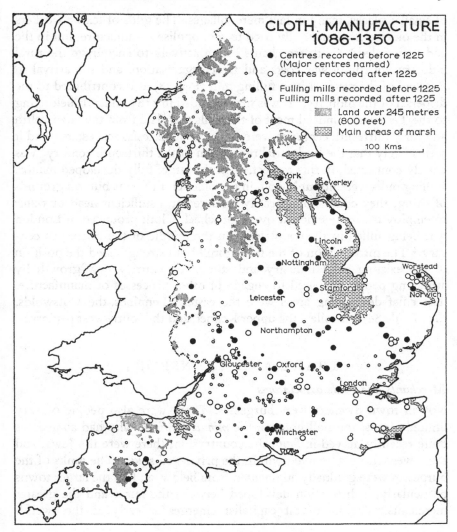

Fig. 28 Cloth manufacture, 1086–1350
Based on miscellaneous sources.

facture were almost certainly increasing their share of the English market.
Home production probably failed to expand very much and may even
have declined around the middle of the century. The most significant
development was the gradual shift in the distribution of cloth-making
away from the old-established towns and towards a much larger number

DHG

of smaller places, many in fact mere villages. The gilds of textile workers in the older centres naturally tried to monopolise manufacture, but in the end they failed.[1] Cloth merchants began actively to encourage the rural industry with its lower costs and freer organisation, and the arrival in England of the water-driven fulling mill (Fig. 28) also contributed to the dispersal of the industry.[2] It was first recorded in 1185 on lands belonging to the Templars;[3] indeed most of the mills that date from the late twelfth and early thirteenth centuries were on royal or ecclesiastical estates, and it seems likely that the 'industrial revolution of the thirteenth century' was closely connected, in this respect, with the large, fully developed manor. Fulling mills were not unknown in the older cloth towns but, on grounds of siting, they could not so easily arrange for a sufficient head of water to employ the more efficient overshot wheel. Cloth produced in London was being fulled in the countryside in the middle of the thirteenth century.[4] The introduction of the fulling mill both strengthened the position of the existing rural industry and stimulated entirely new growth by attracting people prepared to engage in other processes of manufacture. The chief developing areas were the eastern Pennines, the Cotswolds, the middle Severn valley, the upper Kennet, and the south-west peninsula.

TRADE AND TRANSPORT

Merchants and trade associations

Not all town dwellers were burgesses. There were also people of very limited means, the stall holders, the *minuti homines*, who had sometimes quite recently moved in from the countryside; there were the Jews; and there were those who lived within the privileged sokes. The ranks of the burgesses were gradually augmented from below, and, in the larger towns particularly, a distinction developed between the lesser and the greater merchants. The important capitalist emerges as early as the twelfth

[1] See E. E. Hirshler, 'Medieval economic competition', *Jour. Econ. Hist.*, XIV (1954), 52–8.

[2] E. M. Carus-Wilson, 'An industrial revolution of the thirteenth century', *Econ. Hist. Rev.*, XI (1941), 39–60; R. A. Pelham, 'The distribution of early fulling mills in England and Wales', *Geography*, XXIX (1944), 52–6.

[3] B. A. Lees (ed.), *Records of the Templars in England in the twelfth century* (British Academy, London, 1935), 51, 127.

[4] R. R. Sharpe (ed.), *Calendar of letter books of the City of London*, C (London, 1901), 52–3; H. T. Riley (ed.) (1859), 127–9.

century,[1] and wealth derived from commerce was often invested in rural property, thus reinforcing the association between town and country.

A great part of the country's internal trade in the twelfth and thirteenth centuries was in the hands of Englishmen. On the other hand, control of the principal items of international trade – the export of wool and the import of wine – rested largely with the Italians, the Flemings and the Gascons. During the second half of the thirteenth century, the Italians also acted as financiers to the Crown and to monasteries and lay lords. The Jews performed a similar service, and many monasteries were at some time or other in their debt for large sums, the security offered being usually wool or land. Jewries were located almost exclusively in the larger towns. Names mentioned in twelfth-century records indicate important colonies at London, Lincoln, Norwich, Gloucester, Northampton and Winchester. To these should be added Canterbury, Worcester, Oxford and York, which figure prominently in the tallage lists of the thirteenth century.[2] The Jews facilitated the circulation of wealth and the transfer of land; but as usurers they were unpopular, and steps were taken as early as about 1230 to prevent their acquiring further urban property. They were expelled from England in 1290.[3]

The earliest specific reference to the Gild Merchant (*Gilda Mercatoria*) is dated about 1100.[4] Although the association continued to be sanctioned throughout the twelfth and thirteenth centuries, it was not found in the majority of boroughs. The principal advantages of membership were freedom of trade at all times and exemption from the borough tolls; and toll exemptions enjoyed by a borough elsewhere in the country might be confined to gildsmen. The Jews were virtually excluded; they had their counterpart, but it is mentioned only once in England, at Canterbury in 1266.[5] The term 'merchant' appears at first also to have included most, if not all, craftsmen, but weavers and fullers, unlike dyers, were not members in the thirteenth century; on the other hand they had their own craft gilds at this time, whereas the dyers apparently did not. The craft

[1] H. A. Jenkinson, 'Money lenders' bonds of the twelfth century' in H. W. C. Davis (ed.), *Essays in history presented to R. L. Poole* (Oxford, 1927), 190–210.

[2] H. G. Richardson, *The English Jewry under the Angevin kings* (London, 1960), 1–22.

[3] P. Elman, 'The economic causes of the expulsion of the Jews in 1290', *Econ. Hist. Rev.*, VII (1936–7), 145–54.

[4] C. Gross, *The Gild Merchant*, I (Oxford, 1890), 5.

[5] L. Rabinowitz, 'The medieval Jewish counterpart to the Gild Merchant', *Econ. Hist. Rev.*, VIII (1937–8), 180–5.

gilds,[1] first mentioned in the reign of Henry I (1100–35), came into prominence in the late thirteenth century when the Gild Merchant was already beginning to decline as an economic, if not as a social, force. Its influence and power passed, however, not so much to the gilds as to the borough courts and councils which were largely controlled by the same merchant families.

Internal trade

The chief commodities of internal trade were grain, fish and salt, wool and cloth, and metals. Large quantities of grain were moved across country to satisfy manorial needs and obligations.[2] That the trade in grain was also considerable can be inferred from the recurring references to its sale in manorial accounts.[3] Towards the close of the thirteenth century, grain sales commonly accounted for 20 to 30% of manorial profits. The trade was organised primarily on a local and regional basis, with cornmongers, first mentioned in the early thirteenth century, operating between producers and rural markets and the main centres of demand – the towns.[4] There was also some inter-regional movement, particularly coastwise.

Much of the internal trade of the country was carried on at markets and fairs. The right to establish either was a royal prerogative, and during the twelfth and thirteenth centuries approximately 2,500 market charters were granted or confirmed.[5] Every borough eventually had a market, but the larger number of charters were to rural communities. A market lasted a single day and usually took place once a week. The thirteenth-century

[1] E. Power, 'English craft gilds of the Middle Ages', *History*, n.s., IV (1919–20), 211–14.
[2] R. Lennard, 'Manorial traffic and agricultural trade in medieval England', *Proc. and Jour. Agric. Economics Soc.*, V (1938), 259–77.
[3] See, for example, E. M. Halcrow, 'The decline of demesne farming on the estates of Durham cathedral priory', *Econ. Hist. Rev.*, 2nd ser., VII (1954–5), 356.
[4] N. S. B. Gras, *The evolution of the English corn market* (Cambridge, Mass., 1926); E. Kneisel, 'The evolution of the English corn market', *Jour. Econ. Hist.*, XIV (1954), 46–52.
[5] *Report on markets and fairs in England and Wales*, pt 1, Min. of Agric., Econ. Ser., 13 (London, 1927), 7; O. S. Watkins, 'The medieval market and fair in England and Wales', *Y Cymmrodor*, XXV (1915), 21–74; J. L. Cate, 'The Church and market reform in England during the reign of Henry III' in J. L. Cate and E. N. Anderson (eds.), *Medieval historiographical essays in honor of J. W. Thompson* (Chicago, 1938), 27–65; R. A. Donkin, 'The markets and fairs of medieval Cistercian monasteries in England and Wales', *Cistercienser-Chronik*, LXIX (1962), 1–14.

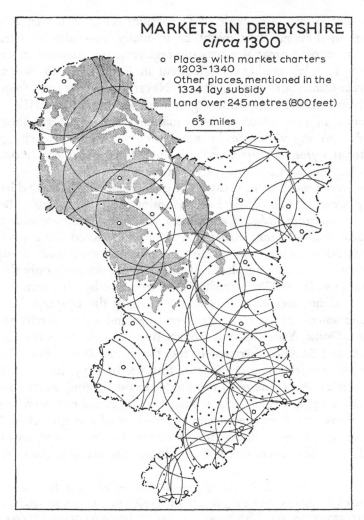

Fig. 29 Markets in Derbyshire *circa* 1300
      Based on B. E. Coates, 'The origin and distribution of markets and fairs in
      medieval Derbyshire', *Derbyshire Archaeol. Jour.*, LXXXV (1965), 92–111.

lawyer Bracton held that markets should not be less than '6 miles and
a half and a third part of a half' (i.e. 6⅔) miles apart, but his argument is
difficult to follow.[1] In fact, markets were often closer than six miles, as

---

[1] L. F. Salzman, 'The legal status of markets', *Cambridge Hist. Jour.*, II (1928), 210.

can be seen from the example of Derbyshire[1] (Fig. 29). Neighbouring
markets on different days might be mutually beneficial by attracting
merchants into an area; this was the argument put forward in 1252 to
justify a Tuesday market at Wingham in Kent, close to Wednesday
markets at Canterbury and Sandwich.[2] Nevertheless, there was frequently
keen rivalry between neighbouring market centres. Edward III's charter
to London in 1327 stated 'that no market from henceforth should be
granted...to any within seven miles in circuit of the said city';[3] and
a market at Lyme in Dorset was condemned because it was 'more than
5 miles but less than 6' from that of Bridport.[4]

Fairs served much wider areas.[5] The greatest fairs in the early thirteenth
century were at Bristol, St Ives (Huntingdon), Winchester, Boston,
Stamford and Stourbridge (Cambridge). As English trade expanded, fairs
became increasingly important; they commonly lasted for a week, but
some extended for a month or more. Booths and stalls were set up and
each trade had its allotted place. Here were to be seen not only tin from
Cornwall, salt from Worcestershire, lead from Derbyshire, iron from the
Forest of Dean, and wool from many parts of the country, but silks,
cloth and spices from abroad. Between 1270 and 1315, merchants from
Cologne, Douai, Ypres, Ghent, Rouen, Bruges, St Omer, Caen, Dinant,
Louvain and Malines frequented the fair at St Ives.[6] But with the
fourteenth century came changes in fortune. In 1335, it was said that
'foreigners do not come to St Botolph's fair (at Boston) as they used to
do'; and in 1363, we hear that the fair of St Ives had not been held 'for
twenty years and more' because of the absence of foreign traders.[7] It is
clear that fairs were declining in importance. As towns grew, they themselves became permanent marts. In the fourteenth century, the provincial

[1] B. E. Coates, 'The origin and distribution of markets and fairs in medieval
Derbyshire', *Derbyshire Archaeol. Jour.*, LXXXV (1965), 92–111. See also R. H. Hilton,
A medieval society: the west Midlands at the end of the thirteenth century (London, 1966),
172.
[2] L. F. Salzman, *English trade in the Middle Ages* (Oxford, 1931), 129; H. S. A.
Fox, 'Going to town in thirteenth-century England', *Geog. Mag.*, XLII (1970), 666.
[3] W. de G. Birch (ed.), *The historical charters and constitutional documents of the
City of London* (London, 1887), 58.
[4] L. F. Salzman (1928), 211.
[5] See, for example, A. H. Thomas (ed.), *Calendar of early mayors' court rolls,
1298–1307* (Cambridge, 1924), 154.
[6] E. Lipson, *The economic history of England*, I (London, 1929), 222.
[7] *Ibid.*, 234.

trade of London merchants 'was by no means confined to the great fairs; they were constantly trading all over the country'.[1]

Goods were moved by pack-horse and by cart.[2] The four-wheeled wagon (*longa caretta*) became common in the thirteenth century. Livestock were sometimes driven several hundred miles between estates or to meet the king's needs.[3] Although not much is heard of the making of roads[4] or even of their maintenance,[5] 'there is little evidence that men complained of bad roads and slow travel in the thirteenth century'.[6] This was partly because rivers were used to convey such bulky commodities as grain, wool, reeds,[7] timber and stone. Roads were also characteristically broad and diversions could easily be arranged. Suitable fording or bridging points, on the other hand, were few and far between, and it is not surprising that there was a considerable amount of bridge-building or rebuilding in stone from the late thirteenth century onwards. Some idea of the main arterial roads and how they focused on London may be obtained from the fourteenth-century Gough map (Fig. 39). As early as the twelfth century, there are references to 'the London Road' in various towns, for example the *Via Londiniensis* in Missenden (Buckinghamshire) and *Londenestret* in Gamlingay (Cambridgeshire).[8] It is also significant that the Gough map, compiled with practical ends in mind, showed a large number of rivers. The most important were those which converged on the Humber estuary and those of the Fenland together with the Thames and the Severn.

[1] S. Thrupp, 'The grocers of London, a study of distributive trade' in E. Power and M. M. Postan (eds.), *Studies in English trade in the fifteenth century* (London, 1933), 273.

[2] J. F. Willard, 'The use of carts in the fourteenth century', *History*, n.s., XVII (1932–3), 246–50.

[3] M. H. Mills (trans.) and R. S. Brown (ed.), *Cheshire in the Pipe Rolls, 1158–1301*, Record Soc. Lancs. and Cheshire, 92 (1938), 174.

[4] F. M. Stenton, 'The road system of medieval England', *Econ. Hist. Rev.*, VII (1936), 6.

[5] D. S. Bland, 'The maintenance of roads in medieval England', *Planning Outlook*, IV (1957), 5–15.

[6] D. M. Stenton, 'Communications' in A. L. Poole (ed.), I, 201.

[7] N. Neilson, *Economic conditions on the manors of Ramsey abbey* (Philadelphia, 1899), 80. See also M. W. Barley, 'Lincolnshire rivers in the Middle Ages', *Rep. and Papers, Lincs. Archaeol. and Architect. Soc.*, n.s., I (1938), 1–22.

[8] F. M. Stenton et al., *Norman London* (Hist. Assoc., London, 1934), 21.

Overseas trade

Imports included woad from Picardy,[1] cloth from Flanders, alum from the Mediterranean lands, iron from Spain and Sweden, and a wide range of luxury goods. In early Norman times, wine appears to have been imported mainly from the Seine Basin, Burgundy and the Rhineland, but after the acquisition of Gascony in 1152, this area provided an ever-increasing quantity.[2] By the early thirteenth century, the greater part of England's supplies came from Gascony. Viticulture in England declined as the trade with Gascony expanded; the latter involved around 20,000 tuns annually in the 1330s, by which time English merchants had captured the lion's share.[3]

Among exports, lead and tin,[4] hides and dairy products[5] were shipped in significant quantities. Newcastle was the leading port for hides; they were probably the basis of its prosperity at the close of the thirteenth century,[6] which suggests that the scattered references to cattle rearing in the Cheviots and along the slopes of the northern Pennines do not reflect its real importance. Grain also was exported. Even as early as 1198, heavy duties on grain sent to Flanders indicate 'an export trade of real magnitude', expecially through East Anglian ports.[7] Lynn lay at the north-eastern end of what was probably the main area of surplus grain, and it was the most important shipping point in the early fourteenth century when port statistics first become available.

The principal item in England's foreign trade in the thirteenth century was wool. In 1273, export licences for nearly 33,000 sacks – perhaps

[1] E. M. Carus-Wilson, 'La guède français en Angleterre: un grand commerce du Moyen Age', *Revue du Nord*, xxxv (1953), 89–105.

[2] E. M. Carus-Wilson, 'The effects of the acquisition and of the loss of Gascony on the English wine trade', *Bull. Inst. Hist. Research*, xxi (1946–8), 145–54; M. K. James, 'The fluctuations of the Anglo-Gascon wine trade during the fourteenth century', *Econ. Hist. Rev.*, 2nd ser., iv (1951), 170–96.

[3] M. K. James, 'The medieval wine dealer', *Explorations in Entrepreneurial History*, x (1957), 45–53; N. S. B. Gras, *The early English customs system* (Cambridge, Mass., 1918), 210. *Studies in the medieval wine trade* (ed. E. M. Veale), Oxford, 1971, brings together M. K. James's work on this subject.

[4] D. M. Stenton (ed.), *Pipe Roll, 10 Richard I, 1198*, Pipe Roll Soc., n.s. (London, 1932), xxxii, 181–2.

[5] H. T. Riley (ed.), *Memorials of London, 1276–1419* (London, 1868), xli.

[6] F. Bradshaw, 273.

[7] D. M. Stenton (ed.), *Pipe Roll, 10 Richard I, 1198*, Pipe Roll Soc., n.s., ix (London, 1932), xiv.

three-fifths of the country's total production – were issued.[1] Between Michaelmas 1304 and Michaelmas 1305 actual shipments reached a record total of over 46,000 sacks. Our knowledge of the organisation of the trade comes largely from the records of transactions between monasteries and other landowners on the one hand, and shippers, or their agents, on the other. The foreign merchants were mainly Flemish or northern French before about 1275, and thereafter Italian.[2] Buyers travelled about inspecting stocks, and producers were encouraged to sort their wool into three grades.[3] Forms of credit – deferred payments for goods sold or advances on future delivery – were used from the second half of the thirteenth century.[4] The Italians also pioneered the use of 'bills of exchange' and 'assignments' or transferred obligations. Although such refinements were much more fully developed in the later fourteenth and in the fifteenth centuries, they were already facilitating overseas financial transactions by 1334.

The bulk of the wool of the larger producers appears to have been dispatched directly to agreed shipping points (if it had already been bought) or to a central depot, rather than disposed of through local markets. By the close of the thirteenth century many Englishmen were involved in the overseas trade, and their position was strengthened in 1303 by the imposition of an additional duty, the 'New Custom', on foreigners importing or exporting goods, including wool. Individual English merchants seem generally to have been associated with particular areas of the country, and they were probably in closer touch with the small and medium producers than were the Italians and the Flemings.

The distribution of the home towns of English merchants licensed to export wool between 1273 and 1278, and of the religious houses included in a late thirteenth-century list as having a surplus of wool, suggest that

[1] A. Schaube, 'Die Wollausfuhr Englands vom Jahr 1273', *Vierteljahrschrift für Sozial- und Wirtschaftsgeschichte*, VI (1908), 68. Export figures are tabulated in E. M. Carus-Wilson and O. Coleman, *England's Export Trade, 1275–1547* (Oxford, 1963).

[2] E. von Roon-Bassermann, 'Die erste Florentiner Handelgesellschaften in England', *Vierteljahrschrift für Sozial- und Wirtschaftsgeschichte*, XXXIX, 1952, 97–128; R. A. Donkin, 'The disposal of Cistercian wool in England and Wales during the twelfth and thirteenth centuries', *Cîteaux in de Nederlanden*, VIII (1957), 109–31, 181–202.

[3] R. A. Donkin, 'Cistercian sheep farming and wool sales in the thirteenth century', *Agric. Hist. Rev.*, VI (1958), 2–8.

[4] M. M. Postan, 'Credit in medieval trade', *Econ. Hist. Rev.*, I (1928–9), 234–61; and 'Private financial instruments in medieval England', *Vierteljahrschrift für Sozial- und Wirtschaftsgeschichte*, XXIII (1930), 26–75.

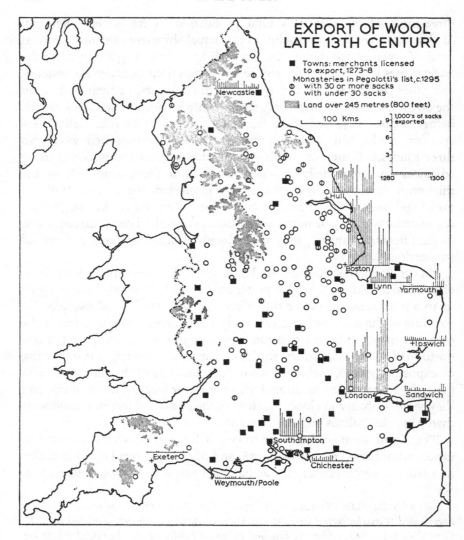

Fig. 30 Export of wool in the late thirteenth century
Based on E. M. Carus-Wilson and O. Coleman, *England's export trade, 1275–1547* (Oxford, 1963). The sources are: (1) *Calendar Patent Rolls, 1272–81* (London, 1901), 13–27; (2) *Calendar of various Chancery Rolls, 1277–1326* (London, 1912), 2–11; (3) A. Evans (ed.), F. B. Pegolotti: *La practica della mercatura* (Cambridge, Mass., 1936).

interest in the export trade lay mainly in the south and east of the country, with an extension into the southern Welsh border (Fig. 30). This cannot generally be explained in terms of the qualities of wool available;[1] it owes more to the siting of the traditional centres of cloth working, and most to the fact that, apart from the Welsh border, which produced the very best wool, no part of the area was inaccessible from the regular shipping points. These stretched from Newcastle round to Exeter, but lay mainly between Hull and Southampton. Boston, serving a large part of eastern and central England, was most important in the 1270s and 1280s,[2] but towards the end of the century it yielded place to London where English shippers were most prominent. Soon after 1300, cloth supplanted raw wool as the country's chief export. In this, as in other ways, the opening decades of the fourteenth century mark the first major watershed in the history of the English landscape and economy since the Norman Conquest.

TOWNS AND CITIES

In England, as in other parts of Europe, the twelfth and thirteenth centuries were characterised by a considerable urban development. Existing centres of trade increased in size; villages grew to urban rank; and entirely new towns were founded. Such growth was promoted by increased commerce, both internal and international, and by a steady expansion of industry. This expansion did not, however, involve any major advance in organisation. Urban industry at the beginning of the fourteenth century was still based on the workshop, a form of enterprise unlikely to attract and to benefit from large amounts of capital. The wealthier citizens were merchants rather than industrialists and they reinvested in trade and property rather than in new production.

Boroughs and towns

The status of borough (*liber burgus*) was either prescriptive (enjoyed 'time out of mind') or conferred by charter granted either by the king for places on the royal demesne or by a feudal lord for other places. The grant

---

[1] For an early list of wool prices, see *Calendar Patent Rolls, 1334–38* (London, 1895), 480–2.

[2] See E. M. Carus-Wilson, 'The medieval trade of the ports of the Wash', *Med. Archaeol.*, VI-VII (1962–3), 182–201; P. Thompson, 'The early commerce of Boston', *Rep. and Papers, Associated Architect. Socs.*, II (1852–3), 362–81.

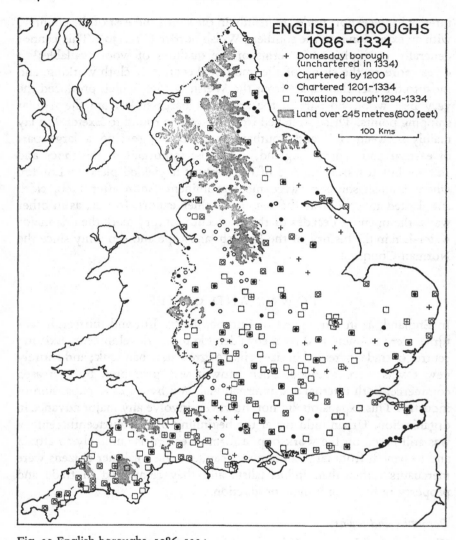

Fig. 31 English boroughs, 1086–1334
  Based on: (1) A. Ballard, *British borough charters, 1042–1216* (Cambridge,
  1913; (2) J. Tait and A. Ballard, *British borough charters, 1216–1307* (Cambridge, 1923); (3) M. Weinbaum, *British borough charters, 1307–1660* (Cambridge, 1943); (4) J. F. Willard, 'Taxation boroughs and parliamentary
  boroughs, 1294–1336' in J. G. Edwards *et al.* (eds.), *Historical essays in honour
  of James Tait* (Manchester, 1933), 417–35.

of a charter had two aspects. On the one hand, it brought privileges of an administrative, tenurial and legal character. The most important of these was the right of burgage or freehold tenure,[1] by which tenants held their lands on payment of fixed money rents and exercised the right of free alienation.

The other aspect was that the burgesses paid for their privileges, and boroughs were therefore sources of profit from rents and tolls; moreover, the burgesses were often ready to pay further for the confirmation and extension of their privileges. The policy of the Crown varied from reign to reign, but the financial exigencies of Richard I (1189–99) and still more of John (1199–1216) led to a readiness to make concessions. Feudal lords, bishops, abbots and laymen alike, also were eager to share in the profits that flowed from the granting of a borough charter. At the death of John, nearly 120 places had secured charters, either to confirm existing rights or to establish new ones;[2] and of these over 60 had not been described as boroughs in the Domesday Book. Between 1216 and 1334, another 90 or so townships received charters[3] (Fig. 31).

Not only did some villages acquire borough status, but some boroughs were created or 'planted' where no previous settlement had existed; their charters, too, came from kings, bishops,[4] abbots and laymen. M. W. Beresford has identified about 160 plantations in England between 1066 and 1334, and another two followed in 1345 and 1368; there were also about 80 plantations in Wales by 1334 (Fig. 32). The rate of founding fell during the decades 1130 to 1150, the years of civil war between Stephen and Matilda, and rose between 1189 and 1230. After 1368, there is no certain record of a new foundation during the remainder of the Middle Ages.

With few exceptions, such as New Salisbury alongside Old Sarum, the new towns did not supplant the old. 'They were planned additions to a stock of towns at a time when the economic situation gave opportunities both to new and to old. Just as agricultural expansion displayed itself

---

[1] M. de W. Hemmeon, *Burgage tenure in medieval England* (Cambridge, Mass., 1914); E. W. W. Veale (ed.), *The Great Red Book of Bristol*, Bristol Record Soc., 2 (1931), Introduction.

[2] A. Ballard, *British borough charters, 1042–1216* (Cambridge, 1913).

[3] A. Ballard and J. Tait, *British borough charters, 1216–1307* (Cambridge, 1923); M. Weinbaum, *British borough charters, 1307–1660* (Cambridge, 1943).

[4] M. Beresford, 'The six new towns of the bishops of Winchester, 1200–1255', *Med. Archaeol.*, III (1959), 187–215; and *History on the ground: six studies in maps and landscape* (London, 1957), 127–49.

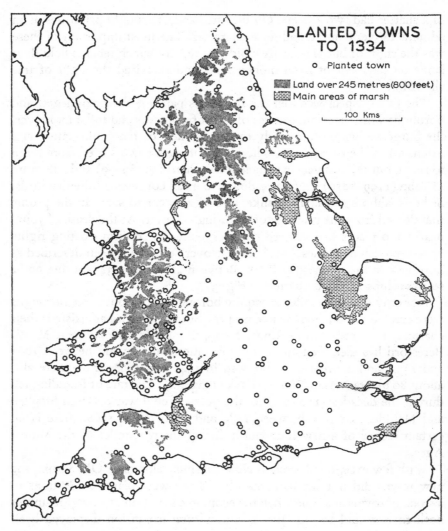

Fig. 32 Planted towns to 1334
    Based on M. W. Beresford, *New towns of the Middle Ages* (London, 1967), 275,
    282, 328, 342.

through an increased number of plough-teams and farm animals, so
urban expansion bore witness to great developments in the warehouse,
in the workshop and on the wharves.'[1] About a third of the plantations

[1] M. Beresford, *New towns of the Middle Ages* (London, 1967), 231.

were by the sea or on estuaries, reflecting the increasing coastwise traffic and the growing export trade in wool, grain and minerals. Some of these places were destined to become prominent seaports – Boston, King's Lynn and Newcastle founded about 1100, Harwich, Liverpool and Portsmouth founded about 1200, and Kingston upon Hull founded about 1300.

On the other hand, many boroughs came to little or nothing, and it would seem that local lords were not always good judges of the economic possibilities of the places to which they granted charters. Thus of the 23 boroughs created in Lancashire, only four – Lancaster, Liverpool, Preston and Wigan – were in any real sense urban at the end of the Middle Ages.[1] Some of the plantations hardly developed beyond mere villages; such was Mitchell in Cornwall. Others like Newtown in Burghclere (Hampshire) started to grow only to decay into nothing. Some declined as a result of physical changes; New Romney and New Winchelsea[2] were deserted by the sea, and Ravenserodd was washed away by the sea. A total of 23 (13%) plantations in England never succeeded, and 18 (21%) of those in Wales. The failures were widely scattered, and they included a high proportion of later foundations.

The boroughs that grew did so largely through immigration from the surrounding countryside. The Domesday village of Stratford upon Avon was granted a charter by the bishop of Worcester in 1196, and by 1252 there were 250 burgages as well as many shops and stalls.[3] About one-third of the burgesses bore the names of towns and villages whence, presumably, they or their ancestors had come (Fig. 33). Interestingly enough, in the late thirteenth or early fourteenth century there was a similar proportion of place-name surnames among the inhabitants of such long-established towns as Bristol, Gloucester and Worcester.[4]

In 1294 differential taxation on moveable property was introduced;[5] rural areas contributed a tenth and the urban centres a sixth. This differential rating, in varying proportions, was continued, and from 1334

[1] A. Ballard and J. Tait, lxxxviii.

[2] W. M. Homan, 'The founding of New Winchelsea', *Sussex Archaeol. Coll.*, LXXXVIII (1949), 22–41.

[3] E. M. Carus-Wilson, 'The first half-century of the borough of Stratford upon Avon', *Econ. Hist. Rev.*, 2nd ser., XVIII (1965), 51.

[4] R. H. Hilton (1966), 183–4.

[5] J. F. Willard, 'Taxation boroughs and parliamentary boroughs, 1294–1336', in J. G. Edwards *et al.* (eds.), *Historical essays in honour of James Tait* (Manchester, 1933), 417–35.

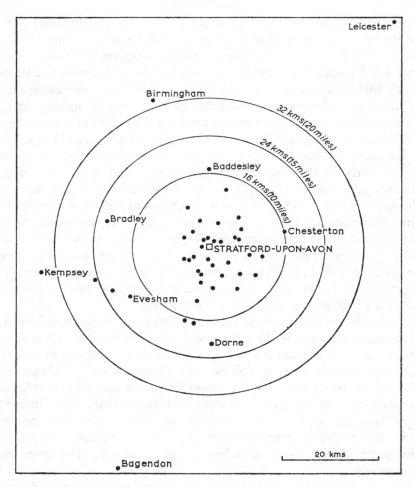

Fig. 33 Immigration into Stratford upon Avon to 1252
        Based on E. M. Carus-Wilson, 'The first half-century of the borough of
        Stratford upon Avon', *Econ. Hist. Rev.*, 2nd ser., XVIII (1965), 51.

onwards it became a fifteenth and a tenth. But the list of 226 places[1]
described as boroughs on the enrolled accounts at various times, and so
taxed at the higher rate, differs from that of the chartered boroughs. On
the one hand, over 40 seignorial boroughs never appeared as taxation
boroughs; thus Bury St Edmunds in Suffolk and Abbots Bromley, New-
borough and Uttoxeter in Staffordshire were always taxed at the rural

[1] M. W. Beresford (1967), 258.

rate; others appeared only intermittently, for example, Morpeth (North-umberland) in 1307 and Burton on Trent in 1307 and 1313. On the other hand, many places without charters were often, and sometimes usually, taxed at the higher rate. They included places on the royal demesne; places designated as boroughs for taxation purposes (there were eleven in Somerset alone); and places described as *villae mercatoriae* (i.e. vills with merchants), such as Old Windsor, Cheltenham and Andover. The taxation boroughs and merchant vills were selected by the 'chief taxers' in each county. No general criterion based on size can be discerned. Thus Yeovil in Devon, a chartered borough with a tax-paying population of 45 in 1327 never appeared as a taxation borough, while Chard in Somerset, also chartered but with only 11 tax payers in 1327, did appear. Probably 'trading activity', measured by local standards, 'had much to do with the selection or omission of seignorial boroughs'.[1]

There is a further complication – that of the parliamentary boroughs. On a number of occasions before 1265, representatives of certain towns had been called before king and council for specific purposes; but in that year, for the first time, such representatives were summoned for a general political purpose.[2] In 1268, men from 27 towns were summoned. In 1275 and again in 1283, *villae mercatoriae* were included, but after the latter date the distinction between boroughs and 'merchant vills' disappeared. From about 1310, the presence of burgesses in parliament was 'coming to be regarded as a matter of course'.[3] The cities and towns represented varied from parliament to parliament, but they usually numbered between 80 and 90 and were chosen for each county by the sheriffs. The places so selected included (i) most of the royal boroughs, (ii) many seignorial boroughs, (iii) a number of unchartered taxation boroughs, and (iv) some towns that were neither chartered nor taxation boroughs. Of the total of about 180 parliamentary boroughs selected at various times between 1295 and 1336, as many as 44 were not taxation boroughs.[4] The history of borough and town in the early Middle Ages presents many puzzling features.

Urban form and planning

A considerable number of towns grew up in the shadow of a castle or a monastery. Further protection was achieved by walling, but this was

[1] J. F. Willard (1933), 424.
[2] M. McKisack, *The parliamentary representation of the English boroughs during the Middle Ages* (Oxford, 1932), 1.
[3] *Ibid.*, 24.   [4] J. F. Willard (1933), 426–7.

an expensive business and the majority of smaller places had either no defence works at all or only ditches and palisades.[1]

There were at least 30 monastic towns – places where an abbot or prior was lord.[2] About three-quarters of these stood immediately adjacent to the abbey, which was usually Benedictine; among the new Orders, only the Austin canons were in control of towns. On the whole, they were small places – Abingdon, Malmesbury, Peterborough and Burton, for example; Coventry was perhaps the largest and the most thriving. In fact, the bulk of the Church's urban property lay elsewhere, in the larger towns, the ports, and above all in London.[3] At the beginning of the fourteenth century, about one-third of the rental of London was in the hands of the Church.[4] Many religious communities had property there or in Southwark – a *hospitium* for use when visiting the capital, warehouses, quays, and land and buildings that simply brought in rent.[5] They were also strongly entrenched in most of the shire towns,[6] and, long before the passing of the Statute of Mortmain in 1279, action had been taken in several places to prevent more land and buildings falling into the hands of the monasteries.[7]

The central areas of most English towns took form during the early Middle Ages, and the street lines and property boundaries then adopted have guided building patterns ever since. The unit of medieval urban building and planning was the burgage plot (*placea*). The buildings on

---

[1] T. D. Hardy (ed.), *Rotuli litterarum clausarum*, I (Record Comm., London, 1833), 193a, 421b; H. M. Colvin, 'Domestic architecture and town planning' in A. L. Poole (ed.), *Medieval England*, I (Oxford, 1958), 65–7. See also H. L. Turner, *Town defences in England and Wales* (London, 1970).

[2] N. M. Trenholme, *The English monastic borough*, University of Missouri Studies, 2 (1927); D. Knowles, *The Monastic Order in England* (Cambridge, 1963), 444–7.

[3] E. Bradley, 'London under the monastic Orders', *Jour. British Archaeol. Assoc.*, 2nd ser., IV (1898), 9–16; M. B. Honeybourne, 'The extent and value of the property in London and Southwark occupied by the religious houses', *Bull. Inst. Hist. Res.*, IX (1931–2), 52–7; H. M. Cam, 'The religious houses of London and the Eyre of 1321' in J. A. Watt (ed.), *Medieval studies presented to Aubrey Gwynn SJ* (Dublin, 1961), 320–9.

[4] H. T. Riley (ed.), *Memorials of London, 1276–1419* (London, 1868), 98.

[5] See, for example, C. D. Ross (ed.), *Cartulary of Cirencester abbey* (London, 1964), I, 173–5, II, 399; A. Watkin (ed.), *The great cartulary of Glastonbury*, Somerset Record Soc., 59 (1947), 128.

[6] R. A. Donkin, 'The urban property of the Cistercians in medieval England', *Analecta Sacri Ordinis Cisterciensis*, XV (1959), 104–31.

[7] C. Gross, 'Mortmain in medieval boroughs', *American Hist. Rev.*, XII (1907), 739.

each plot, house and workshop,[1] clustered, gable-on, towards the front; the area behind was usually open and might be worked as a garden or orchard or be used for livestock. In the planted boroughs, plots were sometimes forty to sixty feet wide,[2] instead of the usual ten to twenty feet around the market places of older centres.[3] Although particular trades tended to be concentrated in certain streets or districts,[4] there is not much evidence of any purely social segregation; large and small properties stood side by side.

By the close of the thirteenth century, buildings had often encroached on streets and on the market place. For example, at Pontefract a line of stalls down the centre of the 'new market' gradually (1236–78) became permanent;[5] and at Gorleston in 1275, the market place had been 'improperly' narrowed by 30 feet all round.[6] A shop carried forward into the street by means of a stall was generally the first step in such encroachment. There is also evidence of the building-up of the backs of burgages, of their lengthwise subdivision, and of much sub-letting.[7] Meanwhile, the town was sometimes spreading beyond its walls or official limits (*in suburbio, extra burgum*),[8] partly because of congestion near the centre, and partly as a consequence of attempts to escape industrial regulations and borough fees. In the second half of the thirteenth century, the market in land and property was as active in the towns as in the country.

[1] See W. O. Hassall, *The cartulary of St Mary Clerkenwell*, Camden Soc., 3rd ser., 71 (1949), 140–1.

[2] E. M. Carus-Wilson (1965), 57.

[3] See W. A. Pantin, 'Medieval English town house plans', *Med. Archaeol.*, VI-VII (1962–3), 202–39.

[4] M. Curtis, 'The London Lay Subsidy of 1332' in G. Unwin (ed.), *Finance and trade under Edward III* (Manchester, 1918), 35–92; W. T. Lancaster (ed.), *Abstract of charters of Bridlington priory* (London, 1912), 322.

[5] R. Holmes (ed.), *The chartulary of St John of Pontefract*, I, Yorks. Archaeol. Soc., Record Series, 25 (1899), 125–6, 182. See also, T. B. Dilks (ed.), *Bridgwater borough archives*, Somerset Record Soc., 48 (1933), liii; H. E. Butler (ed.), *The Chronicle of Jocelin of Brakelond* (London, 1949), 77; E. M. Carus-Wilson (1965), 60.

[6] W. Illingworth and J. Carey (eds.), *Rotuli Hundredorum*, II (Record Comm., London, 1818), 169.

[7] R. H. Hilton, 'Some problems of urban real property in the Middle Ages' in C. H. Feinstein (ed.), *Socialism, capitalism and economic growth: essays presented to Maurice Dobb* (Cambridge, 1967), 329, 334.

[8] S. R. Wigram (ed.), *The cartulary of the monastery of St Frideswide at Oxford*, Oxford Hist. Soc., 28 (1894), 215; M. Bateson, *Records of the borough of Leicester*, I (London, 1899), 389; A. Cronne, *The borough of Warwick in the Middle Ages*, Dugdale Society, Occasional Papers, 10 (1951), 18.

The many planted towns of the twelfth and thirteenth centuries were, almost by definition, planned. Twenty-six or more were organised about a grid,[1] for example Salisbury, Hull, Stratford upon Avon, New Winchelsea and King's Lynn; but in most cases the axis of development was a single, comparatively broad street which also served as a market. New house plots spread over waste, pasture, and even arable holdings. Alnmouth in Northumberland absorbed 296 acres of the common pasture and open fields of neighbouring Lesbury, and at Knutsford in Cheshire the burgages were actually measured in *seliones*.[2] At Eynsham, the abbot planted a new borough in 1215 by enclosing a *cultura* of 20 to 22 acres and dividing it into burgage plots of an acre or less.[3] The whole scheme took the form of a rectangular projection into the open fields.

The changing importance of towns

Any attempt to measure the changes in the relative importance of towns in the early Middle Ages is fraught with difficulty. Table 3.2 attempts to set out the ranking order of the thirty most wealthy provincial towns in the middle of the twelfth century and again in 1334. The earlier list is based on the average of the 'aids' levied during the reign of Henry II (1154–89),[4] but the returns are uneven and some boroughs appear only intermittently in the lists of places taxed. Bristol, being in baronial hands, does not appear at all; nor do Chester and Leicester. The Cinque Ports (Dover, Hastings, Hythe, Romney and Sandwich) were exempted from taxation. But even with its limitations, the list does provide a basis for comparison with the ranking order of towns in the 1334 Subsidy.

By 1334, Exeter, Winchester, Gloucester, Canterbury and Cambridge, all Domesday boroughs, had lost their places among the first twelve towns. Grimsby, Scarborough and Colchester also had declined relatively, and Dunwich had been entirely destroyed by the sea, but other east coast ports had improved their positions – Newcastle, Boston, Lynn, Yarmouth and Ipswich. Southampton was the leading south-coast port in the twelfth century and in 1334, and by the latter date Plymouth also was prominent.

---

[1] M. W. Beresford (1967), 151.

[2] A. Ballard and J. Tait, lxxviii.

[3] H. E. Salter (ed.), *Eynsham cartulary*, Oxford Record Soc., 49 (1906–7), 60; 51 (1908), xli, xliv, 50. See also J. S. Brewer (ed.), *Chronicon monasterii de Bello*, Anglia Christiana Soc. (1946), 12ff.

[4] C. Stephenson, *Borough and town*, Medieval Academy of America (Cambridge, Mass., 1933), 225 (with corrections); and 'The aids of the English boroughs', *Eng. Hist. Rev.*, XXXIV (1919), 457–75.

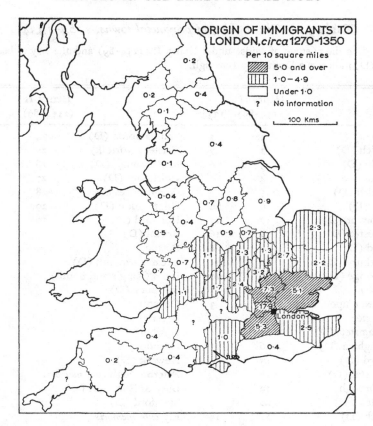

Fig. 34 Origin of immigrants into London *circa* 1270–1350
      Based on E. Ekwall, *Studies in the population of medieval London* (Stockholm, 1956).

York, Norwich and Lincoln remained among the chief inland towns of the realm.

Towering above all was London. From 1183 comes the famous description of the city by William Fitz Stephen.[1] There were then 13 great conventual churches and 126 parish churches. To the port of London 'merchants from every nation that is under heaven' brought 'their trade in ships'. Beyond the ancient walls a 'populous suburb' stretched as far

---

[1] F. M. Stenton *et al.*, *Norman London* (Hist. Assoc., London, 1934). See also U. T. Holmes, *Daily living in the twelfth century, based on observations of Alexander Neckam in London and Paris* (Madison, 1962), 18–43.

Table 3.2 *The ranking of provincial towns, 1154–1334*

Based on the average of 'aids' under Henry II (1154–89) and the Lay Subsidy of 1334. (D) indicates a Domesday Borough.

	Henry II (1154–89)	1334		Henry II (1154–89)	1334
York (D)	1	2	Southampton (D)	24	18
Norwich (D)	2	7	Caister (Norfolk)	25	
Lincoln (D)	3	6	Marlborough (D)	26	
Northampton (D)	4		Colchester (D)	27	
Dunwich (D)	5		Godmanchester	28	
Exeter (D)	6	26	Huntingdon (D)	29	
Winchester (D)	7	17	Hereford (D)	30	14
Gloucester (D)	8	16	Bristol (D)		1
Oxford (D)	9	8	Boston		4
Canterbury (D)	10	15	Great Yarmouth (D)		5
Cambridge (D)	11	20	Lynn (King's and South)		10
Grimsby	12		Salisbury (D)		11
Newcastle upon Tyne	13	3	Coventry		12
Doncaster	14		Beverley		19
Berkhamsted (D)	15		Newbury		21
Nottingham (D)	16	25	Plymouth		22
Bedford (D)	17		Newark (D)		23
Worcester (D)	18		Peterborough *cum membris*		24
Scarborough	19		Bury St Edmunds (D)		27
Carlisle	20		Stamford (D)		28
Ipswich (D)	21	13	Ely *cum membris*		29
Corbridge	22		Luton		30
Shrewsbury (D)	23	9			

as the palace and church of Westminster; and all around lay 'the gardens of the citizens that dwell in the suburbs, planted with trees, spacious and fair, adjoining one another'. There was a wooden bridge on the site of the present London Bridge; a stone bridge, begun in 1176, was not completed until about 1209. By this time, there were also some stone houses; in 1215, stone from the houses of Jews was taken to repair the walls of London.[1] A series of disastrous fires in the middle of the twelfth century was followed by the Building Assize of 1189 (the work of London's first mayor) which stated that party walls should be of stone and that tiles

[1] C. L. Kingsford (ed.), *A survey of London by John Stow*, 1 (Oxford, 1908), 38.

rather than thatch should be used in roofing.[1] Needless to say, the Assize was not fully observed until several centuries later.

If we may trust the taxation figures, London, at the time that Fitz Stephen wrote, was about three times as wealthy as York; and below York came Norwich, Lincoln and Northampton. By 1334, the capital city was about five times as wealthy as Bristol; and next to Bristol came York and Newcastle. London was not only growing absolutely, its relative importance in the urban hierarchy of the country was also increasing. But of its actual size, we are very poorly informed. An intelligent guess might put the population at about 30,000 for the early part of the thirteenth century and at something under 50,000 in 1334.

Whatever its size, immigrants made a considerable contribution to its growth (Fig. 34). E. Ekwall found about 7,000 place-name surnames among Londoners during the period *circa* 1270 to 1350.[2] We cannot be sure that all people with such surnames were themselves immigrants. Some may have been the sons or even the grandsons of immigrants; others may have been apprentices who took the surnames of their masters. On the other hand, we know of some immigrants without local surnames. Allowing for the uncertainty of the evidence, it seems clear that a substantial number of people came from the neighbouring counties and from East Anglia and the east Midlands. As might be expected, the numbers who came from the north, the west and the north-west were very much smaller. Fig. 34 tells its own tale, but were the information available, it might have been better to present the figures in terms of the total population of each county. The names were those of clerks, of merchants, of lawyers, of priests, of physicians and of tradespeople of various occupations – mercers, woolmongers, hatters, goldsmiths, vintners, skinners and the like. Not only was the attraction of London far-reaching, but those who came were making a significant contribution to the growing trade and industry of England's greatest city.

[1] H. T. Riley (ed.) (1859), xxix–xxxiii, 319–36.
[2] E. Ekwall, *Studies on the population of medieval London* (Stockholm, 1956); and *Early London personal names* (Lund, 1947), 119–22. See also J. C. Russell, 'Medieval midland and northern migration to London, 1100–1365', *Speculum*, XXXIV (1959), 641–5; G. A. Williams, *Medieval London* (London, 1963), 18, 138–43; S. L. Thrupp, *The merchant class of medieval London, 1300–1500*, (Chicago, 1948), 206–10, 389–92.

Chapter 4

ENGLAND *circa* 1334

R. E. GLASSCOCK

It is now generally agreed that an age of increasing population and
prosperity had come to an end by the 1330s. The Black Death, once held
responsible for the onset of decline, has been relegated to the less dramatic
role of accelerating forces which had already begun to show in the economy
of some parts of the country well before 1348.[1] M. M. Postan, having
erected in 1950 a general framework for the interpretation of the economy
of the thirteenth and fourteenth centuries, looked forward to the time
when his 'mere outlines' would be filled out by local studies. That is now
happening; twenty years later a number of studies are making us more
aware of the regional differences in early fourteenth-century England.
They show that economic conditions differed widely; evidence of
recession in one place is matched by evidence of expansion elsewhere.
What was true of the Winchester estates did not apply to Cornwall; the
Weald differed greatly from the Midlands; and the north from East
Anglia. Generally speaking it seems that there were fewer signs of
recession in regions where there was a small but vital sector of non-
agricultural activity within the local economy.[2]

[1] M. M. Postan, 'Some economic evidence of declining population in the later
Middle Ages', *Econ. Hist. Rev.*, 2nd ser., II (1950), 221–46; A. R. Bridbury, *Economic
growth: England in the later Middle Ages* (London, 1962); B. H. Slicher Van Bath, *The
agrarian history of western Europe A.D. 500–1850* (London, 1963), 137–44; E. Miller,
'The English economy in the thirteenth century: implications of recent research', *Past
and Present*, XXVIII (1964), 21–40; M. M. Postan in M. M. Postan (ed.), *The Cambridge
economic history of Europe*, I (2nd ed., Cambridge, 1966), 548–91; B. F. Harvey, 'The
population trend in England between 1300 and 1348', *Trans. Roy. Hist. Soc.*, 5th ser.,
XVI (1966), 23–42; J. C. Russell, 'The pre-plague population of England', *Jour. Brit.
Stud.*, V (1966), 1–21; E. J. Nell, 'Economic relationships in the decline of feudalism:
an examination of economic interdependence and social change', *History and Theory*,
VI (1967), 313–50; J. Z. Titow, *English rural society 1200–1350* (London, 1969),
particularly 64–102.
[2] For studies of specific regions see: (1) R. H. Hilton, *A medieval society: the west
Midlands at the end of the thirteenth century* (London, 1967); (2) F. R. H. Du Boulay,

The early mid-fourteenth century is therefore an opportune point in time to pause and to look at the landscape of England, to consider the achievements of the twelfth and thirteenth centuries, and to try to set the back-cloth for events which were to follow. Unfortunately we have no Domesday Book for 1300, possibly the high-water mark of the era of colonisation, nor is there anything so comprehensive that would enable us to take stock of the state of the country in any one year. Only one source, the Lay Subsidy of 1334, although meagre in its detail, gives a fairly complete coverage of the country. This is a valuable date at which to have any kind of survey, as it comes fourteen years before the Black Death, and before the full impact of social and economic disorder. The year 1334 is therefore a convenient date for a description of England; but the lack of sources makes it impossible to reconstruct the geography of England for that single year. It is therefore necessary to examine the state of the country around 1334, that is between say 1320 and 1348.

THE 1334 LAY SUBSIDY

The 1334 Lay Subsidy is a unique source for the early fourteenth century on account of its coverage. The amounts of taxation agreed upon in 1334 were standardised in 1336, and did not finally disappear from use until 1623. This means that the gaps in the 1334 rolls themselves can be filled from later documents which contain the same information.[1] In this way all the counties of England can be covered for 1334, except for the palatine counties of Cheshire and Durham which were not liable for the tax, and for Cumberland, Westmorland and Northumberland which were excused tax in 1334 because of recent damage done by invading Scots; these three latter counties, however, were taxed in 1336 so that their earlier omission can be remedied.

The fifteenth and tenth of 1334 are so called because a fifteenth was asked from rural areas and a tenth from boroughs and ancient demesnes. The tax was one of a series of taxes on movable goods, principally on crops and stock, that had become a standard method of taxation by the

The lordship of Canterbury: an essay on medieval society (London, 1966); (3) J. Hatcher, *Rural economy and society in the duchy of Cornwall, 1300–1500* (Cambridge, 1970).

[1] The rolls of the taxes upon movable goods are all contained within Class E 179, Exchequer Lay Subsidies, at the Public Record Office, London. For studies dealing with the rolls, see p. 138, n. 1.

end of the thirteenth century.[1] The 1334 tax differed in an important way from its immediate predecessors in 1327 and 1332, for whereas those earlier taxations had been based upon the direct assessment of the goods of individuals, the 1334 tax took the form of an agreed sum or quota which each community was expected to pay. This quota was to be not less than the amount paid in 1332, and in most villages the agreed sum turned out to be slightly more.

Before basing any conclusions on the evidence of the 1334 Subsidy we must be under no illusions about some of its weaknesses as a source. Evasion and conventional valuation must have been common. Moreover, fourteenth-century officials of all kinds inherited the knack of lining their pockets by petty extortion. The only consolation is that there is no evidence to suggest that men were more dishonest in some areas than in others, and, therefore, the relative distribution of taxable wealth should not be unduly distorted. Apart from evasion there were several lawful exemptions from the tax.[2] Among the privileged groups so excused were the moneyers of London and Canterbury, the men of the Cinque Ports, and the stannary men of the south-west. Allowance should, therefore, be made for these omissions in Kent, Devon and Cornwall. Then again, a considerable number of people had escaped tax in 1332 because their property was below the taxable minimum, and the 1334 quotas took no account of the small amounts of movables in the hands of a large number of people in every community. A more serious omission was that most of the movable wealth of the Church, separately taxed in Clerical Subsidies, was excluded from the Lay Subsidies.[3] This means that a county with a great amount of clerical property appears less wealthy relative to others than it would have done had a complete assessment been carried out. Nevertheless, even allowing for clerical wealth, it is probable that the pattern of relative prosperity as shown in Fig. 35 would be refined rather than radically altered.[4] It is not likely that counties with low lay assessments would be transformed into counties of great wealth by the inclusion of clerical property.

[1] J. F. Willard, *Parliamentary taxes on personal property, 1290 to 1334* (Cambridge, Mass., 1934); M. W. Beresford, 'The Lay Subsidies', *The Amateur Historian*, III (1958), 325–8, and IV (1959), 101–9. A detailed account of the 1334 Subsidy is contained in R. E. Glasscock, *The Lay Subsidy of 1334* (British Academy, London, 1974).

[2] J. F. Willard (1934), 110–82.

[3] *Ibid.*, 96–109.

[4] As demonstrated in R. S. Schofield, 'The geographical distribution of wealth in England', *Econ. Hist. Rev.*, 2nd ser., XVIII (1965), 483–510.

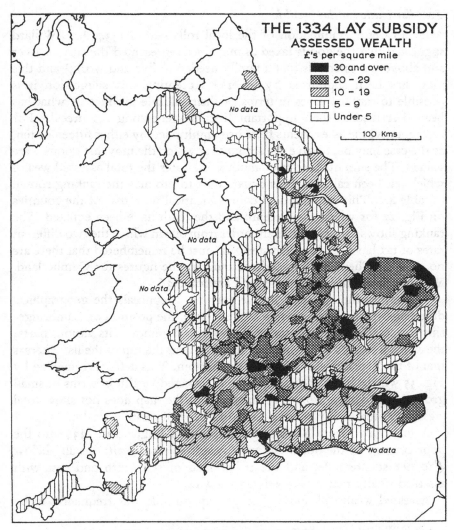

Fig. 35 The 1334 Lay Subsidy: assessed wealth
Based on P.R.O. Exchequer Lay Subsidies, E. 179.

The distribution of lay wealth

After a careful examination of the local rolls before 1334, J. F. Willard suggested that the goods taxed as movables represented the surplus over and above the essentials that a family needed to live and work,[1] and this view has been supported by others.[2] Accepting this suggestion it is possible to rank counties in terms of their saleable produce or what has been described as their important 'income-generating resources'.[3] For each county, the assessments have been multiplied by either fifteen or ten, as the case may be, to give the amounts at which the movable goods were valued. The sum of these two amounts indicates the total assessed wealth which has been calculated per square mile to produce the ranking shown in table 4.1. This ranking is of course identical with that of the counties on Fig. 22 for which the fifteenth and the tenth have been equated. The ranking shows a slightly different order from one in which the two different rates of tax have not been equated.[4] It must be remembered that there are no data for Cheshire and Durham and that the figures for Cumberland, Westmorland and Northumberland are for 1336.

Interesting though such a table may be, it conceals the geographical differences within each county. For example, the position of Cambridgeshire on the list represents an average of the extremes of its various parts; the assessed wealth of the upland would be near the top of the list whereas that of the Fenland would be near the bottom. This defect is remedied by Fig. 35 which shows the distribution of assessed wealth in terms of small groups of parishes. It must be stressed that the map does not show total wealth but that it is probably a fair indicator of relative wealth.

A line drawn from York to Exeter divided England in 1334 into the poor country of the north and west, with assessed wealth usually below £10 per square mile, and the richer land of the south and east with assessed wealth that was mostly above £10.

Assessed wealth of above £20 per square mile was frequently to be

[1] J. F. Willard (1934), 84–5.

[2] A. T. Gaydon, 'The taxation of 1297', *Beds. Hist. Record Soc.*, XXXIX (1959); L. F. Salzman, 'Early taxation in Sussex', *Sussex Archaeol. Coll.*, XCVIII (1960), 29–43, and XCIX (1961), 1–19.

[3] E. J. Buckatzsch, 'The geographical distribution of wealth in England, 1086–1843', *Econ. Hist. Rev.*, 2nd ser., III (1950), 180–202.

[4] As in E. J. Buckatzsch, 187. A table of gross yields which also does not make allowance for the two rates of tax is given in W. G. Hoskins and H. P. R. Finberg, *Devonshire Studies* (London, 1952), 216–17.

Table 4.1   *1334 Lay Subsidy*

Assessed wealth per square mile (by counties)

(The figures for Cumberland, Westmorland and Northumberland are for 1336)

	£		£
Middlesex (including London)	57.7	Somerset	12.0
Oxfordshire	27.2	Hampshire	11.3
Norfolk	24.0	Surrey	11.2
Bedfordshire	21.7	Sussex	11.1
Berkshire	20.8	Essex	11.0
Rutland	20.4	Worcestershire	9.8
Cambridgeshire	18.1	Herefordshire	7.4
Gloucestershire	18.0	Staffordshire	7.2
Huntingdonshire	17.8	Shropshire	7.0
Northamptonshire	17.2	Derbyshire	6.6
Kent	17.1	Devonshire	5.1
Wiltshire	17.1	Cumberland (1336)	4.9
Lincolnshire	16.6	Cornwall	4.9
Hertfordshire	14.8	North Riding	4.4
Warwickshire	14.4	West Riding	4.1
Suffolk	14.1	Westmorland (1336)	3.8
Buckinghamshire	13.8	Lancashire	2.9
Leicestershire	13.7	Northumberland (1336)	2.5
East Riding	13.5	Cheshire	No data
Dorset	12.4	Durham	No data
Nottinghamshire	12.3		

found in a great zone across lowland England, from the Parrett valley of Somerset to the fens and coastlands of Norfolk and Lincolnshire. As the assessments were above average, this must have been an area of considerable agricultural prosperity. That there was a distinct correlation between these areas and the corn growing lands of the country will be shown below. Outside this zone the only areas of comparable wealth were the Thames valley, east Kent, and the coastal plain of Hampshire and Sussex. All these areas carried twice, and in places at least three times as much movable wealth as the west Midlands, the north and the west.

There was also a random scatter of wealthy localities. Values over £30 per square mile were to be found in various parts of Norfolk – around Hunstanton in the north-west, in the silt marshland in the north-east, and to the south-east of Norwich. Upland Cambridgeshire was also very wealthy. So also were mid-Oxfordshire, parts of west Berkshire, parts of the north Kent lowlands, and parts of the Kesteven claylands of Lincolnshire. No common factor can be found to explain such a haphazard distribution. These very high assessments probably resulted from exceptional local agricultural prosperity induced by favourable local economic and social conditions. It is for the local historian to try to unravel the story of each area, and to explain, for example, the curiously high assessment of the north-western corner of Norfolk, as prominent on a map of 1334 as on one of 1086.[1]

By contrast almost all the west and north had assessed wealth below £10 per square mile, but within this area there were variations. The moorlands stand out as blank areas on the map – in the south-west, Exmoor, Dartmoor and Bodmin Moor; and in the north, the Pennines, the North York Moors, the Lake District and the Cheviots. Almost all the settled agricultural lowlands of the west and north were only between £5 and £10 per square mile, and the lowlands of Lancashire, the North Riding and Northumberland were particularly poor. It should be remembered, however, that not only were these areas remote from the centre from which the taxation was administered, but that they had recently been overrun in the war with Scotland.

Almost all the poorer areas of the south-east were also in the £5 to £10 category. These included the infertile sands and the heavy clays of Hampshire, Surrey, and the Weald of Sussex and Kent, and also parts of Essex and Suffolk. Furthermore, parts of the chalkland had similarly low assessments, for example, in Hampshire, and in the Chilterns. In some areas the assessed wealth fell below £5 per square mile, as, for example, in the New Forest, north-west Surrey and parts of the Weald.

Fig. 35 represents, so to speak, a framework into which might be fitted the mass of evidence for population and agriculture for the different regions of medieval England. One thing, however, is abundantly clear: different amounts of movable wealth were to be found in areas of similar physical conditions, and, on the other hand, similar amounts were found in contrasting areas. For example, values of between £20 and £30 per square mile were to be found on the silts of the Lincolnshire fens, on the clays of western Cambridgeshire, and on the medium soils of much of Norfolk. While climate, relief and soils exerted broad controls over differences in farming, and underlay the strong contrast between the

[1] H. C. Darby, *The Domesday geography of eastern England* (Cambridge, 3rd ed., 1971), 117.

north-and-west and the south-and-east, the wealth of any particular locality can only be explained in terms of its local economic and social conditions. To this end a number of studies of the 1334 assessment for smaller areas and particular counties have already been produced.[1]

POPULATION

While the 1334 Subsidy allows us to grasp some essentials about the distribution of lay wealth over the country it does not enable us to estimate the total population. Estimates of the population of England just before the Black Death vary between $2\frac{1}{2}$ million and just over 6 million.[2] If demographers cannot agree on the population of the country in 1377, when the Poll Tax at least gives some quantitative basis for calculations, it is unlikely that they will agree on the population forty years earlier. In 1948 J. C. Russell estimated the total in 1377 as 2,332,373[3] and, although this figure has been criticised as too low by Stengers[4] and Krause,[5] his recently revised estimate is even lower at 2,199,916.[6] Assuming that the Black Death carried off around 40% of the population, Russell arrived at a pre-plague total of about 3.7 million. While many regard this as far too low, there is little hope of reaching a definite figure from the fragmentary sources available; and as the accepted method is to work backwards from the supposed totals of 1377, estimates for the pre-plague population will

---

[1] F. W. Morgan, 'The Domesday geography of Devon', *Trans. Devon Assoc.*, LXXII (1940), 321; B. Reynolds, 'Late medieval Dorset; three essays in historical geography', unpublished M.A. thesis, University of London, 1958; C. T. Smith, in W. G. Hoskins and R. A. McKinley (eds.), *V.C.H. Leicestershire*, III (1955), 134; H. C. Darby, *The medieval Fenland* (Cambridge, 1940), 134–5; W. G. Hoskins and E. M. Jope, in A. F. Martin and R. W. Steel (eds.) *The Oxford region* (Oxford, 1954), 109; R. E. Glasscock, 'The distribution of wealth in East Anglia in the early fourteenth century', *Trans. and Papers, Inst. Brit. Geog.*, XXXII (1963), 113–23; and 'The distribution of lay wealth in Kent, Surrey and Sussex, in the early fourteenth century', *Archaeol. Cantiana*, LXXX (1965), 61–8; W. G. Hoskins and H. P. R. Finberg, *Devonshire studies* (London, 1952), 212–49.

[2] M. M. Postan (1966), 561–2. A summary of the controversy over the pre-Black Death population of England is given by J. Z. Titow (1969), 66–73.

[3] J. C. Russell, *British medieval population* (Albuquerque, New Mexico, 1948), 146.

[4] J. Stengers, in a review in *Revue Belge de philologie et d'histoire*, XXVIII (1950), 600–6.

[5] J. Krause, 'The medieval household: large or small?', *Econ. Hist. Rev.*, 2nd ser., IX (1957), 420–32.

[6] J. C. Russell (1966), 14.

continue to vary according to the weight given to the mortality of the Black Death. For various reasons very large numbers of people were not recorded on the more detailed Lay Subsidy rolls of 1327 and 1332, and any calculation of total population based on these rolls is bound to be 'purely conjectural and probably erroneous'.[1] Recent work on the local rolls of Sussex puts this beyond any doubt.[2] At best the rolls may be used to infer the relative size of places, such has been done for Warwickshire.[3]

A figure midway between the various estimates gives a population of between say 4 and 4½ million in England before the Black Death. It may therefore have been about two and a half times what it had been in 1086, which indicates the striking growth during the intervening 250 years. Unfortunately, except for some classes of people, for example the clergy,[4] we have no idea of the breakdown of this number, nor is there direct evidence of how this population was distributed. It is reasonable, however, to suppose that the distribution of population bore a fairly close relationship to that of lay wealth (Fig. 35).

By modern standards England in the early fourteenth century might appear to have been sparsely peopled, but it should be remembered that most people lived directly off the land. The early fourteenth century constituted a crucial phase in the relationship between a rapidly growing population and its ability to produce food. Much land had been taken into cultivation since the Norman Conquest, but grain yields remained low and there were extra mouths to feed. As Miller has pointed out, in the absence of any great new technological breakthrough, parts of the country became overpopulated relative to food supply, with the result that bad harvests and famines, when they came, were progressively harder to bear.[5] The great European famine of 1315–17, in England as elsewhere, was a critical point in the population cycle. Undoubtedly its severity varied from place to place. Figures for some Winchester manors suggest that the death rate was twice as high as usual during these years.[6] Despite Russell's view that the population loss of the decade 1310–19 was only a few per

---

[1] J. F. Willard (1934), 181.

[2] L. F. Salzman (1960), 40–3.

[3] J. B. Harley, 'The settlement geography of early medieval Warwickshire', *Trans. and Papers, Inst. Brit. Geog.*, XXXIV (1964), 115–30.

[4] J. C. Russell, 'The clerical population of medieval England', *Traditio*, II (1944), 177–212.

[5] E. Miller (1964), 37–9.

[6] M. M. Postan and J. Titow, 'Heriots and prices on Winchester manors', *Econ. Hist. Rev.*, 2nd ser., XI (1959), 392–411.

cent and that the population went on increasing until 1347,[1] there is, at present, much more evidence to support the belief that, in the period under consideration, namely between 1320 and the Black Death, the total population was at best static, at worst declining. It is no coincidence that the founding of new towns, representative of an expanding population and flourishing commercial activity, fell rapidly away in the early fourteenth century.[2] As Miller has put it, 'the first half of the fourteenth century was a time of waiting for the blow to fall'.[3]

THE COUNTRYSIDE

Villages and houses

By the fourteenth century almost all the villages known to us today were in existence, together with many more which have long since disappeared.[4] Fortunately we know for almost all of England which villages were in existence from the *Nomina Villarum* of 1316.[5] Added to this is the almost complete coverage of the Lay Subsidy of 1334 with its long lists of vills. Few places of any size can have escaped both lists: 13,089 places are named on the records of the 1334 Lay Subsidy,[6] and in addition the rolls include many unspecified hamlets and other places.

Yet we know relatively little about the size and shape of the villages or about the dwellings of the villagers. Later layouts have replaced the medieval ones, and newer houses have obliterated the traces of their predecessors. Even the shape of a village in the fifteenth or sixteenth century may be no guide to its plan two or three centuries earlier. This has been shown at Wharram Percy in the East Riding, where excavations suggest three distinct stages of growth between the eleventh and fourteenth centuries, and at Wawne in the same county where the village was laid out on a completely new alignment sometime in the fourteenth century.[7] These examples sound a useful cautionary note against the habit of discussing settlement types and origins from the viewpoint of the

---

[1] J. C. Russell (1966), 21.

[2] M. W. Beresford, *New towns of the Middle Ages* (1967), 319–38.

[3] E. Miller (1964), 39.

[4] M. W. Beresford, *The lost villages of England* (London, 1954).

[5] Published in *Feudal aids*, 6 vols. (H.M.S.O., 1899–1920).

[6] This figure excludes places in Durham and Cheshire, palatine counties which were exempt from tax, but includes places in Cumberland, Westmorland and Northumberland taxed in 1336.

[7] M. W. Beresford and J. G. Hurst (eds.), *Deserted medieval villages* (London, 1971), 125–6.

present form of villages or at best from their form in first map evidence. In much the same way as many place-names have changed beyond recognition, so too have villages. No doubt most of the 'green' villages, whose distribution presents such a fascinating problem of interpretation,[1] existed in the early fourteenth century, but we know very little about their organisation. Planned villages were certainly a feature of the early fourteenth-century landscape, as there is increasing evidence, especially from excavation, that many villages and their surrounding fields were laid out anew in the thirteenth century.[2]

The best hope of discovering the form of English villages just before the Black Death lies in the excavation of those villages that were depopulated either in, or soon after, the early fourteenth century, and whose only traces lie 'fossilised' under the soil. Such desertions are not very numerous but there are several examples of local decline and abandonment in the early fourteenth century, for reasons unknown, for example at Beere in Devon,[3] at Pickwick in Somerset,[4] and at Heythrop, Shelswell and Langley in Oxfordshire.[5] Elsewhere the Black Death hastened the decline of many settlements that were already in difficulties; excavated examples are Hangleton in Sussex, and Seacourt in Berkshire.[6] Only a handful of villages seem to have disappeared as a direct result of the plague[7] and as yet none of these has been excavated. Examples of this type are Tilgarsley and Tusmore in Oxfordshire,[8] Middle Carlton in Lincolnshire, and Ambion in Leicestershire.[9]

The only village buildings of the fourteenth century that survive today are those which were built in stone, namely churches, manor houses and

---

[1] L. Dudley Stamp, 'The common lands and village greens of England and Wales', *Geog. Jour.*, cxxx (1964), 465; and L. D. Stamp and W. G. Hoskins, *The common lands of England and Wales* (London, 1963), 28–34.

[2] See, for example, P. Allerston, 'English village development: findings from the Pickering district of north Yorkshire', *Trans. and Papers, Inst. Brit. Geog.*, LI (1970), 95–109.

[3] E. M. Jope and R. I. Threlfall, 'Excavation of a medieval settlement at Beere, North Tawton, Devon', *Med. Archaeol.*, II (1958), 115.

[4] *Med. Archaeol.*, IV (1960), 156.

[5] K. J. Allison et al., *The deserted villages of Oxfordshire* (Leicester, 1965), 5–6.

[6] E. W. Holden, 'Excavations at the deserted medieval village of Hangleton, Part I', *Sussex Archaeol. Coll.*, CI (1963), 72; M. Biddle, 'The deserted village of Seacourt, Berkshire', *Oxoniensia*, XXVI–XXVII (1961–2), 70–201.

[7] M. W. Beresford (1954), 158–62.

[8] K. J. Allison et al., 44–5.

[9] W. G. Hoskins, *The making of the English landscape* (London, 1955), 93.

an occasional barn. While the churches survive in great numbers there are comparatively few surviving manor houses. There are, however, enough to show that the manor houses of the early fourteenth century were either of the simple, rectangular, first-floor hall type,[1] dating from the twelfth and thirteenth centuries, or of the newer end-hall types.[2] In addition, the moated manorial homestead was in fashion and was especially characteristic of late-colonised woodland areas such as Essex, Suffolk and the Arden of Warwickshire where land had been sub-infeudated to new freeholders.[3] In all manor houses the emphasis was on the hall, the home of the lord and the place of the manorial court. As the nearest equivalent to a public building it usually warranted the use of stone. By 1334 most castles had also become centres of local administration where courts were held, accounts audited and records kept.[4] Only a few, such as those on the borders with Wales and Scotland, retained their importance as military strongholds.

In contrast, while many of the peasant houses had stone wall-footings they were frail structures, and none has withstood the ravages of time. Our knowledge of them depends on archaeological rather than architectural investigation, and, while this has increased in the last twenty years with the new interest in medieval archaeology, there are still relatively few excavated peasant houses of the period. Excavation has shown that the peasant houses of the fourteenth century were of three types, the one-roomed cot, the long-house, and the farm complex.[5]

At Wharram Percy on the chalk wolds of the East Riding, deserted just after 1500, excavations have shown that the houses in use in the fourteenth century were of the long-house type, sometimes up to 90 feet in length, and with the living end separated from the lower end by a cross passage with opposite doorways.[6] Whether, in the early fourteenth century,

---

[1] As illustrated in M. Wood, *The English medieval house* (London, 1965), plate v and frontispiece.

[2] P. A. Faulkner, 'Domestic planning from the twelfth to the fourteenth centuries ', *Archaeol. Jour.*, CXV (1958), 164.

[3] F. V. Emery, 'Moated settlements in England', *Geography*, XLVII (1962), 378–88; and B. K. Roberts, 'A study of medieval colonization in the Forest of Arden, Warwick-shire', *Agric. Hist. Rev.*, XVI (1968), 101–13.

[4] N. Denholm-Young, *The country gentry in the fourteenth century* (Oxford, 1969), 32–3.

[5] M. W. Beresford and J. G. Hurst (1971), 104–12.

[6] For example house B1 in the plan of Wharram Percy, Area 10 in *Med. Archaeol.*, VIII (1964), Fig. 95; dated to the fourteenth century in *Med. Archaeol.*, IV (1960), 161.

the byre ends were used for animals, or whether they had been taken over for human use is not always clear. The presence of animals can only be certain where there is evidence of actual mangers, pens, or drains. Undoubtedly land had become precious by this period for at Wharram Percy chalk pits had been filled and the ground levelled off, and even the undercroft of the twelfth-century manor house was filled with rubbish in the first half of the fourteenth century to provide a level surface upon which to build houses.[1]

An idea of the character of a medieval village may be seen from the reconstruction drawing of Wharram Percy.[2] It shows the village as it might have looked in the early fifteenth century, but excavation has revealed that the shape of the houses was much the same a century earlier. The half-timbered dwellings of the fourteenth century rested on low stone foundations. On one house-site a complete reconstruction of the house seems to have taken place every thirty or forty years, and on occasions the alignment of the house was turned through 90 degrees from being long-side to the street to being gable-end to the street; a fourteenth-century house was overlain by three fifteenth-century houses on different alignments.[3] Similar changes took place on other house sites.[4]

One most important fact to emerge through the air survey and excavation of deserted medieval villages[5] is that the long-house was widespread over lowland England and not restricted to the highland zone as was formerly thought.[6] Excavations in many parts of the country, for instance at West Whelpington (Northumberland), Hangleton (Sussex), Holworth (Dorset), Houndtor (Devon), Gomeldon (Wiltshire), Garrow and Treworld (Cornwall),[7] have produced long-houses similar to the classic

[1] Med. Archaeol., II (1958), 206.

[2] Illustrated London News, No. 6485, 16 November 1963, 816–17.

[3] Med. Archaeol., I (1957), Fig. 34; IV (1960), 161; and VIII (1964), Fig. 95.

[4] Med. Archaeol., VIII (1964), 291. For a recent summary of changes in house alignments see M. W. Beresford and J. G. Hurst (1971), 122–3.

[5] Listed in Villages désertés et histoire économique (Ecole Pratique des Hautes Etudes – VIe. Section, S.E.V.P.E.N. Paris, 1965), 573–6.

[6] J. G. Hurst, 'The medieval peasant house', in A. Small (ed.), The Fourth Viking Congress (Edinburgh, 1965), 191.

[7] M. G. Jarrett, 'The deserted village of West Whelpington, Northumberland', Arch. Aeliana, 4th ser., XL (1962), 189–225; E. W. Holden, 73, Fig. 5; and J. G. Hurst and D. G. Hurst, 'Excavations at the deserted medieval village of Hangleton, Part II', Sussex Arch. Coll., CII (1964), 112, Fig. 6; P. A. Rahtz, 'Holworth medieval village excavations', Proc. Dorset Nat. Hist. and Archaeol. Soc., LXXX (1959), 127–47. For Houndtor, see Med. Archaeol., VI–VII (1962–3), Fig. 102 and plate XXXIIB, and

example excavated at Beere (Devon) in 1938–9,[1] mainly of thirteenth-century date, but used into the fourteenth century. The foundations of similar houses show up from the air on deserted village sites in areas of stone building, for example, in Lincolnshire and the Cotswolds. The presence of timber-built long-houses in the Midlands is highly probable but difficult to prove as they have left little trace. While the long-house was obviously not the only kind of medieval peasant house it must have been a common element in the rural landscape of the early fourteenth century. But our knowledge of this and other house types, and of the most interesting problems of the use of wood and stone in peasant house building, await the excavation of more fourteenth-century sites.

The agrarian landscape

Most peasant houses stood within small enclosures. At Wharram Percy all the houses stood in small enclosures which were used for pasture, vegetables, pigs or poultry. Behind these, and stretching back to the edge of the open fields, were the crofts, larger areas held in severalty and used either for arable or pasture, as they were in some Leicestershire villages.[2] On Midland deserted village sites, where timber houses have left no trace, it is often the small enclosures and crofts which show up most clearly from the air.[3]

To present any picture of the common fields in the early fourteenth century is much more difficult. It is hazardous to generalise, if only because we now know that 'field' arrangements varied enormously from place to place according to social organisation, physical conditions and types of farming.[4] While there is evidence for the existence of some kind of open-field cultivation in every English county at some time,[5] the

VIII (1964), Figs. 90 and 91; for Gomeldon, see *Med. Archaeol.*, VIII (1964), Fig. 94; D. Dudley and E. M. Minter, 'The medieval village at Garrow Tor, Bodmin Moor, Cornwall', *Med. Archaeol.*, VI–VII (1962–3), 272–94, esp. Fig. 88; for Treworld see *Med. Archaeol.*, VIII (1964), 282.

[1] E. M. Jope and R. I. Threlfall, Fig. 26.

[2] R. H. Hilton in W. G. Hoskins and R. A. McKinley (eds.), *V.C.H. Leicestershire*, II (1954), 166.

[3] E.g. Cestersover, Warwickshire, in M. W. Beresford (1954), plate 15, and Burston, Buckinghamshire, M. W. Beresford and J. K. S. St Joseph, *Medieval England: an aerial survey* (Cambridge, 1958), Fig. 44.

[4] A. R. H. Baker, 'Howard Levi Gray and *English field systems*: an evaluation', *Agric. Hist.*, XXXIX (1965), 86–91.

[5] J. Thirsk, *Tudor enclosures* (London, 1959), 4, footnote.

overall situation as it was at any one time, at say about 1330, cannot be pictured. We do not yet know, for example, the extent of enclosed land at this time.

Clearly in the early fourteenth century most of the villages in H. L. Gray's 'Midland Zone' were surrounded by their open fields (Fig. 23). The details of field arrangement varied with local conditions, and in some districts two- and three-field systems existed side by side, as for example in Wiltshire,[1] and Leicestershire. Yet in some Leicestershire villages it was often the furlong and not the open field which was the main cropping unit.[2] This implies that even for the Midlands we can no longer envisage, without qualification, villages surrounded by large open fields each given over to one crop or lying fallow. Invariably some arable land was held in the separate closes of holdings created as a result of thirteenth-century assarting.[3] We must therefore allow for considerable flexibility in the system by 1330, and picture some villages with different furlongs under various crops.[4] Relatively advanced crop rotations must have developed earlier than we have hitherto thought. By the early fourteenth century, three-course rotations were practised on both demesne and tenant holdings, as for example in the Chilterns,[5] and there is increasing evidence that a more refined four-course rotation, which included legumes, was becoming increasingly common, for example on the estates of Ramsey Abbey,[6] and at Westerham.[7]

Even within the furlongs, rigidity was further broken down as the holdings of landlord and tenant, perhaps once arranged in a particular order, became scattered irregularly as a result of exchange and amalgamation in the previous two centuries. In addition, by the early fourteenth century much demesne land had been enclosed and was valued more highly than demesne land cultivated in the common fields. For example,

---

[1] W. G. Hoskins in E. Crittall (ed.), *V.C.H. Wiltshire*, IV (1959), 14.

[2] R. H. Hilton in W. G. Hoskins and R. A. McKinley (eds.), *V.C.H. Leicestershire*, II (1954), 159–61.

[3] Well illustrated in C. C. Taylor, 'Whiteparish: a study of the development of a forest-edge parish', *Wilts. Archaeol. and Nat. Hist. Mag.*, LXII (1967), 90–1.

[4] V. H. T. Skipp, and R. P. Hastings, *Discovering Bickenhill* (Birmingham, 1963), 15–18.

[5] D. Roden, 'Demesne farming in the Chiltern Hills', *Agric. Hist. Rev.*, XVII (1969), 9–23.

[6] J. A. Raftis, *The estates of Ramsey Abbey* (Toronto, 1957), 220.

[7] T. A. M. Bishop, 'The rotation of crops at Westerham, 1297–1350', *Econ. Hist. Rev.*, IX (1934), 38–44.

in Wiltshire, at Wootton Bassett 126 acres of demesne lay in severalty in 1334, and at Draycote Cerne 66 of the 240 acres of demesne had also been enclosed by 1344.[1]

H. L. Gray's recognition that the field systems of the Midlands differed from those of the west and east still holds good, although his ethnic reasons for the origin of these differences are slowly being revised. It has recently been argued that the differences between the areas of regular two- or three-field systems and areas of irregular field arrangements stem from a contrast between arable and pasture farming;[2] and that many pastoral areas of west and east which have hitherto been thought of as 'early enclosed' may be so called not because enclosure proceeded early in these areas but because they had never known a fully developed common-field system. In addition it is now clear that systems of infield and outfield were much more frequent throughout England than was formerly supposed, especially on light and less fertile land.

While there are differences of opinion on field systems[3] (and on their terminology), most authorities would agree that the agricultural landscapes of much of East Anglia, Kent, the Welsh borderlands, Somerset, and the south-west, must have looked very different from those of the Midlands in the early fourteenth century. In both west and east much land was parcelled up into numerous small enclosures; hedgerows were more important in the landscape, and field and tenurial arrangements were more flexible.

Arable farming

The available evidence suggests that the early fourteenth century, a time of rapid inflation, was a period of instability in agriculture. The first phase, before about 1315 was, as it were, a left-over from the thirteenth century, a period of fifteen years in which the overall picture was still one of prosperity. After about 1320, conditions became more confused. We must presume that the effects of the famine of 1315–17 varied from place to place, and it is therefore not surprising that evidence from different estates suggests that agricultural fortunes varied considerably in the period 1320–45. In some places there appears to have been only temporary

[1] W. G. Hoskins in E. Crittall (ed.), *V.C.H. Wiltshire*, IV (1959), 13.

[2] J. Thirsk, 'The common fields', *Past and Present*, XXIX (1964), 3–25.

[3] See J. Z. Titow, 'Medieval England and the open field system', *Past and Present*, XXXII (1965), and the reply by J. Thirsk, *Ibid.*, XXXIII (1966); G. C. Homans, 'The explanation of English regional differences', *Past and Present*, XLII (1969), 18–34.

dislocation and a fairly quick return to reasonable agricultural productivity, sometimes accompanied by changes of land use. In others the effect of the famine seems to have been more prolonged and there was no return to early fourteenth-century levels of production. To suggest this contrast before and after the 1320s is to make a hazardous generalisation, and what might be true for one area was not true for another. Moreover because peasant agriculture has left little written record, the evidence must be based on the accounts of a few large estates, mainly in the south-east of the country, where records of profit and loss were meticulously kept. For it was on these estates, where capital and efficiency in administration overcame the difficulties of scattered demesne production, that the main commercial agricultural enterprise of the early fourteenth century was to be found. The monastic houses, in particular, were best suited to efficient production not only because of their better administration,[1] but because in some cases their estates were more compact, for example, those of Battle abbey in Sussex and of St Augustine's priory in Canterbury.

Whereas demesne land gave scope for specialised production we know very little of the fortunes of the peasants tied down to customary practice on small and scattered holdings, and tilling the land for their own needs. Not that we should underestimate the role of the peasant in the food supply of medieval England. On the bishop of Winchester's estates in the thirteenth century, for example, a considerable amount of the town demand was met by local peasant produce.[2] This is likely to have been even more true in the second quarter of the fourteenth century when more demesne was being let out by landlords, and when peasants had need of a higher income to meet increasing rents. Consequently there was an increase in the number of land transactions among peasants, as for example on the estates of Titchfield abbey in Hampshire.[3] The network of local markets in early fourteenth-century England, inherited from the previous century, was partly a response to peasant production.

The first two decades of the century were years of prosperous demesne farming in the best thirteenth-century tradition.[4] On estates as far apart as Durham and Canterbury, Peterborough and Glastonbury, more land

---

[1] D. Knowles, *The Religious Orders in England*, I (Cambridge, 1956), 32.

[2] E. Kosminsky, 'The evolution of feudal rent in England from the xith to the xvth Centuries', *Past and Present*, VII (1955), 20.

[3] D. G. Watts, 'A model for the early fourteenth century', *Econ. Hist. Rev.*, 2nd ser., XX (1967), 543–4.

[4] E. Miller (1964), 32.

had come under cultivation, and administration and husbandry were more efficient.[1] The Benedictine houses in particular, by a careful division of the types of farming on their estates, were able to specialise in cereal farming in contrast to the specialisation in wool that was characteristic of the Cistercians.[2] A good example of a commercialised monastic estate was that of Canterbury cathedral priory, midway between the markets of London and the Continent, where grain production reached a peak in the period 1306–24.[3] The agricultural achievement of this priory helped to make north-east Kent one of the most prosperous parts of England in 1334.[4] On the priory estates increasingly efficient commercial agriculture was partly achieved by a drastic change in the system of the food rents, whereby distant manors which had formerly supplied food to the priory took advantage of rising grain prices to sell their produce at favourable prices in local markets.

The cash crop of midland and southern England was wheat. Barley, although more widely grown, fetched lower prices; so did oats and also rye, the crop of the light soils. The market for grain, especially for wheat, in the early fourteenth century was an incentive to improve yields, both of seed and of land. Not that any dramatic overall results were achieved. Yields of grain remained low by modern standards throughout the thirteenth and fourteenth centuries.[5] The statistics are full of difficulties,[6] but a probable average yield of 8–9 bushels of wheat per acre represents little over a quarter of an expected yield today.[7] Similarly, there has been roughly a threefold increase in the yield of seed since the medieval period.[8] Despite the efforts on the estates of Canterbury priory to sow seed more intensively and to act on the advice of Walter of Henley to get their seed from outside, the yields showed no great improvement and remained

[1] D. Knowles, 44–7.

[2] R. A. L. Smith, 'The Benedictine contribution to medieval agriculture', in *Collected Papers* (London, 1947), 103–16.

[3] R. A. L. Smith, *Canterbury cathedral priory* (Cambridge, 1943), 128–45.

[4] R. E. Glasscock (1965), 62–8.

[5] Lord Beveridge, 'The yield and price of corn in the Middle Ages', *Economic History*, I (1927), 155–67, reprinted in E. M. Carus-Wilson (ed.), *Essays in economic history* (London, 1954).

[6] R. Lennard, 'Statistics of corn yields in medieval England', *Economic History*, III (1934–7), 173–92 and 325–49.

[7] M. K. Bennett, 'British wheat yield per acre for seven centuries', *Economic History*, III (1934–7), 21.

[8] Lord Beveridge, 158–9.

below expectations.[1] As on many other estates, attempts were also made to improve the soil by draining, dunging and marling. In coastal areas where marl was not available, seaweed and sand were used to sweeten the land.[2]

In the drive to increase cereal production and take advantage of rising prices not only was agricultural efficiency stepped up, but the acreage under crops was increased. Tillage was increased by the conversion of pasture land, and on the Canterbury priory manors by the reclamation of marshland which was particularly suited to oats. The result of all these efforts at Canterbury may be seen in Prior Henry of Eastry's remarkable survey of 1322,[3] which shows the great area under wheat and barley in east Kent, under oats on the marshland manors, and the importance of legumes in the various rotations (Fig. 36).

The Canterbury priory estates enjoyed of course special advantages, firstly in possessing the redoubtable Prior Henry, one of the most enlightened landowners of his time,[4] and secondly in their nearness to the markets of London and the Continent. But at the other end of the country on the estates of Durham cathedral priory, similar improvements were taking place, and there was a corresponding peak in grain production at the beginning of the fourteenth century. Increased grain production was not merely the result of the cultivation of more land, but also of improved techniques.[5] Elsewhere also the monastic houses gave a lead in agricultural improvement at this period, for experiment was only possible where capital was available and where demesne land was held in severalty. On some estates, enclosure and improvement went hand in hand, for example on the lands of the abbey of Evesham.[6]

While improvement may be detected it is not surprising that there was no great breakthrough in medieval agriculture because the level of investment was extremely low. Even on the best run monastic and lay estates gross investment was usually under 5%.[7] The likelihood is that it was lower on smaller estates, although it must be admitted that the role of

[1] R. A. L. Smith (1943), 133–5.

[2] H. P R. Finberg, *Tavistock Abbey* (Cambridge, 1951), 89.

[3] R. A. L. Smith (1943), 140–1.

[4] D. Knowles, 49–54.

[5] E. M. Halcrow, 'The decline of demesne farming on the estates of Durham cathedral priory', *Econ. Hist. Rev.*, 2nd ser., VII (1955), 345–56.

[6] R. A. L. Smith (1947), 110.

[7] M. M. Postan, 'Investment in medieval agriculture', *Jour. Econ. Hist.*, XXVII (1967), 579.

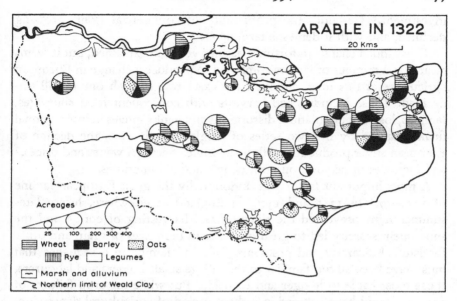

Fig. 36 Arable in 1322 on Canterbury cathedral priory estates in Kent
Based on R. A. L. Smith, *Canterbury cathedral priory* (Cambridge, 1943),
140–1.

medium-size estates within medieval agriculture is, as yet, largely un-
explored. Capital invariably went into the purchase of more land or the
improvement of buildings, neither of which did anything to improve
overall economic development or to increase crop yields. Profits went
everywhere except into improving agricultural production. And if a low
level of investment was characteristic of the lords then it was even more
true of the peasants, three-quarters of whom must have lived at near
subsistence level with little chance to save or to invest anything in their
agriculture.

By about 1330 there were unmistakable signs that agriculture was
less prosperous, for not only were the prices of cereals falling and rents
and wages rising, but in a few places land was lying uncultivated, all
symptoms of a decreasing demand.[1] In addition, after 1330, landlord
interest in acquiring new property slackened off (partly because the
squeeze between wages and prices reduced profits), and sometimes ceased
entirely, a sure sign that land was no longer thought of as a good

[1] M. M. Postan (1950), 225–40.

investment.[1] Certainly the profitability of commercial grain growing declined, and with it demesne farming.[2]

This sudden change in fortunes has not yet been explained, but it seems clear that a number of factors combined to produce a change in European agriculture at this time, although the exact role of each one is still undecided. These included bad harvests with consequent food shortages, famines, losses of population, disruptions and bankruptcies in international finance caused by the two series of Anglo-French wars, the decline of European silver production with its possible effects on wages and prices,[3] and increases in taxation on both people and commodities.

A most important factor was undoubtedly the great European famine of 1315–17.[4] In 1315 and 1316, in England as elsewhere, heavy late-summer rains prevented the ripening and harvesting of corn,[5] and the subsequent scarcity led to great increases in the prices of grain and other foodstuffs.[6] Scarcity and pestilence followed famine in a sequence that must have been all too familiar to the village small-holders and labourers, those most liable to hunger and hardship. But while the crop failures of 1315–16 could have ushered in a short period of agricultural depression, they can hardly be held responsible for a depression that was to last a century and a half. It is not yet known whether the bad weather, and in particular, the heavy summer and autumn rainfall of these years was a local intensification of a wider climatic deterioration at this period. There is some evidence to suggest that it might have been. The severity of the climate at this period is well known,[7] and climatic change may well have

---

[1] G. A. Holmes, *The estates of the higher nobility in fourteenth-century England* (Cambridge, 1957), 113–14.

[2] For a case study illustrating this see I. Keil, 'Farming on the Dorset estates of Glastonbury abbey in the fourteenth century', *Proc. Dorset Nat. Hist. and Archaeol. Soc.*, LXXXVII (1965), 234–50.

[3] J. Schreiner, 'Wages and prices in England in the later Middle Ages', *Scand. Ec.H.R.*, II (1954), 61–73.

[4] H. S. Lucas, 'The great European famine of 1315, 1316 and 1317', *Speculum*, V (1930), 343–77.

[5] J. Titow, 'Evidence of weather in the account rolls of the bishopric of Winchester, 1209–1350', *Econ. Hist. Rev.*, 2nd ser., XII (1960), 385–6.

[6] J. E. T. Rogers, *A history of agriculture and prices in England*, 7 vols. (Oxford, 1866–1902), I, 230; N. S. B. Gras, *The evolution of the English corn market* (Cambridge, Mass., 1915), 47.

[7] E. Huntington and S. S. Visher, 'The climatic stress of the fourteenth century', being ch. 6 of *Climatic changes* (New Haven, 1922), 98–109; H. H. Lamb, 'Britain's changing climate', *Geog. Jour.*, CXXXIII (1967), 454–60.

been important in the economic life of Europe in the fourteenth and fifteenth centuries.[1] Certainly we cannot dismiss the possibility that it was a climatic deterioration, and especially heavier summer rainfall, that triggered off the chain reaction of poor harvests, lack of seed corn for the following year, famine and pestilence, economic dislocation, a declining population, the reduction of the cultivated area, and political unrest. All these appeared in England well before 1348 when the Black Death caught the country at such a low ebb.

Prices of agricultural produce, especially of wheat and barley, began to decline steadily after about 1320, although there was a temporary recovery in some years, for example in 1330 and 1331.[2] The problem of declining prices and what it meant to an estate dependent upon selling agricultural produce is well illustrated on the manors of the bishop of Ely at Great Shelford and at Wisbech in Cambridgeshire.[3] At Shelford, wheat which had sold at between 10s. 6d. and 14s. a quarter in the early twenties, sold at an average of 5s. 6d. in the period 1325–33, and only just over 4s. in the years around 1340. The revenue on both manors dropped drastically between 1325 and 1348, a drop that was due for the most part, though not entirely, to the great fall in prices. A consequence of this fall was that less grain was marketed locally and more was sent instead direct to the bishop's household. Moreover, less capital was available for the maintenance and improvement of the manors. Higher wages and declining income from sales had to be offset by increasing income from rents, both by letting out more demesne and by commuting labour services. The bishop of Ely benefited from the new high rents which thirteenth-century inflation had made possible, and the letting out of demesne continued during the period before the Black Death. On his Cambridgeshire estates, at Downham, for example, 264 acres were let out during 1299–1337, at Stretham 135 acres during 1316–46, and at Linden End 362 acres during 1316–45. At Wisbech, the income from demesne leaseholds rose from 53s. 4d. in 1320 to £48 10s. in 1345.[4]

The same was happening on the estates of Ramsey abbey,[5] and towards

[1] G. Utterstrom, 'Climate fluctuations and population problems in early modern history', *Scand. Ec.H.R.*, III (1955), 3–47.

[2] J. E. T. Rogers, I, 230–2, and N. S. B. Gras, 60. D. L. Farmer, 'Some grain price movement in thirteenth century England', *Econ. Hist. Rev.*, 2nd ser., X (1957), 207–20.

[3] E. Miller, *The abbey and bishopric of Ely* (Cambridge, 1951), 105–12.

[4] E. Miller (1951), 100.

[5] J. A. Raftis, 241.

the end of the thirteenth century Cistercian holdings began to be regularly leased.[1] Sometimes, as on the abbot of Westminster's estates, leasing involved just a few acres piecemeal, sometimes whole demesnes.[2] Commutation of labour services increased; thus on 15 of 81 manors in twenty counties in 1350, labour services had either entirely disappeared or were insignificant.[3] Both leasing and commutation increased the freedom of the tenantry and in the long term contributed to the break-up of the manorial economy.

While these were general characteristics of the period, the pace of change varied from place to place. Thus on many of the Durham manors the sales of grain increased after 1340, and the decline of demesne farming did not take place until the second half of the fourteenth century,[4] nor did it on the Berkeley estates.[5] On the estates of the archbishopric of Canterbury there was slow contraction of demesne land under cultivation throughout the fourteenth century.[6] In Leicestershire, demesne farming flourished well into the fifteenth century.[7] Generally speaking, however, the 1330s and 1340s were times when landlords, lay and ecclesiastic, found money was much easier to come by through rents than through sales.

References to *terra frisca* (uncultivated land) in the *Nonarum Inquisitiones* of 1342 show that much land was lying uncultivated in various parts of the country in 1341. The record of uncultivated land is very uneven, and there is difficulty in knowing what exactly the term means from one place to another.[8] In some cases it may have implied merely a change in land use or the temporary abandonment of outfield. Nevertheless it seems clear that much land which was formerly cultivated was no longer so, and the same impression is gained from *Inquisitiones Post Mortem* of the period. In counties that have been studied in some detail the contraction

[1] R. A. Donkin, 'Cattle on the estates of medieval Cistercian monasteries in England and Wales', *Econ. Hist. Rev.*, 2nd ser., xv (1962), 44.

[2] B. Harvey, 'The leasing of the abbot of Westminster's demesnes in the later Middle Ages', *Econ. Hist. Rev.*, 2nd ser., xxii (1969), 17–27.

[3] E. Lipson, *The economic history of England* (12th ed. London, 1959), I, 95–7.

[4] E. M. Halcrow, 347.

[5] Lord Ernle, *English farming past and present* (6th ed. London, 1961), 46.

[6] F. R. H. Du Boulay, 218–19.

[7] R. H. Hilton, *The economic development of some Leicestershire estates in the fourteenth and fifteenth centuries* (London, 1947).

[8] A. R. H. Baker, 'Evidence in the *Nonarum Inquisitiones* of contracting arable lands in England during the early fourteenth century', *Econ. Hist. Rev.*, 2nd ser., xix (1966), 518–32.

of arable land was, as we might expect, more marked on the uplands than the lowlands. In some places waste land was attributed to poverty; for other places the entry was more specific, as for example at Stockland and Compton abbey in Dorset where the tenants had departed and their land lay uncultivated.[1] On the other hand, uncultivated land in seven Wealden parishes amounted to over 2,000 acres yet only for two parishes was it specifically attributed to poverty, and some of the parishes concerned were more prosperous in 1341 than they had been fifty years before.[2] Cambridgeshire had almost 5,000 acres out of cultivation, yet this was exactly the period when the nearby Huntingdonshire estates of Ramsey abbey were enjoying a temporary revival of the prosperity of earlier years.[3] It would seem that the references to uncultivated land in the *Nonarum Inquisitiones* cannot be regarded as indicators of widespread retreat but they may be seen as symptomatic of places which were falling on hard times.

Pasture farming

Livestock formed an integral part of medieval agriculture not only in the pastoral west of the country but in the drier east where the emphasis was on grain production. Oxen for ploughing, plough and cart horses, and cattle for milk and hides were essential to every village. Pasture farming was a useful stand-by in years of bad harvests, and the prices of animals fluctuated less than those for grain which varied sharply from year to year.[4] Sheep were valued throughout the country for their wool, mutton, and skins, as is clear from the frequency with which animals are listed on the local rolls of taxation of the early fourteenth century.[5]

There may have been about eight million sheep in England in the mid-fourteenth century.[6] Peasant flocks often exceeded those of demesne but we still know very little about peasant wool production.[7] In addition to

[1] B. Reynolds, 'Late medieval Dorset; three essays in historical geography', unpublished M.A. thesis, University of London, 1958, 166–7.

[2] J. L. M. Gulley, 'The Wealden landscape in the early seventeenth century and its antecedents', unpublished Ph.D. thesis, University of London, 1960, 345–8.

[3] J. A. Raftis, 241.

[4] D. L. Farmer, 'Some livestock price movements in thirteenth-century England', *Econ. Hist. Rev.*, 2nd ser., XXII (1969), 1–16.

[5] J. F. Willard (1934), 73.

[6] R. A. Pelham in H. C. Darby (ed.), *An historical geography of England before A.D. 1800* (Cambridge, 1936), 240.

[7] E. Power, *The wool trade in English medieval history* (Oxford, 1941), 29–31.

the many thousands of animals in the hands of the peasants there were the large flocks of the lay and ecclesiastical landlords. Sheep-farming as a source of wealth was the oldest of all forms of commercial farming, and retained its importance to such great monastic houses as those of Ely, Peterborough, Glastonbury, Crowland, and Canterbury cathedral priory.[1] The Cistercians, having access to great areas of moorland sheep-walk, specialised in wool production, and on their largest estates, such as Fountains in the West Riding, flocks numbered anything up to 15,000. The limestone uplands of the Pennine flanks, the Peak District, the Cotswolds, Lincolnshire and the chalklands of southern England were the great sheep runs of fourteenth-century England. In the south, the flocks of the bishop of Winchester, and of the priory of St Swithun in Winchester compared in number with those of the great Cistercian houses in the north. Battle abbey ran flocks of up to 3,000 sheep on the Sussex downs but this was only a small proportion of the total in the county. It would seem that in 1340–1 there were about 110,000 sheep in Sussex;[2] while they were most numerous on the chalk downs there were very large numbers on the grain-growing land of south-west Sussex owned by tenant farmers who grazed their animals on open-field fallow and waste. Even on the lands of Canterbury cathedral priory, the greatest of the grain-producing estates, wool production and dairy farming were just as important as cereals, especially on manors with extensive marshland pastures. Here, the peak years of wool production were 1319–21,[3] so that the survey of 1322 comes just at the right time to show the importance of sheep especially on Thanet (Fig. 37).

Numbers of sheep varied, of course, from year to year according to the incidence of disease, but on many Cistercian estates the numbers of sheep began to drop steadily well before the middle of the fourteenth century.[4] At Canterbury also the flocks never recovered from losses by floods and disease in 1324–6. While it is difficult to isolate trends in pasture farming in the early fourteenth century it is clear that the decline in arable farming was not due to expansion in pasture and livestock. With sheep so widely distributed it is not surprising that wool varied greatly in quality from the

[1] D. Knowles, I, 41–2.

[2] R. A. Pelham, 'The distribution of sheep in Sussex in the early fourteenth century', *Sussex Archaeol. Coll.*, CXXV (1934), 128–35.

[3] R. A. L. Smith (1943), 156.

[4] R. A. Donkin, 'The Cistercian Order in medieval England: some conclusions', *Trans. and Papers, Inst. Brit. Geog.*, XXXIII (1963), 191.

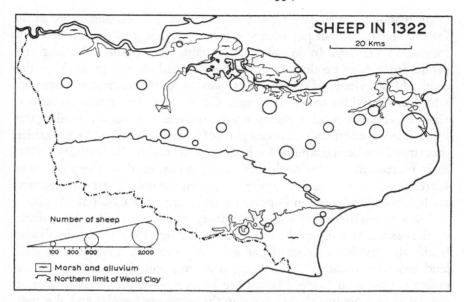

Fig. 37 Sheep in 1322 on Canterbury cathedral priory estates in Kent
        Based on R. A. L. Smith, *Canterbury cathedral priory* (Cambridge, 1943), 156.

poor, coarse wools of the south-west to the fine wools of Shropshire and Lincolnshire. To judge from the prices in 1343 the best wools were the short wools of the Ryeland sheep of the Welsh border country and the west Midlands, and the long wools of the Leicester and the Lincoln sheep. Pennine wool, although not highly priced, was nevertheless far more in demand than the medium-quality wools of the chalklands, East Anglia, and the marshlands of south-east England.[1]

While oxen were kept for ploughing and carting, cattle met the need for milk and hides. The scarcity of winter feed no doubt placed restrictions on the number of cattle, for whereas pigs could root around on any waste ground and sheep exist on the poorest grazing, cattle could survive the winter only with sufficient hay, together with legumes when available. Generally, the animals supplied only the needs of the manor and village, but where there was extensive marshland grazing such as on Romney Marsh, the Fens, and the Somerset Levels, some specialised dairying and cheesemaking took place. The Canterbury manors of Romney and east Kent provide examples of efficient stock rearing and dairying.[2] On some

[1] R. A. Pelham (1936), 245.   [2] R. A. L. Smith (1943), 157–8.

of the manors, where dairying was only a sideline, the leasing out of cattle for milking for a cash payment was a custom that grew up in the early fourteenth century. In keeping with arable and pasture farming the profits from dairy produce at Canterbury waned after 1327. Perhaps the death of Prior Henry of Eastry in 1331 was one of the contributory reasons why these activities never recovered. Cattle were also important on the Cistercian estates.[1] Cattle grazing was an important means of extending the limits of occupation, and the upper parts of Nidderdale and other Yorkshire dales may have been colonised for pasture in the late twelfth century. In the early fourteenth century cattle were more important than sheep on some Cistercian estates in central Wales, and from the mid-fourteenth century cattle farming increased in importance on many other Cistercian lands.[2]

Throughout the early fourteenth century inundations had serious effects on the pasture farming of the coastal areas through the loss of animals and land. On the marsh manors of Canterbury, protection of newly reclaimed land required constant embanking, draining, and the building of sea defences.[3] Around Canvey Island the fourteenth century was a time of continual fight against the sea,[4] and in the Somerset Levels[5] and the fenlands of Norfolk and Lincolnshire,[6] floods and devastation were frequent.

Lay wealth and agriculture

If goods taxed in 1334 represent the surplus that was for sale, then we might expect to find some relationship between areas with high assessments and those of the greatest grain and wool production. The year 1334 was a good year for the sale of both wool and grain. The export of wool was rising following the abolition of the home staples and the restoration of free trade in 1328. Although grain exports were lower in the reign of Edward III (1327–77) than earlier, they were still considerable, and the year 1332, on which the tax quotas of 1334 were based, was a good year and grain prices were low.[7] In 1334 we hear of seven merchants being licensed to export 52,000 quarters of wheat to Gascony.[8]

[1] R. A. Donkin (1962), 31–53.   [2] R. A. Donkin (1963), 191–2.
[3] R. A. L. Smith (1943), 186–9.
[4] B. E. Cracknell, *Canvey Island: the history of a marshland community* (Leicester, 1959), 12.
[5] P. J. Helm, 'The Somerset Levels in the Middle Ages (1086–1539)', *Jour. Brit. Archaeol. Assoc.*, XII (1949), 37–52.
[6] H. C. Darby, *The medieval Fenland* (Cambridge, 1940), 58–9.
[7] J. Titow (1960), 363; and N. S. B. Gras, 60.
[8] M. McKisack, *The fourteenth century, 1307–1399* (Oxford, 1959), 349.

Wheat and barley were the preferred cereals over most of south-east England, although oats were very widely grown and rye was important on the lighter soils. On the basis of average wheat prices between 1259 and 1500 N. S. B. Gras divided the country into somewhat arbitrary price regions, in six of which the price of wheat did not exceed 6*s*. per quarter.[1] These areas of low wheat prices, all in lowland England, were almost certainly areas of high wheat production. Three of his six regions, the Upper Thames basin, the Cambridge region (apart from the peat fens), and the Norwich region, coincided with the areas of the highest assessments in 1334. The fourth region, the Lower Severn, was an area of medium-to-high assessments, while the remaining two, the Bristol area and east Suffolk, were, on the whole, areas of low assessments. South-east Lincolnshire, north Kent, the Thames valley and the south coast, however, were areas of high assessments in 1334 but not among the lowest-priced wheat areas. There is little doubt that wheat was the main crop in these regions, and the price may have been kept above 6*s*. a quarter by the demand, especially in Kent and in Sussex from the Continent,[2] and in the Thames valley from London.[2] On the whole there is a definite coincidence between areas of high assessment in 1334 and those of considerable wheat production.

No such relationship existed between the 1334 tax and the important wool-producing areas. A valuation of 1343 (nine years later) shows that the highest priced wools were those of the Welsh border, the west Midlands, Leicestershire, and Lincolnshire. All of these, with the exception of eastern Lincolnshire, were relatively poor areas in 1334, with assessed wealth mostly below £10 per square mile. Even the chalklands, where a greater number of sheep might be expected, had low assessments. Nor was there any coincidence between low wool price areas and high assessments, not that this would be expected because wool was priced on quality not quantity. Some low-price counties, for example Norfolk and Berkshire, had a high 1334 tax; others such as Surrey, Hampshire and Suffolk had a low tax.

The pattern was very irregular and no conclusions can be drawn at present in the face of so many unknown quantities. Certain surviving local taxation rolls of before 1334 show that sheep and lambs were fairly numerous in some districts, yet they were 'far from being as plentiful as

[1] N. S. B. Gras, 41 and 47. See also E. Kneisel, 'The evolution of the English corn market', *Jour. Econ. Hist.*, XIV (1954), 46–52.

[2] R. A. Pelham (1936), 238.

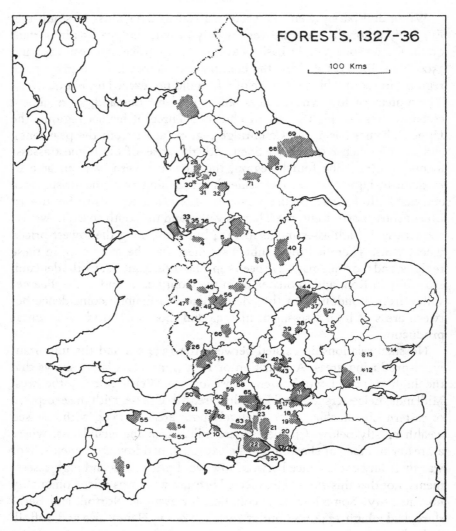

Fig. 38 Forests, 1327–36
      Based on N. Neilson, 'The forests', being ch. 9 of J. F. Willard and W. A.
      Morris (eds.), *The English government at work, 1327–1336*, I (Cambridge,
      Mass., 1940), map v.
      Names of forests are given opposite.

Key to Fig. 38

Berkshire
1 Windsor
Buckinghamshire
2 Bernwood
Cheshire
3 Wirral
4 Delamere
5 Macclesfield
Cumberland
6 Inglewood
Derbyshire
7 High Peak
8 Duffield
Devonshire
9 Dartmoor
Dorset
10 Gillingham
Essex
11 Essex
12 Writtle
13 Hatfield
14 Kingswood
Gloucestershire
15 Dean
Hampshire
16 Pamber
17 Bagshot
18 Alice Holt
19 Woolmer
20 Bere by Porchester
21 Bere by Winchester
22 New Forest
23 Buckholt
24 Chute
25 Isle of Wight
Herefordshire
26 Hereford Hay
Huntingdonshire
27 Wauberghe
Lancashire
28 Quernmore
29 Bleasdale
30 Myerscough
31 Fulwood
32 Blackburn
33 Simonswood
34 West Derby

35 Toxteth
36 Croxteth
Northamptonshire
37 Rockingham
38 Salcey
39 Whittlewood
Nottinghamshire
40 Sherwood
Oxfordshire
41 Wychwood
42 Shotover
43 Stowood
Rutland
44 Rutland (or Leighfield)
Shropshire
45 Lithewood
46 Wellington
47 Stretton
48 Shirlet
49 Morfe
Somerset
50 Kingswood
51 Mendip
52 Selwood
53 Neroche
54 North Petherton
55 Exmoor
Staffordshire
56 Kinver
57 Cannock
Wiltshire
58 Braden
59 Chippenham
60 Pewsham
61 Melksham
62 Selwood
63 Clarendon
64 Chute
65 Savernake
Worcestershire
66 Feckenham
Yorkshire
67 Galtres
68 Spaunton
69 Pickering
70 Bowland
71 Knaresborough

the amount of wool raised in medieval England, and the descriptions of large herds would render necessary'.[1] The low numbers would be understandable if they represented a surplus over and above the essentials needed by a family. Nor do we know whether the valuation of sheep varied from place to place. Were the Welsh border sheep, for example, valued at the same sum as marshland sheep? Were they valued on their size or fleece? In addition to these unanswered questions, another serious drawback is the fact that sheep on monastic demesne land were excluded from the subsidy.

In conclusion, therefore, the assessments suggest that the wealthiest part of England in 1334 was the grain country of the south and east Midlands, long settled and intensively farmed. The wealth of coastal Sussex rested on a sheep-and-corn husbandry in exceptionally favourable physical conditions.[2] With the exception of Norfolk and north Kent, these rich lands were situated within H. L. Gray's zone of the Midland field system. In contrast, there was little movable wealth in those parts of the country famous for their wool. It must be remembered, however, that a low assessment did not necessarily mean that an area was poor. While its resources of movable goods (grain and stock) may have been low in comparison with those of other places, it may have had non-movable resources, for example, in timber. Almost all the main forests were areas of very low assessments, e.g. Windsor, Epping, the New Forest and the Forest of Dean. The same was true of the woodlands of Essex, the Weald, the Chilterns and the west Midlands.

Forests, woodland and parks

A considerable area of England was still subject to forest law in the early fourteenth century despite the fact that the extent of royal forest had diminished by about a third since about 1250. There seem to have been at least 71 forests in the period 1327–36.[3] The largest were the New Forest, and the forests of Dean, Sherwood, Essex, Windsor, Inglewood and Pickering. Their distribution was very uneven; whereas one half of Hampshire and about a third of Wiltshire, for example, were subject to forest law, East Anglia was entirely free from it (Fig. 38).

---

[1] J. F. Willard (1934), 73.

[2] R. E. Glasscock (1965), 61–8; and P. F. Brandon, 'Demesne arable farming in coastal Sussex during the later Middle Ages', Agric. Hist. Rev., XIX (1971), 113–34.

[3] N. Neilson, 'The forests', being ch. 9 of J. F. Willard and W. A. Morris (eds.), The English government at work, 1327–1336, I (Cambridge, Mass., 1940), 394–467.

The forests, which varied in character from open ground to dense woodland, were not only timber reserves but often important grazing areas for deer, cattle, pigs and horses, as at Duffield Frith in Derbyshire and Needwood in Staffordshire.[1] Forest timber served to augment the patches of woodland that still remained after centuries of clearing and colonisation. The value of timber and its general use meant that charcoal-burners, carpenters, smiths and other craftsmen lived and worked in and around almost all the remaining stands of woodland.[2] Local shortage of timber was an additional incentive for landlords to enclose land for small parks in the early fourteenth century; thus in Wiltshire, we know that there were two enclosed parks in Wootton Bassett by 1334, a 200-acre park at Colerne in 1311, a 95-acre park in Oaksey in 1347, and another park in Castle Combe by 1328.[3] Emparking of this kind was also taking place on the Evesham estates, and from such enclosures coppice wood and loppings were sold for firewood. In the Weald the number of parks increased from 46 in about 1300 to 68 in 1350,[4] and they varied in size from 60 to 4,000 acres. While they were mainly recreational their assets were not wasted, and, in addition to their use as timber reserves, they also provided grazing. In some there was assarting and intermittent cropping, as at Westerham in Kent.[5]

Agricultural regions?

Before concluding this section on the countryside we might ask whether there were signs in the early fourteenth century of the distinctive local economies that were evident by the sixteenth. Can their origins be seen before the Black Death or did they develop only after 1350, when the emergence of regional farming was strengthened by the break up of highly organised demesne farming?[6] Technology and exchange had not progressed far enough by the early fourteenth century to allow much specialisation. While it is true that certain parts of the country came under the strong influence of the monastic houses this does not mean that agriculture

---

[1] J. R. Birrell, 'The forest economy of the Honour of Tutbury in the fourteenth and fifteenth centuries', *Univ. of Birmingham Hist. Jour.*, VIII (1962), 117–19.

[2] J. Birrell, 'Peasant craftsmen in the medieval forest', *Agric. Hist. Rev.*, XVII (1969), 91–107.

[3] W. G. Hoskins in E. Crittall (ed.), *V.C.H. Wiltshire*, IV (1959), 18.

[4] J. L. M. Gulley, 312.

[5] R. A. L. Smith (1943), 186–9.

[6] W. G. Hoskins, 'Regional farming in England', *Agric. Hist. Rev.*, II (1954), 7.

was one-sided.[1] Livestock were essential for the fertility of arable land. On the great grain-growing estates, such as those of Canterbury cathedral priory, cattle formed a vital element in the economy, and conversely even the Cistercian granges, with which we usually associate wool production, could be regarded primarily as arable holdings.[2]

Nevertheless the pattern of agriculture in early fourteenth-century England reflected the broad physical controls of soils, climate and topography. Climatically the south-west and the northern uplands were more suited to pasture, while the drier lowlands of the Midlands and the east were well suited to grain. The chalk and limestone hills were sheep country; the forest and woodland provided grazing for cattle, swine and deer, and the marshland pastures for cattle and sheep. Within this framework the specialisations of the monastic houses were a natural outcome of the areas which they settled, the Benedictines in the south and east, and the Cistercians in the north and west. By 1300 their enterprise in seeing the potential of the land they settled had already underlined the fundamental distinction between the pastoral west and the arable east. Beyond this broad division it was only in areas of very distinctive physical environment, where soils and climate made the land especially suitable to a particular use, that some specialisation in agriculture first emerged, for example on the Essex marshlands (as early as 1086?), on the Fenland pastures, on those of Sedgemoor, and on the limestone uplands. Or again, the sensitivity of farm practice to soil conditions may be seen from details for Kent about 1300. Here, oats predominated on the marshland manors while elsewhere wheat and barley were the predominant crops (Fig. 36).[3] Barley was the dominant crop of the high wolds of Yorkshire where an agriculture based upon barley and sheep had already emerged by the mid-fourteenth century.[4] But until more studies are available, we cannot be clear about the exact degree of specialisation in the fourteenth century over England as a whole.

[1] The diversity of production is well illustrated in B. Waites, *Moorland and vale-land farming in north-east Yorkshire: the monastic contribution in the thirteenth and fourteenth centuries* (York, 1967).

[2] R. A. Donkin (1963), 187.

[3] R. A. Pelham (1936), 241. Ann Smith, 'Regional differences in crop production in medieval Kent', *Archaeol. Cantiana*, LXXVIII (1963), 147–60.

[4] B. Waites, 'Aspects of thirteenth and fourteenth century arable farming on the Yorkshire Wolds', *Yorks. Archaeol. Jour.*, XLII (1967), 136–42.

INDUSTRY

England in 1334 was overwhelmingly rural. The majority of people were engaged in tilling the land, in looking after livestock, and in meeting their everyday needs from whatever materials were near at hand. This is not to say that villages were entirely self-sufficient. Peasants had to sell produce in order to get cash for rents and taxes, and to buy some goods which were not available locally such as cloth, leather, salt, pottery and metal goods. Of the various industries that catered for more than local needs, the following were of especial importance – cloth manufacture, mining and quarrying and the making of salt.

The cloth industry

The 1330s were midway in time between the 'urban' cloth industry of the thirteenth century and the 'country' cloth industry of the fifteenth. Throughout the thirteenth century, England had been exporting wool to the Low Countries and importing finished cloth. The export of raw wool reached a peak of 46,000 sacks in 1304–5[1] and although we do not have an exact figure it is likely that cloth imports reached a peak at about the same time.[2] After this date the output of home-produced cloth undoubtedly increased, paradoxically at a time which most scholars see as a period of contraction. Yet both propositions are tenable; at times when all the outward signs point to general economic difficulty there can often be found expanding sectors in an economy. The difficulty lies in knowing where the increase in production took place. The respective roles of town and country in the making and marketing of cloth in the early mid-fourteenth century is a matter of debate. The new growth seems to have been principally in the rural areas, with the towns, which had been the backbone of the thirteenth-century industry, fluctuating in their fortunes. In the 1330s the difficulties of the weavers in towns had intensified. By 1334 the weavers of Northampton, formerly about 300, had apparently disappeared completely, and their numbers were also dwindling in London, Winchester, Oxford, and Lincoln.[3] Yet we know from the 1334 Lay Subsidy that these same towns and other centres of cloth-making were still among the

[1] E. M. Carus-Wilson and O. Coleman, 1963, *England's export trade, 1275–1547* (Oxford, 1963), 41–122.

[2] E. M. Carus-Wilson, *Medieval merchant venturers* (2nd ed. London, 1967), 242–5.

[3] E. Miller, 'The fortunes of the English textile industry during the thirteenth century', *Econ. Hist. Rev.*, XVIII (1965), 70.

wealthiest towns in the country. In 1334, London, York, Newcastle, Bristol, Lincoln, Norwich and Oxford were among the wealthiest ten, and Salisbury, Coventry and Beverley were also wealthy towns. Clearly the wealth that was built on wool and cloth was still evident despite a recession in some places. On the other hand, cloth-making was spreading rapidly in the rural areas, made possible on the streams of the upland valleys by the spread of the fulling-mill,[1] and encouraged elsewhere by urban entre-preneurs who, free from the restrictions of the town gilds, could produce cloth more cheaply in rural areas.[2] By so doing, English cloth-makers could compete with the Flemish who had dominated the trade in 1300. An additional stimulus to home production was provided by the war-time conditions and policies under which the export of raw wool to Flanders was temporarily prohibited and the import of cloth virtually ceased. The Flemish industry never fully recovered from these measures, and, even after their withdrawal, increasingly heavy export duties on raw wool pushed up the cost for the Flemish cloth-makers. As the Flemish industry declined so the English cloth industry, in the late 1330s and 1340s, developed under economic protection and captured the home market. English industry also benefited from Edward III's policy of encouraging foreign cloth-workers to come to England. Many Flemings left the unrest and disturbances of the Low Countries. Alien weavers, dyers and fullers seem to have settled mainly in such centres as London, York, Winchester, Norwich, Bristol and Abingdon, but we also hear of them in the West Riding and in the West Country.[3] But we must not exaggerate their importance; their coming was a symptom rather than 'a cause of the progress of English enterprise'.[4] By the eve of the Black Death England was exporting a considerable amount of cloth each year.[5]

Sustained demand led to the expansion of rural cloth manufacture in the West Riding, in the Lake District, in Wiltshire and the West Country, in the Mendips, the Cotswolds, the Kennet valley and East Anglia. As the market for home produced cloth grew, so the demand for 'Ludlows' and 'Cotswolds', 'Stroudwaters', 'Westerns' and 'Worsteds' began to replace that for 'Lincoln Scarlets', 'Beverley Blues' and 'Stamfords'. Although

[1] E. M. Carus-Wilson (1967), 183–210.   [2] E. Miller (1965), 73–4.
[3] E. Lipson, *The economic history of England: the Middle Ages* (5th ed., London, 1929), 399–400.
[4] E. M. Carus-Wilson in M. Postan and E. E. Rich (eds.), *The Cambridge economic history* II (Cambridge, 1952), 415.
[5] E. M. Carus-Wilson (1967), 245.

the industry was growing it is impossible to estimate the number employed in woollen manufacture in the 1330s; we might guess at something of the order of 20–25,000, having regard to the estimates which have been made for the industry later in the century.[1]

Mining and quarrying

Mining played a small but important part in the economy of the early fourteenth century. Coal, iron, tin, lead and silver were all worked along the fringes of the uplands of the north and west, but except for tin mining in the south-west the numbers employed must have been very small (Fig. 26).

Coal, which was still insignificant as a fuel in comparison with charcoal and wood, was mined in surface workings along the Pennine flanks, in South Wales, and in the north-east, where it achieved its greatest local importance.[2] By the early fourteenth century coal was being exported from Newcastle, and the traffic in 'sea-coal' from the Tyne to London was sufficiently well established for a tax to be levied on it.[3] Coal, however, was not very popular with Londoners who in 1307 complained about the smell,[4] having earlier said that 'the air is infected and corrupted to the peril of those frequenting and dwelling in those parts'.[5] As charcoal was the first fuel for smelting, and as domestic coal burning did not appear until the late fourteenth century, the coal was mainly used in kilns for lime-burning, brewing, baking, and metal work.

The Forest of Dean, where there were still plentiful supplies of wood, was the principal centre of the iron industry, although the developing Wealden industry was already supplying the London market. Iron was also mined in the Cleveland Hills, along the Pennines, and in Furness, where a large quantity of iron formed part of the booty of a Scots raid in 1316.[6] But in many parts of the country fuel supplies for iron-working were becoming very short and for the first time the depletion of woodland resources began to cause anxiety. The iron industry, in the region of Skipton in Craven, closed down; and iron-working in the Forest of Knaresborough, in other parts of the West Riding and in Duffield Frith in

[1] See note in E. M. Carus-Wilson (1967), 261, n. 3.
[2] J. U. Nef, *The rise of the British coal industry*, 2 vols. (London, 1933), I, 9.
[3] R. A. Mott, 'The London and Newcastle chaldrons for measuring coal', *Archaeol. Aeliana*, XL (1962), 228; R. Smith, *Sea-coal for London* (1961), 2.
[4] *Calendar of the Close Rolls, 1302–7* (H.M.S.O., 1908), 537.
[5] *Calendar of the Patent Rolls, 1281–92* (H.M.S.O., 1893), 207 and 296.
[6] L. F. Salzman, *English industries of the Middle Ages* (Oxford, 1923), 26.

Derbyshire virtually ceased.[1] The early fourteenth century was a period of slump in the industry which resulted in its disappearance from some districts and its increasing concentration in well-wooded areas. Even so, England could not meet the demand for iron for the making of arms and agricultural implements, and ore was being imported from Spain in the early fourteenth century.

Tin production was localised in Devon and Cornwall, where it was obtained by streaming. The early fourteenth century was a period of exceptional activity in tinning, and the charters of Edward I in 1305 had the desired effect of encouraging the industry and confirming the ancient privileges of the stannary men, which included exemption from ordinary taxation.[2] The main producing areas lay in Cornwall where production rose to high levels in the 1330s and 1340s.[3] Production was severely curtailed by the Black Death, but it recovered again by the late fourteenth century.[4] In Devon also the mines were prospering, although administration was chaotic, and there were complaints that the tinners were destroying good farm land at the rate of over 300 acres a year.[5] There is hardly any evidence of the working of the copper deposits of Cornwall and Devon at this time. Most of the copper used in fourteenth-century England must have been imported.

Silver and lead were precious commodities, the one needed for coin and ornament, the other for roofing and piping. The two minerals were mined together in the Mendips,[6] around Bere Alston in Devon,[7] at Alston in Cumberland,[8] in the Pennine dales of Yorkshire[9] and Durham, and in Derbyshire.[10]

Mention must also be made of stone quarrying and the building

[1] H. R. Schubert, *History of the British iron and steel industry* (London, 1957), 111–15.

[2] G. R. Lewis, *The stannaries* (London, 1908), 39.

[3] J. Hatcher, 'A diversified economy: later medieval Cornwall', *Econ. Hist. Rev.*, 2nd ser, XXII (1969), 208–27.

[4] A. R. Bridbury (1962), 25–6.

[5] H. P. R. Finberg, 175–81.

[6] J. W. Gough, *The mines of Mendip* (Oxford, 1930).

[7] L. F. Salzman, 'Mines and stannaries', in J. F. Willard *et al.* (eds.), *The English government at work, 1327–1336*, III (Cambridge, Mass., 1950), 67–104.

[8] J. Walton, 'The medieval mines of Alston', *Cumberland and Westmorland, Antiq. Soc.*, XLV (1946), 22–33.

[9] A. Raistrick and B. Jennings, *A history of lead mining in the Pennines* (London, 1965).

[10] L. F. Salzman (1923), 41–3.

industry. Although building activity was much less than in the preceding century, the early fourteenth century was nevertheless a period of great architectural achievement in the Decorated Gothic style. The west front of York Minster, the choir at Gloucester Cathedral and the nave at Exeter all belong to this period; so does the central tower at Wells, supported on crossed arches that are one of the most spectacular achievements of Gothic building. Quarrying stone was an important local industry, e.g. at Barnack in Northamptonshire; and boatloads of stone moved many miles by river, and around the coast. Thus, in 1317 Kentish rag was used for work on the Tower of London; Ramsgate stone was taken to Westminster in 1324 and 1333, Yorkshire stone to Westminster in 1343, Purbeck marble to Exeter in 1309, and Portland stone to Exeter in 1303, and to London and Westminster in 1349. Caen stone was still being brought across the Channel, for example to Exeter and Norwich.[1]

Salt making

Salt was an essential commodity in fourteenth-century England as in earlier times. By the 1330s, the main coastal producing area was in Lincolnshire, but substantial amounts of salt were also produced elsewhere along the south and east coasts, notably in Norfolk, Kent and Sussex.[2] Inland, salt was produced from the brine springs of Worcestershire and Cheshire whence it was exported through Chester to Ireland. In 1334, export to the Continent was less than it had been early in the century, and, as salt was greatly in demand, England imported some, mainly from Bourgneuf Bay south of the estuary of the Loire, and, to a lesser extent, from Spain and Portugal. In the 1320s and 1330s Scarborough, Lynn and Hull handled the bulk of the imports, and only later did London, Bristol and Yarmouth take over as the main points of distribution. Much of the home-produced and imported salt was used at the coast for salting fish; the rest went inland where it had a multitude of uses in every home including the salting of butter and cheese, and the preserving of meat.

[1] All examples from L. F. Salzman, *Building in England down to 1540* (Oxford, 1967), 119–39.
[2] A. R. Bridbury, *England and the salt trade in the later Middle Ages* (Oxford, 1955), 16–39.

174   R. E. GLASSCOCK

TRADE AND TRANSPORT

Most places were within three or four hours' journey of one or more market centres. The markets ranged from small weekly gatherings in villages to the specialised commodity markets of the larger towns. Many places also had annual fairs, but the great international fairs of the thirteenth century had declined in importance; their functions had been assumed by the permanent and increasingly complex trading arrangements of the towns.

Any consideration of roads in the early fourteenth century must be based upon the Gough map, formerly thought to date from about 1300 but now assigned to *circa* 1360.[1] The original is on a scale of approximately 16 miles to one inch, and the more it is studied the more remarkable appears the knowledge of towns, roads, and routes, of the unknown map-maker (Fig. 39). The essence of the modern road pattern existed in the early fourteenth century, except that very few roads crossed the country from south-west to north-east. London was at the hub as it had long been, and Coventry was the great crossing point in the Midlands. On the roads we must picture travellers on foot and on horseback, merchants with pack-horses, carts, and occasionally with great four-wheeled wagons.[2] Carts were used to transport goods such as fish, grain, flour, wine, salt, cloth, hay, faggots and brushwood, peat and stone, and less frequent loads of iron, tin, and military weapons.[3] The carriage of Exchequer goods from Westminster to York in the early fourteenth century took between 10 to 14 days;[4] the journey from Malmesbury to Carlisle in 1318 took 12 days;[5] London to Gloucester was an eight-day return journey.[6] Travel and methods of transport by road formed one of the few aspects of the

[1] E. J. S. Parsons, *The map of Great Britain circa A.D. 1360, known as the Gough Map* (1958), with facsimile. A photograph of the Gough map is included in Lady Stenton's chapter on Communications in A. L. Poole (ed.), *Medieval England*, 2 vols. (Oxford, 1958), I, 208.

[2] J. F. Willard, 'Inland transportation in England during the fourteenth century,' *Speculum*, I (1926), 361–74.

[3] J. F. Willard, 'The use of carts in the fourteenth century', *History*, XVII (1932), 246–50; R. A. Pelham, 'Studies in the historical geography of medieval Sussex', *Sussex Archaeol. Coll.*, LXXII (1931), 167–75.

[4] D. M. Broome, 'Exchequer migrations to York in the thirteenth and fourteenth centuries', in A. G. Little and F. M. Powicke (eds.), *Essays presented to T. F. Tout* (Manchester, 1925), 298.

[5] *Calendar of the Close Rolls, 1313–18* (H.M.S.O., 1893), 548.

[6] J. F. Willard (1926), 367.

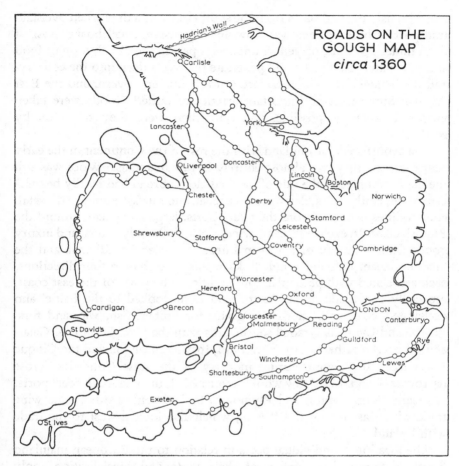

Fig. 39 Roads on the Gough map *circa* 1360
Based on E. J. S. Parsons, *The map of Great Britain circa A.D. 1360, known as the Gough map* (Oxford, 1958).
It is impossible to provide an accurate scale.

geography of England in 1334 that the Black Death and the next century would scarcely change.

As well as transport by road there was movement along the navigable rivers and also coast-wise traffic. Thus, boats plied to Southampton with herrings and stockfish from the east coast, with coal from the north-east, with wheat, malt, and iron from nearby Kent and Sussex, with ropes, sails, and cordage from Bridport, with Purbeck stone and marble from Poole, and with fish, tin, and slates from the south-west. At a time when overland transport of bulky goods was slow and expensive, small boats, most of them under 100 tons, played an important part in the distribution of food and raw materials. They could penetrate the rivers deep into the country, and the Thames, Trent, Yorkshire Ouse, Humber, Severn, and the East Anglian Stour were all important arteries of trade.[1] Goods were taken upstream as far as possible, and only then were they taken on by road.

The export trade of England with the rest of the Continent in the early fourteenth century was almost entirely in raw materials.[2] Wool was still the main export, and the sale of finished cloth abroad had hardly begun.[3] Grain, tin, cloth, hides, dairy produce, coal, and salt were among the main commodities exported from the many ports, large and small, around the English coast. In exchange, came wine, cloth, timber, dyestuffs, and luxury goods from a variety of places in Europe between Scandinavia and the Mediterranean. England's trade links at this time were in four directions, each associated with particular commodities. The ports of the east coast, especially Newcastle, Hull, Boston and Lynn looked to the Baltic, and exchanged coal, salt, cloth and grain, for timber, fish, wax and furs. Boston and Lynn also traded southwards with the Low Countries, Calais and France.[4] Farther south, Yarmouth, Ipswich, London and the Cinque Ports faced the manufacturing industry of the Low Countries across narrow seas, and traded raw wool for finished cloth. The south coast ports, especially Southampton and Plymouth, looked southwards to the wine trade with Gascony, as did Bristol, which also looked west to the trade with Ireland.

The location of individual ports in relation to the European mainland largely determined the nature of their trade and specialisation; only London handled almost every commodity. By the early fourteenth century many ports were associated with a particular trade: Lynn, for example, with wool and grain, Boston with wool and salt, Yarmouth with herring and salt, Newcastle with coal and hides, and Hull with salt. The fortunes of the ports were closely linked with the prosperity of the cities which

[1] R. A. Pelham (1936), 264.
[2] M. M. Postan, 'The trade of medieval Europe: the north', being ch. 4 of M. M. Postan and E. E. Rich (eds.), *The Cambridge economic history of Europe*, II (Cambridge, 1952).
[3] E. M. Carus-Wilson and O. Coleman, 41–122.
[4] E. M. Carus-Wilson, 'The medieval trade of the ports of the Wash', *Med. Archaeol.*, VI–VII (1962–3), 182–201.

they served, Boston with Lincoln, Yarmouth with Norwich, Hull with York, Southampton with Winchester and Salisbury. Wool, the 'sovereine marchandise', was exported in the early fourteenth century at an average rate of 30,000 sacks a year, equivalent to just over 4,000 tons.[1] London handled almost half of this, followed in order of importance by Boston, Hull, Southampton, Ipswich, Lynn, Newcastle and Yarmouth.[2] Cloth was exported in small quantities; Bristol exported from the growing Cotswold woollen industry, and London and Yarmouth from the worsted industry of East Anglia.

Among imports, wine featured prominently, and the early fourteenth century was the peak period of the Anglo-Gascon wine trade. Between 1305 and 1336, when the Gascon trade was relatively stable, the average annual export of wine from the Bordeaux ports was about 83,000 tuns, with up to 100,000 tuns in peak years such as 1308–9. From the evidence of wine imports contained in the English customs accounts, the reliability of which has been debated, it seems that England probably took between a fifth and a quarter of this amount.[3] On the outbreak of the Hundred Years' War in 1337 wine exports dropped, and prices rose on account of limited supplies and the increased cost of escorting the wine vessels safely across the Channel. After a partial recovery the trade was curtailed in 1345 on the resumption of hostilities, and again in 1348 when the Black Death swept through Europe.

TOWNS AND CITIES

Under the double rating plan used in 1334 boroughs were taxed at the higher rate of the tenth, but the selection of boroughs by the taxers does not seem to have conformed to any consistent rules. Consequently some places which were selected as taxation boroughs could hardly have been thought of as towns, either by their size or their economic activity. Conversely some very large towns, for example Boston and Coventry, were not considered taxation boroughs. No doubt some towns were anxious not to be regarded as boroughs so that they might escape the higher rate

[1] One sack = 364 lb. E. M. Carus-Wilson and O. Coleman, 13.
[2] E. M. Carus-Wilson and O. Coleman, 122–37. A detailed study of the business of a leading wool merchant in 1337 is E. B. Fryde, *The wool accounts of William de la Pole* (York, 1964).
[3] M. K. James, 'The fluctuations of the Anglo-Gascon wine trade during the fourteenth century', *Econ. Hist. Rev.*, 2nd ser., IV (1951), 170–96. See also E. M. Carus-Wilson and O. Coleman, 201–7.

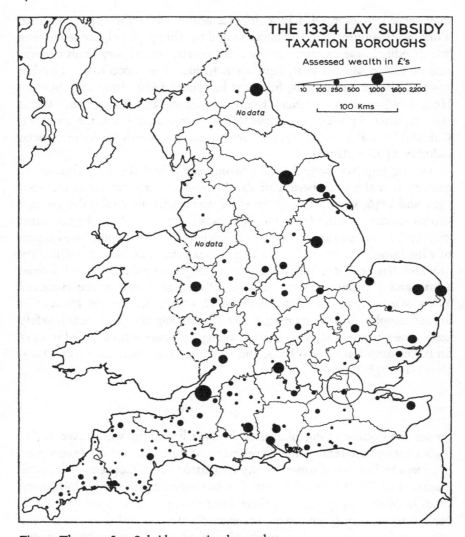

Fig. 40 The 1334 Lay Subsidy: taxation boroughs
      Sources as for Fig. 35.
      London taxed at a fifteenth is shown by an outline circle for comparative
      purposes in view of its great size.

of tax. Coventry's protest in the early fourteenth century that it was not a borough[1] was upheld, and in consequence it paid tax at a fifteenth in 1334.

Fig. 40 shows the taxation boroughs in 1334 and their respective wealth. The Leicestershire and Shropshire boroughs are those taxed in 1332 and 1336, and it is presumed that they are the unspecified boroughs mentioned in the 1334 Accounts for these counties.[2] Carlisle, Corbridge, Bamburgh and Appleby were taxation boroughs in 1336 having escaped tax in 1334. The city of London, not strictly a borough as it was taxed at a fifteenth, is included on the map for comparative purposes.

Taxation boroughs

There were surprisingly few taxation boroughs in the main zone of wealth. Most of the Midland counties had only one each. Rutland, Bedfordshire and Buckinghamshire had none in 1334. Of the wealthy counties, Oxfordshire had three boroughs and Norfolk four. Taxation boroughs were far more numerous in the west and south. Shropshire, Staffordshire and Herefordshire had nine between them, Hampshire, Surrey and Sussex had 18; Somerset, Devon and Cornwall had no less than 57 – a number which probably reflects the multiplication of small boroughs as a result of late colonisation. The north had hardly any. The 1334 taxers for the most part followed the choices of previous taxers, and most taxation boroughs had been selected before 1306.[3] A comparison of taxation and parliamentary boroughs for the period 1240–1336 shows that taxers and sheriffs used much the same criteria for selection; but there were differences. Some parliamentary boroughs were omitted by the 1334 taxers, and conversely some taxation boroughs were not parliamentary boroughs. This may have been due to the fact that the meaning of 'market town' (a term used in the instructions to the taxers between 1290 and 1297) varied from place to place. Many small settlements, some hardly more than villages, qualified as market towns in the eyes of the taxers of Devon and Cornwall, but would no doubt have failed in the eyes of Norfolk men.

The choice of some places and not others is frequently most puzzling, and is exemplified in Hampshire where, of the 20 chartered boroughs in

---

[1] P.R.O., E. 368, L.T.R. Memoranda Roll 77, m. 8.

[2] P.R.O., E. 179/133/3 and E. 179/166/4.

[3] J. F. Willard, 'Taxation boroughs and parliamentary boroughs, 1294–1336', in J. G. Edwards *et al.* (eds.), *Historical essays in honour of J. Tait* (Manchester, 1933), 417–35.

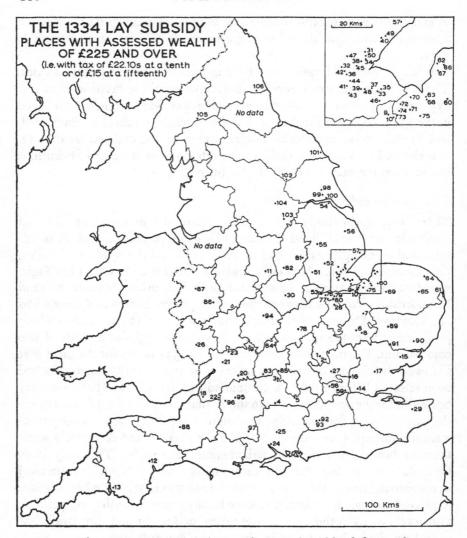

Fig. 41 The 1334 Lay Subsidy: places with assessed wealth of £225 and over
       Sources as for Fig. 35.
       Names of places are given on pp. 181–2.

Table 4.2   *Places with assessed wealth of £225 and over*

(i.e. with tax of £22 10s. and over at a tenth or of £15 and over at a fifteenth)
NOTE: The numbers in brackets indicate the places on Fig. 41.

	£		£
Bedfordshire		Kent	
Leighton Buzzard (1)	249	Canterbury (29)	599
Luton (2)	349	Leicestershire	
Berkshire		Leicester (30)	267
Abingdon (3)	269	Lincolnshire (Holland)	
Newbury (4)	412	Boston (31)	1,100
Reading (5)	293	Donington (32)	250
Cambridgeshire		Fleet (33)	270
Cambridge (6)	466	Frampton (34)	255
Ely (7)	358	Gedney (35)	233
Fulbourn (8)	293	Gosberton (36)	450
Leverington (9)	360	Holbeach (37)	495
Wisbech (10)	410	Kirton (38)	413
Derbyshire		Moulton (39)	465
Derby (11)	300	Old Leake (40)	315
Devonshire		Pinchbeck (41)	675
Exeter (12)	366	Quadring (42)	225
Plymouth (13)	400	Spalding (43)	630
Essex		Surfleet (44)	315
Barking (14)	341	Sutterton (45)	320
Colchester (15)	261	Sutton St James (46)	375
Waltham Holy Cross (16)	262	Swineshead (47)	285
Writtle (17)	267	Whaplode (48)	480
Gloucestershire		Wrangle (49)	235
Bristol (18)	2,200	Wyberton (50)	240
Campden (19)	255	Lincolnshire (Kesteven)	
Cirencester (20)	250	Grantham (51)	293
Gloucester (21)	541	New Sleaford (52)	241
Marshfield (22)	270	Stamford (53)	359
Tewkesbury (23)	243	Lincolnshire (Lindsey)	
Hampshire		Barton upon Humber (54)	246
Southampton (24)	511	Lincoln (55)	1,000
Winchester (25)	515	Louth (56)	227
Herefordshire		Wainfleet (57)	233
Hereford (26)	605	Middlesex	
Hertfordshire		Harrow (58)	257
St Albans (27)	265	London (59)	11,000
Huntingdonshire		Norfolk	
Yaxley (28)	227	Gayton (60)	225

Table 4.2 (*cont.*)

	£		£
Norfolk (cont.)		Shrewsbury (87)	800
Great Yarmouth (61)	1,000	Somerset	
Heacham (62)	248	Bridgwater (88)	260
King's Lynn (63)	500	Suffolk	
North Walsham (64)	225	Bury St Edmunds (89)	360
Norwich (65)	946	Ipswich (90)	645
Sedgeford (66)	233	Sudbury (91)	281
Snettisham (67)	285	Surrey	
South Lynn (68)	270	Bramley (92)	298
Swaffham (69)	300	Godalming (93)	247
Terrington (70)	607	Warwickshire	
Tilney (71)	450	Coventry (94)	750
Walpole (72)	533	Wiltshire	
Walsoken (73)	396	Bremhill (95)	233
West Walton (74)	345	Corsham (96)	225
Wiggenhall (75)	555	Salisbury (97)	750
Northamptonshire		Yorkshire (East Riding)	
Barnack (76)	269	Beverley (98)	500
Castor (77)	276	Cottingham (99)	330
Northampton (78)	270	Hull (100)	333
Paston (79)	251	Yorkshire (North Riding)	
Peterborough (80)	383	Scarborough (101)	333
Nottinghamshire		York (102)	1,620
Newark on Trent (81)	390	Yorkshire (West Riding)	
Nottingham (82)	371	Doncaster (103)	255
Oxfordshire		Pontefract (104)	270
Bampton (83)	969	Cumberland (1336)	
Banbury (84)	267	Penrith (105)	398
Oxford (85)	914	Northumberland	
Shropshire		Newcastle upon Tyne (106)	1,333
Bridgnorth (86)	244		

the county, only five were considered taxation boroughs in 1334;[1] to these must be added the soke of Winchester which was listed separately as a taxation borough in addition to the city itself. Of the remaining 14, four were treated as ancient demesnes and taxed at a tenth, ten were taxed

[1] M. W. Beresford, 'The six new towns of the bishops of Winchester 1200–55', *Med. Archaeol.*, III (1959), 213.

at a fifteenth, and one was not mentioned. The 1334 taxers must have been as puzzled as we are why Porchester was a taxation borough and not Overton, and why New Alresford and not New Lymington. Whatever the criteria of selection, perhaps a combination of commercial activity, population, or most probably, local reputation, they were not uniformly applied. Many of the taxation boroughs of the south-west were taxed at only a fraction of the amounts paid by some large villages in eastern England.

Fig. 41 shows those places which had assessed wealth of £225 and over, i.e. with quotas of either £15 and over at a fifteenth or with £22 10s. and over at a tenth. Comparison with Fig. 40 shows that only about one-fifth of the taxation boroughs were in this category. Only 3 out of 57 in the west country qualified, only 2 of the 18 in Hampshire, Surrey and Sussex, and only 3 of the 11 in the west Midlands. Even if they were locally important, almost all the boroughs of the west and south had much less wealth than many places in midland and eastern England that were not considered boroughs. Thus almost every fenland township in the Holland division of Lincolnshire and in Norfolk marshland had assessed wealth of over £225. Clearly most of these were not towns. They were wealthy because they included extensive and rich areas of agricultural land.[1] With these exceptions the wealthier places were the large trading centres. Some were taxation boroughs as noted above, others were towns that had somehow escaped the higher rate of tax, for example Coventry, Boston, Bury St Edmunds, and Newbury. A few were probably only large market villages such as Fulbourn (Cambs), Marshfield (Gloucs), Yaxley (Hunts). Most of the larger trading centres were situated in the east Midlands and East Anglia. They were conspicuously lacking in the west Midlands and the border counties, in the south-west, in the north, and on the south coast. The Cinque Ports were excluded from the Lay Subsidy, but even so, the impression given by the map may be correct, for, with the exception of Southampton and Plymouth, the south coast trade was shared between many small ports rather than concentrated in any large one.

The chief towns

Table 4.3 shows all towns with assessed wealth of £300 and over, ranked according to their values. Allowance has been made for the different rates of tax; the yields for taxation boroughs have been multiplied by ten; and

[1] R. E. Glasscock (1963), 120.

Table 4.3   *Ranking list of chief towns in 1334*

It is difficult to construct a satisfactory ranking list of towns in 1334. Many other settlements would qualify for inclusion in the list below, on the basis of their wealth alone. Thus there were nineteen fenland townships with valuations of £300 or more; the largest of these were Pinchbeck (£675) and Spalding (£630). Another addition would be Bampton *cum membris* (in Oxfordshire) with a valuation of £969. Penrith, too, with £398 in 1336 would be included, but it was taxed at a tenth as Ancient Demesne. Furthermore, Chester and Durham were not taxed, and they would certainly come fairly high on the list. Even so, the table may serve to indicate some, at any rate, of the main facts about the relative sizes of English towns in 1334. (F) denotes the lower rate of the fifteenth and so indicates the towns that were not considered to be taxation boroughs in 1334, although some of them were at other dates.

	Assessed wealth £		Assessed wealth £
London (F)	11,000	Beverley	500
Bristol	2,200	Cambridge	466
York	1,620	Newbury (F)	412
Newcastle upon Tyne	1,333	Plymouth	400
Boston (F)	1,100	Newark on Trent (F)	390
Great Yarmouth	1,000	Peterborough *cum membris* (F)	383
Lincoln	1,000	Nottingham	371½
Norwich	946	Exeter	366
Oxford	914	Bury St Edmunds (F)	360
Shrewsbury	800	Stamford	359
Lynn (King's and South)	770	Ely *cum membris* (F)	358
Salisbury	750	Luton (F)	349
Coventry (F)	750	Barking (F)	341
Ipswich	645	Hull	333
Hereford	605	Scarborough (F)	333
Canterbury	599	Cottingham, Yorks. E.R. (F)	330
Gloucester	541	Derby	300
Winchester	515	Swaffham (F)	300
Southampton	511		

those for places such as Boston, Coventry, and Spalding have been multiplied by fifteen.[1] Some totals such as those for Ely and Peterborough include suburbs. The figure for Plymouth covers Sutton Prior and Sutton Vautort. Chester and Durham, not taxed, are omitted. So is Bampton, in Oxfordshire, with a huge quota that included many nearby places. Moreover, £300 is an arbitrary minimum and some towns of local importance were not far below this figure. They included Reading, Northampton, Leicester, Colchester, Bridgwater, Cirencester and Bridgnorth.

In spite of these difficulties, the list provides a general indication of the most important towns in England in 1334 and of their relative wealth. Some generalisations emerge from the list. Firstly, most of the towns listed lay south of the Trent and the Severn. In the north, only Newcastle, York, Beverley, Nottingham, Hull and Scarborough were of comparable importance to the towns of the south; and in the west and south-west only Bristol, Shrewsbury, Hereford, Gloucester, Exeter and Plymouth. Secondly, all the leading towns were either ports or centres of cloth manufacture. Thirdly, there was the overwhelming predominance of London, a city of perhaps 50,000 people, which had more wealth than the three leading provincial cities, Bristol, Newcastle and York, combined. It was 'a metropolis, bearing far more resemblance to the great cities of northern Continental Europe than to any other English town'.[2]

[1] The ranking shows a different order from that in which the two different rates have not been equated as in W. G. Hoskins, *Local history in England* (London, 1959), 176.

[2] S. L. Thrupp, *The merchant class of medieval London, 1300–1500* (Chicago, 1948), 1.

Chapter 5

CHANGES IN THE LATER MIDDLE AGES

ALAN R. H. BAKER

The geographical changes in England between 1334 and 1600 formed part of a far wider transformation. From Norman times English kings had been pre-occupied with France, but after the so-called Hundred Years' War (1338–1453) nothing was left of English territory on the Continent except Calais, and that was lost in 1558. Factional struggles of the Wars of the Roses (1455–85) had brought a new dynasty to the throne; and the crowning of Henry Tudor in 1485, we can see in retrospect, was the beginning of a new age. Savage and ruthless though the Wars of the Roses had been, they had not inflicted such damage on the countryside as, say, the disorders of the twelfth century. Domestic events within the British Isles were overshadowed by changes of a world-wide character. When Columbus crossed the Atlantic in 1492, and when Portuguese mariners rounded the Cape of Good Hope to India in 1497–9, a new epoch had been inaugurated. The trade of the marginal seas of Europe was now to be extended to the great oceans beyond. England at first played but little part in the maritime explorations and expansion of the sixteenth century. Its searches for a north-west and a north-east passage to the East came to relatively little. But other routes were open, even if they involved conflict with the Spanish and the Portuguese. Drake's voyage around the world in 1577–80, Raleigh's settlement in Virginia in 1558–95, the foundation of the East India Company in 1600 – these were among the symptoms of the change in England's position from that of an off-shore island of a continent to the centre of world trade routes. Clearly, the England of 1600 was very different from that of the early fourteenth century.

Not only was the geography of England changing, but men in general were becoming more self-conscious, or at any rate more vocal, about it. The invention of printing in the fifteenth century and its introduction to England by William Caxton in 1476 resulted in an explosion of information. The Tudors, in the words of Charles Whibley, 'recognised that the most brilliant discovery of a brilliant age was the discovery of their country'.[1] At any rate, a rich tradition of topographical writing was launched upon its course. John Leland's 'Itinerary' of the 1530s and 1540s was not printed until many years after his death, but William Harrison's *Historical description of the island of Britain* appeared in 1577, and William Camden's *Britannia* in 1586. Then, too, came the printed maps of Christopher Saxton in the 1570s and 1580s and those of John Norden in the 1590s. There were also other more detailed descriptions – William Lambarde's *Perambulation of Kent* (1576) and John Stow's *Survey of London* (1598) were the precursors of a long line of local studies.

POPULATION

Evidence about population during the later Middle Ages is usually either direct but incomplete, or indirect and controversial, and trends and changing distributions are more easily established than absolute numbers. By about 1330 growth had ceased, and population was possibly declining; the rapid rate of increase in the thirteenth century had certainly been retarded.[2] Population growth had outstripped the means of subsistence, producing widespread malnutrition and increasing susceptibility to famine and disease. Harvest failures and years of summer epidemics were accompanied by exceptionally high death rates on some Winchester manors,[3] and generally rising wage rates and falling food prices suggest a declining population.[4] If the first symptoms of decline appeared before 1348, it was nevertheless the Black Death of 1348–50 which decimated an already vulnerable and unstable population.

The invasion of the British Isles by bubonic plague in 1348 was only an incident in a great epidemic outburst of the disease from its Indian home. Between about 1340 and 1352 this outburst involved most of Asia Minor, much of North Africa, the whole of Europe and some of the islands lying

---

[1] C. Whibley, 'Chronicles and antiquarians', being ch. 15 of *The Cambridge history of English literature* (Cambridge, 1908), III, 313.

[2] J. E. T. Rogers, *Six centuries of work and wages* (London, 1884), I, 217; J. C. Russell, *British medieval population* (Albuquerque, New Mexico, 1948), 246–60; B. F. Harvey, 'The population trend in England, 1300–1348', *Trans. Roy. Hist. Soc.*, 5th ser., XVI (1966), 23–42.

[3] M. M. Postan and J. Titow, 'Heriots and prices on Winchester manors', *Econ. Hist. Rev.*, 2nd ser., II (1958–9), 392–411.

[4] M. M. Postan, 'Some economic evidence of the declining population of the later Middle Ages', *Econ. Hist. Rev.*, 2nd ser., II (1949–50), 221–46.

off that continent such as the Channel Islands, the British Isles, and Greenland. The arrival of the plague added to the pattern of mortality; diseases such as smallpox, measles, typhus fever and dysentery were repeatedly epidemic in Britain during the later Middle Ages; pneumonia undoubtedly occurred in epidemic form in the winter months, and whooping cough, the enteric fevers and influenza in all probability were also epidemic at times.[1] After the plague subsided in England in 1350, the country seems to have been free from a major eruption of epidemic disease for ten or eleven years; the epidemics of 1361–2, of 1369 and of 1374 may have been of bubonic plague or another deadly disease.

It has been variously estimated that the Black Death of 1348–50 carried off between one-third and one-half of the population. Studies of particular manors and of particular districts, however, reveal widely differing death rates, and the incidence of the plague was extremely irregular. Plague, both the bubonic and the more deadly pneumonic type, appeared in England in August 1348, entering through the south-western ports; it spread to Bristol and thence by way of Oxford to London, which it reached by the end of October or the beginning of November. In early 1349 it spread northwards, reaching Yorkshire in March, by which time it had almost ceased in London, although it raged in York until the end of the summer.[2] Its toll was heaviest in crowded towns, especially in ports. At Bristol, between 35% and 40% of the population were victims.[3] Mortality was heavier among clergy than laity; by the very nature of their profession priests were exposed to contagion, and, as a group, their average age was higher than that of the community as a whole.[4] Mortality rates among parish clergy were highest in the dioceses of Exeter, Winchester and Norwich; the deanery of Kenn, to the south of Exeter, was the worst hit deanery in all England, and lost 86 incumbents from its 17

[1] J. F. D. Shrewsbury, *A history of bubonic plague in the British Isles* (Cambridge, 1970), 37–263; P. Ziegler, *The Black Death* (London, 1969).

[2] C. Creighton, *A history of epidemics in Britain from A.D. 664 to the extinction of plague* (Cambridge, 1891), 116–18; J. F. D. Shrewsbury, 37–53; E. Miller, 'Medieval York' in P. M. Tillott (ed.), *V.C.H. Yorkshire: The city of York* (1961), 85; J. M. W. Bean, 'Plague, population and economic decline in England in the later Middle Ages', *Econ. Hist. Rev.*, 2nd ser., xv (1962–3), 422–37.

[3] C. E. Boucher, 'The Black Death in Bristol', *Trans. Bristol and Glos. Archaeol. Soc.*, 60 (1938), 31–46.

[4] G. G. Coulton, *Medieval panorama. The English scene from Conquest to Reformation* (Cambridge, 1938), 495–503; J. C. Russell, 218 and 230; Y. Renouard, 'Conséquences et intérêt démographiques de la Peste Noire de 1348', *Population*, III (1948), 459–66.

churches between 1349 and 1351.[1] On manors of the see of Winchester, in central southern England, about a third of the population died, but mortality varied widely, not only from manor to manor but from tithing to tithing. At Witney in Oxfordshire and at Downton in Wiltshire, A. Ballard estimated a mortality rate of about 66% in 1349; at Brightwell in Berkshire, on the other hand, the mortality was less than 30%.[2] In the hundred of Farnham, in Hampshire, 344 heads of households died within three years (185 in 1348–9, 101 in 1349–50 and 58 in 1350–1), representing between one-third and one-half of the total.[3]

In Essex, 70 tenants appear to have died from plague on the manor of Fingreth in Chelmsford hundred during the first six months of 1349, but in the adjacent hundred of Ongar the effects were slight and only two places out of 25 received any tax relief in 1352.[4] Tax reliefs given in 1352, 1353 and 1354 to villages in Norfolk hard-hit by plague show an interesting distribution of stricken communities; the plague appears to have entered by the ports of Yarmouth and Lowestoft and by the smaller Norfolk ports, and it was severe around these places.[5] The first victims on manors of Crowland abbey in Cambridgeshire were reported in October 1348; between mid-May and July 1349 plague was rife; and by January 1350 it was passing away. At Dry Drayton 20 out of 42 tenants (47%) died, at Cottenham 33 out of 58 (57%), and at Oakington 35 out of 50 (70%).[6]

To the west and north, in the diocese of Lincoln (which then stretched to Bedfordshire and Oxfordshire) 40% of all benefices became vacant by death between Lady Day (25 March) 1349 and Lady Day 1350. Mortality was highest in Lincolnshire itself (48% in the archdeaconry of Lincoln and 57% in the archdeaconry of Stow) and lowest in Oxfordshire (34% in the archdeaconry of Oxford).[7] In the large diocese of York (which

[1] G. G. Coulton, 496; W. G. Hoskins, *Devon* (London, 1954), 169–70.

[2] A. Ballard, 'The manors of Witney, Brightwell and Downton' in A. E. Levett (ed.), *The Black Death on the estates of the see of Winchester* (Oxford, 1916), 181–216.

[3] E. Robo, 'The Black Death in the hundred of Farnham', *Eng. Hist. Rev.*, XLIV (1929), 560–72.

[4] J. L. Fisher, 'The Black Death in Essex', *Essex Review*, LII (1943), 13–20; M. W. Beresford, 'Analysis of some medieval tax assessments: Ongar Hundred' in W. R. Powell (ed.), *V.C.H. Essex*, IV (1956), 296–302.

[5] K. J. Allison, 'The lost villages of Norfolk', *Norfolk Archaeol.*, XXXI (1957), 116–62; G. G. Coulton, 496.

[6] F. M. Page, *The estates of Crowland abbey* (Cambridge, 1934), 120–5.

[7] A. H. Thompson, 'The registers of John Gynewell, bishop of Lincoln, for the years 1349–1350', *Archaeol. Jour.*, LXVIII (1911), 301–60.

included all or parts of Nottinghamshire, Lancashire, Westmorland and Cumberland as well as Yorkshire) mortality rates also varied considerably. In the city of York it was 32%, and in the diocese as a whole it was 39%. A. H. Thompson concluded that 'mountainous country on the one hand and marshland on the other were comparatively immune from pestilence, while normal agricultural country and the lower highlands suffered most heavily'; thus in the moorland deanery of Cleveland, mortality was only 21% compared with 61% in that of Dickering, on the Wolds. Mortality was highest where population was most thickly settled.[1]

Although mortality was high there are few unequivocal instances of the total depopulation of villages; the last recorded reference to Ambion in Leicestershire was in 1346, and it may have been completely depopulated by the plague; so, too, Tilgarsley and Tusmore in Oxfordshire and Middle Carton in Lincolnshire.[2] Such phenomena were rare; settlements shrank rather than disappeared.[3] One further point must be made. Although population was generally declining during the third quarter of the fourteenth century, some localities and towns witnessed an increase; the population of York, for example, has been estimated to have been 50% higher in 1377 than it had been just before 1348.[4]

Some idea of the distribution of population during the fourteenth century can be derived from the Poll Tax returns of 1377. This tax was imposed at the flat rate of a groat (4d.) a head on the lay population; only those under 14 years old and those who regularly begged for a living were exempted. Various attempts have been made to calculate the total population of the country from these figures, notably by J. C. Russell, but all involve varying degrees of conjecture and controversy.[5] The Poll Tax is a better guide to relative densities of population than to absolute numbers. But even then, assumptions are involved: that there were no significant differences from one county to another in the proportion of the population

[1] A. H. Thompson, 'The pestilences of the fourteenth century in the diocese of York', *Archaeol. Jour.*, LXXI (1914), 97–154.

[2] K. J. Allison, M. W. Beresford and J. G. Hurst, 'The deserted villages of Oxfordshire' (Leicester, 1965), 44–5; W. G. Hoskins, *Essays in Leicestershire history* (Liverpool, 1950), 104, and *The making of the English landscape* (London, 1955), 93.

[3] M. W. Beresford, *The lost villages of England* (London, 1954), 159, 269, 286 and 289.

[4] E. Miller, 84.

[5] J. C. Russell, 132–46; J. Krause, 'The medieval household: large or small?', *Econ. Hist. Rev.*, 2nd ser., IX (1956–7), 420–32; J. Stengers, review note, *Revue Belge de Philologie et d'Histoire*, XXVIII (1950), 600–6.

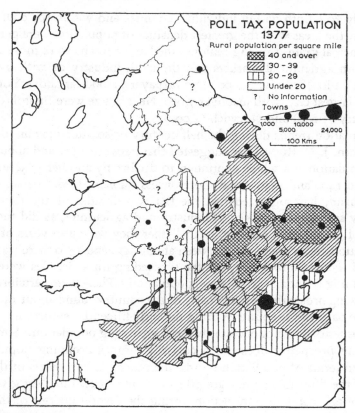

Fig. 42 Poll Tax population, 1377
   Based on J. C. Russell, *British medieval population* (Alberquerque, 1948),
   132–3, 142–3 (P.R.O. Exchequer Lay Subsidies E. 179).
   The Poll Tax figures refer to lay population over 14 years of age.

who were under 14 years old, or in the proportion who were mendicants,
or in the proportion who evaded the tax.[1] These assumptions must be
borne in mind when looking at any map based on these returns (Fig. 42).
There was a sharp contrast between highland and lowland zones; no
county to the north of a line joining the Severn and Humber estuaries had
thirty or more taxpayers to the square mile. Averages, however, conceal
wide variations within individual counties; the average density for Sussex
was 25 taxpayers per square mile, but along the coastal plain it was almost

[1] M. W. Beresford, 'The Poll Taxes of 1377, 1379 and 1381', *Amateur Historian*,
III (1956–8), 271–8.

certainly 40.[1] Both within individual counties and within the country as a whole, the areas with the greatest densities of population were generally the principal grain-growing regions; the largest towns were those associated with agricultural markets and the cloth industry or with overseas trade. London, with its 23,000 or so taxpayers, stood far above York, the second largest city, with just over 7,000. Next in size were Bristol (6,000), Coventry (5,000) and Norwich (4,000).

Population trends during the half-century or so after 1377 are difficult to discern. J. C. Russell has suggested that between 1377 and about 1400 the population as a whole continued to decline by another 5%, and that between 1400 and 1430 it was more or less stable.[2] M. M. Postan, on the other hand, has suggested that the last two decades of the fourteenth century saw some recovery in industry and agriculture, as did the early years of the fifteenth century, but whether they were also years of rising population seems uncertain. Between about 1410 and 1430 there was some general economic recovery, but the succeeding three decades were years of declining economic activity and population.[3] There were certainly signs of growing prosperity in many parts of the country round about 1400, and in some parts of the country during the early fifteenth century, but growing prosperity does not necessarily mean a growing population.[4] Similarly, declining prosperity does not necessarily mean a declining population; hence in terms of population trends, the numerous tax reliefs of the first half of the fifteenth century are difficult to interpret. But they give clear signs of an economic contraction during the second quarter of the century. In Leicestershire, one of the more densely populated counties in 1377, the tax reduction of 16% in 1445 was distributed unevenly throughout the county; some of the larger settlements of the county appear to have suffered a decline in wealth, and possibly numbers, that was greater than average. The 1334 assessment at Melton Mowbray was cut in 1445

[1] R. A. Pelham, 'Fourteenth-century England' in H. C. Darby (ed.), *An historical geography of England before A.D. 1800* (Cambridge, 1936), 230–65. See also C. T. Smith, 'Population' in W. G. Hoskins and R. A. McKinley (eds.), *V.C.H. Leicestershire*, III (1955), 133–6.

[2] J. C. Russell, 269.

[3] M. M. Postan (1949–50), 245, and 'The fifteenth century', *Econ. Hist. Rev.*, IX (1938–9), 160–7.

[4] P. F. Brandon, 'Arable farming in a Sussex scarp-foot parish during the late Middle Ages', *Sussex Archaeol. Coll.*, C (1962), 60–72; J. A. Raftis, *The estates of Ramsey abbey* (Toronto, 1957), 264–5; R. A. L. Smith, *Canterbury cathedral priory: a study in monastic administration* (Cambridge, 1943), ix.

by 38%, at Wigston Magna by 40% and at Barrow upon Soar by 47%, and the largest reduction of all was one of 60% at Humberstone.[1] Six villages which received tax cuts of between 30% and 40%, and two with cuts of more than 40%, were totally depopulated and deserted later in the century. In these villages a notable falling-off in population may have been produced by successive pestilences.

During the economic contraction between 1350 and 1450, some villages were deserted. There was in particular a retreat from marginal soils as the pressure of population upon land was relaxed. On the Lincolnshire Wolds, for example, large reliefs from tax were granted in 1352–4, and the amalgamation of parishes which had already begun in 1428 suggests a retreat of settlement there.[2] Most deserted villages in Norfolk were situated in the west on the light marginal soils, and even in south and east Norfolk the deserted villages tended to lie on areas of lighter soils, such as the plateau gravels north of Norwich and the sands and gravels of the Wensum valley. Many of these desertions represented a true retreat of settlement from marginal soils.[3] Something similar may have taken place on the Wolds of the East Riding of Yorkshire.[4] But many more villages were deserted later for quite other reasons.[5]

Towards the middle of the fifteenth century, population apparently began to increase again; at the beginning of the next century it was increasing rapidly, and it probably continued to do so throughout the sixteenth century. But growth was selective rather than general. In southeast Lancashire there was a marked expansion in the number of chapelries dependent upon parish churches between 1470 and 1548, more so than elsewhere in the county because of the growing textile industries.[6] Expansion of the cloth industry in England saw the rise of new centres of population, while some of the older centres stagnated or declined. Growth was associated more with rural industry and less with agriculture. The

[1] C. T. Smith, 137; W. G. Hoskins, 'The population of an English village, 1086–1801: a study of Wigston Magna', *Trans. Leics. Archaeol. and Hist. Soc.*, XXXIII (1957), 15–35.

[2] M. W. Beresford (1954), 164 and 170–2.

[3] J. Saltmarsh, 'Plague and economic decline in England in the later Middle Ages', *Cambridge Hist. Jour.*, VII (1941–3), 23–41; K. J. Allison, 138–40.

[4] M. W. Beresford, 'The lost villages of Yorkshire. Part II', *Yorks. Archaeol. Jour.*, XXXVIII (1952–5), 44–70; M. W. Beresford (1954), 150, 170 and 241.

[5] M. W. Beresford (1954), 164 and 170–2.

[6] G. H. Tupling, 'The pre-Reformation parishes and chapelries of Lancashire', *Trans. Lancs. and Cheshire Antiq. Soc.*, LXVII (1957), 1–16.

Weald of Kent, for example, sparsely peopled in the fourteenth century had, by the middle of the sixteenth century and in consequence of the growth of the cloth and iron industries, joined the arable and sheep lands of north-east Kent as the most densely peopled parts of the county.[1]

Migration probably contributed as much to changing population distributions as differential birth and death rates. During the sixteenth century there was increasing seasonal migration, principally of rural land-owners and their families, to and from London, as well as more permanent migrations. The population of London and its immediate suburbs grew more rapidly than the population of the country as a whole.[2] Population in the provinces was also becoming increasingly mobile, especially in the Midlands and south-eastern England.[3] A comparison of the names of 1544 subsidy payers in the North Clay division of Bassetlaw wapentake, Nottinghamshire, with those of 1557 reveals that 24% of the names in the later list were new to the district.[4] In the hundreds of Godalming, Farnham and Godley in Surrey, more than 50% of the men who answered the muster of 1575 did not answer that of 1583. Some of this change was due to old men passing beyond military age and to young men growing into it, but much of it was a consequence of emigration. Moreover, people were not only leaving the district but also arriving and almost a third of those registered in 1583 bore family names not included in the earlier list.[5]

But in some places natural increase was of paramount importance. For example, the population of Wigston Magna (Leicestershire) increased dramatically between 1563 and 1603 by about 50%, as a result partly of immigration of new families but largely of natural increase consequent upon a rising birth rate (itself due in part to earlier marriages) and a diminishing death rate (especially a fall in infant mortality).[6] In Leicestershire as

---

[1] H. A. Hanley and C. W. Chalkin, 'The Kent Lay Subsidy of 1334/5', *Kent Records*, XVIII (1964), 58–172. See also: E. M. Yates, 'A contribution to the historical geography of north-west Staffordshire', *Geog. Studies*, II (1955), 39–52; J. M. W. Bean (1962–3), 435.

[2] F. J. Fisher, 'The development of London as a centre of conspicuous consumption in the sixteenth and seventeenth centuries', *Trans. Roy. Hist. Soc.*, 4th ser., XXX (1940), 37–50.

[3] E. J. Buckatzsch, 'The constancy of local populations and migration in England before 1800', *Population Studies*, V (1951–2), 62–9.

[4] S. A. Peyton, 'The village population in the Tudor Lay Subsidy rolls', *Eng. Hist. Rev.*, XXX (1915), 234–50.

[5] E. E. Rich, 'The population of Elizabethan England', *Econ. Hist. Rev.*, 2nd ser., II (1949–50), 247–65.

[6] W. G. Hoskins (1957), 18–19 and 32.

a whole the population in 1563 was still far below its level in 1334, and was substantially smaller than it had been in 1377 (as it also was in many other counties). But the county's population was increasing; coalmining was beginning to add to the population of the north-west, and market towns now had relatively larger populations compared with the purely agricultural settlements. Parish registers from many counties indicate a rapid population increase during the second half of the sixteenth century, but with sharp, localised setbacks in years of pestilence.[1]

In some localities, plague and other diseases reversed the general population trend; at Crediton, in Devon, 551 people died during 1571. The average number of burials for preceding normal years was 40 to 45, so that nearly 500 people must have died of plague in one year in this small town – possibly a third of its population. Between the autumn of 1590 and that of 1592 another 535 people also died here. Thus Crediton lost over 1,000 people by pestilence in the space of 21 years.[2] During the later Middle Ages, plague occurred intermittently but with a generally decreasing vehemence nationally, and it became increasingly a regional, particularly an urban (especially a London), phenomenon.[3]

The distribution of wealth in England in the early sixteenth century as reflected in the Lay Subsidies of 1524–5 (Fig. 43) may be taken as an approximate summation of the economic changes, including the population changes, of the later Middle Ages; but it must be borne in mind that wealth was often concentrated in the towns and in the hands of a few individuals.[4] Estimates of the total population vary greatly both for 1334 and for 1600. Whatever the uncertainty about these figures, it may be reasonably assumed that resurgence and growth during the late fifteenth and sixteenth centuries had brought the total population of England back to its pre-Black Death figure of about 4½ million or so.

[1] C. T. Smith, 137–41; J. W. F. Hill, *Tudor and Stuart Lincoln* (Cambridge, 1956), 88; J. Cornwall, 'An Elizabethan census', *Records of Bucks.*, XVI (1953–60), 258–73.

[2] W. G. Hoskins (1954), 171.

[3] J. Saltmarsh, 32–40, and J. M. W. Bean (1962–3), 428–32.

[4] J. Sheail, 'The distribution of taxable population and wealth in England during the early sixteenth century', *Trans. and Papers, Inst. Brit. Geog.*, LV (1972), 111–26.

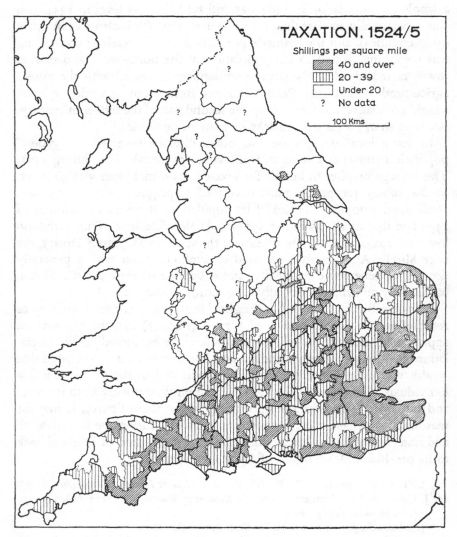

Fig. 43 Taxation, 1524–5
        Based on J. Sheail, 'The distribution of taxable population and wealth in
        England during the early sixteenth century', *Trans. and Papers, Inst. Brit.
        Geog.*, LV (1972), 120, and on additional information provided by Dr Sheail.
        (P.R.O. Exchequer Lay Subsidies E. 179.)

THE COUNTRYSIDE

Demographic changes were linked with price changes. As the supply of labour and the demand for produce changed, so did prices of products and the profitability of different economic enterprises. Price changes encouraged fundamental changes in both agriculture and industry during the later Middle Ages. A brief descriptive account of these changes therefore precedes discussion of agricultural and industrial developments.

The construction of accurate, meaningful price indices for this period is extremely difficult. Isolated references to prices are found in a host of documentary sources, but their isolation detracts from their utility. Most useful are the records of institutions with regular purchases or regular sales of goods, and so with continuous, or near continuous, series of prices. Moreover, comparability of data from different sources is essential; the use of local measures often bedevils direct comparison, and prices are affected by the time, place and conditions of sale.[1] The index of wool prices compiled by J. E. T. Rogers, who himself admitted its defects, has been described by P. J. Bowden as worthless, for two reasons. In the first place, the number of items utilised was very small and sometimes only two or three price quotations represented an entire decade; secondly, no differentiation was made between different qualities of wool.[2] Despite difficulties of this nature, trends of some prices during the later Middle Ages can be discerned (Fig. 44). Those of wages, consumables and land will be discussed briefly.

By about 1340, *wages* for agricultural workers were rising and continued to rise steeply in the 1340s and 1350s, indicative of a growing scarcity of labour, consequent upon a declining population. The rise both preceded and succeeded the Black Death and cannot be attributed solely to it. Nevertheless, the decades after 1350 saw a sharp rise in wages, a rise that became permanent and that levelled out after about 1370. W. Beveridge's study of wages on eleven manors of the see of Winchester, spread over seven counties in southern England, noted the rise and fall of threshing and winnowing costs between 1362 and 1368, followed by the establishment of a new high level about 1374. These fluctuations were not simultaneous on all manors, and Beveridge regarded them as a delayed

[1] E. V. Morgan, *The study of prices and the value of money*, Helps for History Students, 53 (London, 1950).

[2] P. J. Bowden, 'Movements in wool prices, 1490–1610', *Yorks. Bull. Econ. and Soc. Research*, IV (1952), 109–24.

198 ALAN R. H. BAKER

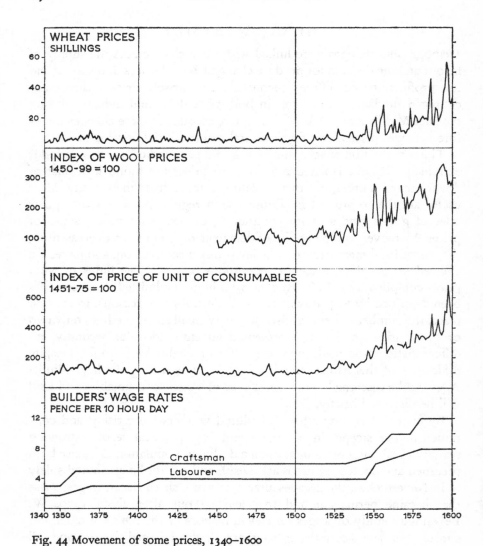

Fig. 44 Movement of some prices, 1340–1600
Based on: (1) J. E. T. Rogers, *A history of agriculture and prices in England*,
7 vols. (Oxford, 1866–1902), I, 228–34; IV, 282–90; VI, 268; (2) P. J. Bowden,
The wool trade in Tudor and Stuart England (London, 1962), 219–20; (3) E. H. P.
Brown and S. V. Hopkins, 'Seven centuries of the prices of consumables
compared with builders' wage-rates', *Economica*, n.s., XXIII (1956), 311–12.
Breaks in the graphs are the result of no data.

reaction to the change of labour conditions begun by the Black Death, although it was the pestilence of 1361 which finally broke down resistance to a change of wage-rates.[1] According to M. M. Postan, wages rose throughout the greater part of the fourteenth and fifteenth centuries.[2] Builders' wages during the later Middle Ages certainly rose sharply during 1350–65, 1400–12 and 1545–80, and showed remarkably level trends during the late fourteenth century and throughout most of the fifteenth century.[3] In both agricultural and industrial occupations, the rise in the wages of unskilled labourers was relatively greater than in those of skilled craftsmen. And, until the sixteenth century, the rise was a considerable one in real wages, for the prices of consumables rose relatively less than wage-rates. After about 1510 there was a considerable fall of wage-rates in relation to prices, indicative of a growing population.[4]

A price index of a composite unit of *consumables*, constructed by E. H. P. Brown and S. V. Hopkins, shows a remarkably level but slightly falling trend from 1350 or so to about 1500. Prices began to creep upwards after about 1510, and climbed steeply during the 1540s.[5] Wheat prices showed a very similar trend: they were generally level in the fourteenth and fifteenth centuries, and increased greatly during the sixteenth century, especially in the second half of the century. The trend of wool prices was somewhat different, the upward movement beginning earlier. The trend of wheat prices was eventually steeper than that of wool prices and the two crossed in the middle of the sixteenth century. Between 1490 and 1552 there were only two transitory periods in which the price of wheat was high relative to the price of wool. It was not until well after 1552 that prices moved relatively in favour of wheat for any length of time. The period 1552–81 was one of transition, in which, on balance, relative prices appear to have slightly favoured wool. From 1581 to 1609 the price of wheat

[1] W. Beveridge, 'Wages in the Winchester manors', *Econ. Hist. Rev.*, VII (1936–7), 22–43, and 'Westminster wages in the manorial era', *Econ. Hist. Rev.*, 2nd ser., VIII (1955–6), 18–35.

[2] M. M. Postan (1949–50), 226–33.

[3] E. H. P. Brown and S. V. Hopkins, 'Seven centuries of building wages', *Economica*, n.s., XXII (1955), 195–206.

[4] W. Beveridge (1936–7), 31–5, and (1955–6), 26; E. H. P. Brown and S. V. Hopkins, 'Wage-rates and prices: evidence for population pressure in the sixteenth century', *Economica*, n.s., XXIV (1957), 289–306; M. M. Postan (1938–9), 166, and (1949–50), 226–7.

[5] E. H. P. Brown and S. V. Hopkins, 'Seven centuries of the prices of consumables compared with builders' wage-rates', *Economica*, n.s., XXIII (1956), 296–314.

moved steeply upwards and except for two short periods was markedly high in relation to the price of wool.[1]

For a century after the Black Death, *land values* were at a low and often falling level. There was in general terms a land surplus, as vacant holdings and lapsed rents testified. According to M. M. Postan, the agricultural slump, which began in the early fourteenth century (in some places before the Black Death), continued, with only a slight halt around 1410, until the late 1470s and 1480s.[2] Towards 1500, however, the picture changed and rents rose as demand for land increased, and the sixteenth century witnessed a considerable increase in land values.[3]

An acceptable explanation of these price changes remains to be firmly established. M. M. Postan has related the low and falling prices of consumables, the high and rising wages, and the low land values to reduced population pressure during the fourteenth and most of the fifteenth century. This view has, for the most part, been accepted by English although not by some foreign scholars.[4] The inflationary price rises of the sixteenth century are now similarly being attributed to increased population pressure, and to 'the fact that population outdistanced supplies'.[5] The view that the price changes of the sixteenth century were caused by a large influx of precious metals from the New World has been challenged, principally on two grounds: (1) that a fall in the value of money began to be marked, in some instances, several decades before the influx from the New World became appreciable, and (2) that not all commodities were similarly affected as might be expected with a change in the value of money brought about purely by an expansion in the circulating medium. Similarly, Henry VIII's devaluation and debasement of the coinage after 1543 certainly aggravated, but did not initiate, the inflationary processes. When the currency was restored in 1561, inflation continued, although at a slower rate.[6]

[1] P. J. Bowden (1952), 116–23; W. G. Hoskins, 'Harvest fluctuations and English economic history, 1480–1619', *Agric. Hist. Rev.*, XII (1964), 28–46.
[2] M. M. Postan (1938–9), *passim*, and (1949–50), 236–9.
[3] A. Simpson, *The wealth of the gentry, 1540–1660* (Cambridge, 1961).
[4] M. M. Postan (1938–9) and (1949–50), *passim*. For dissenting views, see for example: J. Schreiner, 'Wages and prices in England in the later Middle Ages', *Scand. Econ. Hist. Rev.*, II (1954), 61–73; E. Kosminsky, 'The evolution of feudal rent in England from the XIth to the XVth centuries', *Past and Present*, VII (1955), 12–36; W. C. Robinson, 'Money, population and economic change in late medieval Europe', *Econ. Hist. Rev.*, 2nd ser., XII (1959–60), 63–76.
[5] J. Hurstfield, *The Elizabethan nation* (London, 1964), 15.
[6] J. Hurstfield, 13–15. The older view stemmed especially from E. J. Hamilton,

A further factor inducing short-term fluctuations in agricultural prices and in population levels were climatic changes. Patterns of alternating good and bad runs of harvests have been identified in England between 1480 and 1619, and on the estates of the abbey of Battle in particular between 1340 and 1444. The years of harvest failure were associated with periods of excessive rainfall. But there is also increasing evidence for a possible long-term deterioration in the climate during the later Middle Ages at certain periods, notably between about 1420 and 1445 and between 1555 and 1600.[1]

Whatever the exact chronology and causes of changes in prices and population levels, their consequences for agriculture were enormous. The later Middle Ages witnessed three fundamental agricultural changes; first, the decline of demesne farming; secondly, the growth of peasant farming; and thirdly, changes in the use of land, notably enclosure for livestock farming.

The decline of demesne farming

By 1330 or so the age of medieval high farming had ended – in some places long ended – and during the century, as costs rose with wage-rates and as receipts fell with prices, profits of large-scale production declined. At dates differing according to the locality and the policy of a particular lord, grain production was curtailed and demesnes were increasingly leased, at first in portions and then in their entirety, on ever longer leases to decreasing numbers of aspiring tenants. The usual reaction of lords to declining profits was an attempt to reduce costs and to stabilise revenues. The establishment of rentier economies was a major economic change of the later Middle Ages.[2]

Even when demesnes were being cultivated directly for profit by lords

'American treasure and the rise of capitalism', *Economica*, XXVII (1929), 338–57. Current objections to this view are summarised in J. D. Gould, 'The price revolution reconsidered', *Econ. Hist. Rev.*, 2nd ser., XVII (1964–5), 249–66.

[1] W. G. Hoskins (1964), 28–46; H. H. Lamb, 'Britain's changing climate', *Geog. Jour.*, CXXXIII (1967), 444–66; G. Manley, 'Climate in Britain over 1,000 years', *Geog. Mag.*, XLIII (1970–1), 100–7; P. F. Brandon, 'Late-medieval weather in Sussex and its agricultural significance', *Trans. and Papers, Inst. Brit. Geog.*, LIV (1971), 1–17; G. Utterström, 'Climatic fluctuations and population problems in early modern history', *Scandinavian Econ. Hist. Rev.*, III (1955), 3–47.

[2] J. A. Raftis, 217–50; E. M. Halcrow, 'The decline of demesne farming on the estates of Durham cathedral priory', *Econ. Hist. Rev.*, 2nd ser., VII (1954–5), 345–56; A. R. Bridbury, *Economic growth. England in the later Middle Ages* (London, 1962), 18.

or their officials, rents of various kinds provided a substantial proportion of the total manorial income. A major part of the archbishop of Canterbury's income from land, throughout the medieval period, was from rents, as it was on other estates, both lay and ecclesiastical.[1] It was not the introduction but the growth of a rentier economy (i.e. the renting of demesne lands), and the almost total disappearance of direct demesne farming, that was so momentous. The cultivated acreages of demesnes diminished, partly because some arable land was abandoned and more especially because lands were increasingly leased to tenants.

Contraction of demesne farming was apparent some decades before the Black Death, which for many landlords came as the culmination of a period of shaken prosperity. This retrenchment was to continue, with occasional and local revivals, until about 1450 or so. Marginal lands brought into cultivation at the height of the medieval arable expansion now tumbled to grass.[2] Decreasing money profits were a disincentive to arable farming. Thus in the years immediately after the Black Death total grain production was cut by about one half on the Huntingdonshire manors of Ramsey abbey, and the acreage of crops sown on the demesne at Hutton (Essex) in 1389 was not one half that in 1342, having especially diminished since 1369.[3] On many manors, however, reduction in demesne cultivation immediately after the Black Death was only temporary, most of the loss being made up again in the following decade or so.[4] On other manors, land lay waste because the lord considered it unprofitable to cultivate, and because he could find no tenant willing to lease it.[5] Some lords turned more towards pasture farming to reduce their costs.[6]

[1] F. R. H. Du Boulay, 'A rentier economy in the later Middle Ages: the Archbishopric of Canterbury', *Econ. Hist. Rev.*, 2nd ser., XVI (1963–4), 427–38; G. A. Holmes, *The estates of the higher nobility in fourteenth-century England* (Cambridge, 1957), 88–9 and 109–13; J. S. Donnelly, 'Changes in the grange economy of English and Welsh Cistercian abbeys', *Traditio*, X (1954), 399–458.

[2] P. F. Brandon (1962), 71; M. M. Postan (1938–9), 161; F. R. H. Du Boulay, 'Late-continued demesne farming at Otford', *Archaeol. Cantiana.*, LXXIII (1959), 116–24; M. Morgan, *The English lands of the abbey of Bec* (London, 1946), 98–104.

[3] J. A. Raftis, 253; K. G. Feiling, 'An Essex manor in the fourteenth century', *Eng. Hist. Rev.*, XXVI (1911), 333–8.

[4] For example, on the East Anglian manors of the bailiwick of Clare: G. A. Holmes, 91.

[5] F. R. H. Du Boulay (1959), 121.

[6] K. G. Feiling, 336.

Diminishing grain production and arable cultivation on demesnes was accompanied by a movement to lease them, at first piecemeal for short periods to a few tenants, sometimes to a single tenant. Leasing was an attempt to maximise and stabilise manorial incomes. It was not an innovation of the later Middle Ages, but the ubiquity and the finality of the process during that period was new. On some manors leasing of parcels of demesne for short periods in the two decades before and after the Black Death suggests that the process was intended to be temporary, an expedient method of raising ready cash.[1] But with continued economic stagnation, more landlords adopted a policy of leasing out whole demesnes.[2] From the 1360s and 1370s onwards, the process quickened, until by the mid-fifteenth century it was generally completed. The forty demesnes of the see of Canterbury, widely distributed over south-east England, began to be leased out permanently between the 1380s and 1420s, and most had been leased by 1422; the manors of Durham cathedral priory were all leased by 1451; the leasehold system was established on almost all the estates of Canterbury cathedral priory by 1396; most Cistercian granges had been broken up into tenant holdings and leased either in part or as units by 1410 or so.[3] The process was equally frequent on lay estates; a rentier economy was established on the estates of the Percy family in Yorkshire, Northumberland, Cumberland and Sussex by 1416; and all the demesne lands of the bailiwick of Clare, a group of manors in East Anglia, passed out of the lord's hands between 1360 and 1400.[4] Short-term leases, taken up by tenants anxious for quick profits and unconcerned with the state of buildings or with maintaining soil fertility, were gradually replaced by long-term leases incorporating conditions for the maintenance of buildings and soils; thus it was hoped to stabilise rent incomes and to preserve property. With this type of leasing, the early

[1] K. G. Feiling, 335; F. M. Page, 114.

[2] R. H. Hilton, *The economic development of some Leicestershire estates in the 14th and 15th centuries* (London, 1947), 105; J. A. Raftis, 281–301.

[3] F. R. H. Du Boulay (1959), 116, and (1963–4), 426; E. M. Halcrow, 355–66; R. A. L. Smith, 192; J. S. Donnelly, 451. See also: J. S. Drew, 'Manorial accounts of St Swithun's priory, Winchester', *Eng. Hist. Rev.*, LXII (1947), 20–41; R. H. Hilton (1947), 85–91; F. W. Maitland, 'The history of a Cambridgeshire manor', *Eng. Hist. Rev.*, IX (1894), 417–39; M. Morgan, 113–24; J. A. Raftis, 251–80; N. S. B. and E. Gras, *The economic and social history of an English village (Crawley, Hampshire), A.D. 909–1928* (Cambridge, Mass., 1930), 80–3; F. G. Davenport, *The economic development of a Norfolk manor, 1086–1565* (Cambridge, 1906), 49–55.

[4] J. M. W. Bean, *The estates of the Percy family 1416–1537* (Oxford, 1958), 13; G. A. Holmes, 92.

medieval relationship of lord and villein changed to that of landowner and tenant farmer.[1]

Leasing increased the proportion of a landlord's income derived from rent without necessarily increasing his total income. In fact, for most lords the century after 1334 was a period of declining incomes from land. Incomes from the customary lands of tenants as well as from leased demesnes showed an overall decline; rents per acre fell, the demand for tenements which lapsed into the lord's hands was slack, and arrears of rent accumulated. The rural landscape assumed in many parts a neglected aspect as buildings decayed, as tenements were abandoned and as land reverted to rough pasture. Landlords and landscape were the twin sufferers from the agricultural slump which characterised the period 1350–1450. It was not until after about 1460 that rents again began to rise, and on some estates the rise did not come until much later.[2]

This general trend (contracting demesne cultivation, spread of leasing, declining incomes from land) was typical, but it was not universal, and the effects of the retrenchment were not felt with equal severity throughout the country. The experiences of lay and ecclesiastical estates were essentially similar, but it seems that small estates were more able than large ones to weather the economic storm.[3] On large estates, conversion to wholesale demesne leasing took place earlier on distant manors than on 'home-farms' retained in cultivation to produce not for sale but for consumption.[4] Nor was the general trend continuous; the years around 1400 witnessed a measure of economic recovery throughout England, more especially in the south-east. The incomes of Canterbury cathedral priory and of the see of Canterbury, both of whose estates were widely spread over the south-east, were apparently more stable during the later Middle Ages than those of most other large estates, and this owed something to an expansion of the cultivated areas. But the peculiarity of these estates was one of degree not

[1] E. M. Halcrow, 352–4; M. M. Postan, 'The rise of a money economy', *Econ. Hist. Rev.*, XIV (1944–5), 123–34; J. M. W. Bean (1958), 55–6; R. H. Hilton (1947), 126–8; R. A. L. Smith, 200; R. Scott, 'Medieval agriculture' in E. Crittall (ed.), *V.C.H. Wiltshire*, IV (1959), 7–42.

[2] J. M. W. Bean (1958), 17–48; R. H. Hilton (1947), 85–8; G. A. Holmes, 114–20; W. G. Hoskins, *The Midland peasant: the economic and social history of a Leicestershire village* (London, 1957), 84–5; F. M. Page, 147–9; M. M. Postan (1938–9), 161–2; F. G. Davenport, 56–9; J. A. Raftis (1957), 285–94.

[3] R. H. Hilton (1947), 117–21.

[4] R. A. L. Smith, 200–1; R. H. Hilton (1947), 88, 91 and 132; P. F. Brandon (1962), 69.

of kind, for both witnessed a decline of demesne farming. Both prior and archbishop became landlords pure and simple.[1]

The growth of peasant farming

The emergence of rentier landlords during the later Middle Ages was paralleled by the appearance in increased numbers of both rich peasant farmers and poor landless labourers. Throughout this period land became available on terms freer than the customary terms governing traditional holdings. Demesnes were increasingly leased for money rents; land taken in from the waste was almost always leased out for a money rent only; and peasant tenements were being divided up even before the Black Death, and taken piecemeal by other peasant lessees at money rents.[2]

Peasants who leased demesnes often found themselves in possession of potentially prosperous farms because demesnes were often situated on the best soils in a locality and had often been meticulously manured and cultivated.[3] When worked by lessee peasants utilising little hired labour, incurring no managerial expenses, and working for their own profit, former demesne lands again became viable economic units, operated now by a prosperous section of the peasantry. Occasionally, the lessees were newcomers to a district, but more often they were enterprising local men, sometimes former manorial officials, who had built up substantial holdings by purchasing and leasing usually small plots, and who now augmented their holdings further by leasing part or all of a demesne.[4] The process of demesne leasing gave added momentum to an already fluid land market and accentuated the growing inequality in the sizes of peasant holdings. Holdings were far from being equal in size at the beginning of the fourteenth century, and they became decreasingly so during the later Middle Ages. From an economic point of view, the commutation of villein services was one of many ways of lightening the terms on which customary lands were let, and as the demands of demesnes declined so lords could dispense with some of the services previously required. The leasing of demesnes was essentially complete by about 1450 and the exaction of labour services had all but ceased by 1500.[5]

[1] F. R. H. Du Boulay (1963–4), 438.

[2] R. H. Hilton, 'Medieval agrarian history' in W. G. Hoskins (ed.), *V.C.H. Leicestershire*, II (1954), 145–98.

[3] E. M. Halcrow, 349; A. R. Bridbury (1962), 91.

[4] R. H. Hilton (1954), 94 and 157–62; M. Morgan, 111–12 and 115–18.

[5] M. M. Postan (1949–50), 238, and 'The chronology of labour services', *Trans. Roy. Hist. Soc.*, 4th ser., XX (1937), 169–93; L. C. Latham, 'The decay of the manorial

Social status came to be determined more by the amount of land occupied than by the nature of its tenures.[1] One direct consequence of the decline in population during the fourteenth century was an increase in the average size of peasant holdings. Demesne land and vacant tenant land provided ample opportunity for the enterprising and industrious peasant to build up an estate. Lands which became vacant during and after the Black Death were frequently taken up by land-hungry tenants.[2] The land market became increasingly active, and holdings were frequently subdivided and let to new tenants. In Leicestershire, the disintegration of tenements and regrouping of lands produced a sort of polarisation process in which on the one hand richer peasants built up farms above the size of the 'normal' virgate holding of 20–30 acres and, on the other hand, the poorer lost what land they had and so became labourers. This stratification of the peasantry was one of the most important developments in the English countryside during the fourteenth and fifteenth centuries and was accompanied by the rise of yeomen and husbandmen from bondage.[3]

By the first quarter of the sixteenth century, as evidenced in Leicestershire assessments of 1524 and 1525, the rough economic equality of the early medieval community had been shattered beyond recognition. Approximately one man in five was assessed on his wages and could be regarded as being dependent on them as a source of income rather than on income from land or possessions, so that he belonged undoubtedly to the labouring classes.[4] Economic inequality was well-marked in Leicestershire in 1524–5, for (omitting the squirearchy, who were less wealthy than many a yeoman, in personal estate at least) 4% of the rural population owned a quarter of the personal estate and $15\frac{1}{2}\%$ owned half of it. And even in villages where personal estate was more evenly distributed, as at Wigston Magna, 20% of the taxpayers owned half of the personal estate.[5]

system during the first half of the fifteenth century, with special reference to manorial jurisdiction and to the decay of villeinage, as exemplified in the records of twenty-six manors, in the counties of Berkshire, Hampshire and Wiltshire', *Bull. Inst. Hist. Research*, VII (1929–30), 113–16.

[1] K. C. Newton, *Thaxted in the fourteenth century* (Chelmsford, 1960), 20–32; R. H. Hilton (1954), 96.

[2] G. A. Holmes, 90–2; F. M. Page, 123–4 and 152; J. A. Raftis, 252–3.

[3] R. H. Hilton (1947), 79–105, and (1954) 185–8; W. G. Hoskins (1957), 39–52; N. S. B. and E. Gras, 95–8; F. G. Davenport, 70–97.

[4] R. H. Hilton (1954), 94–5; A. R. Bridbury (1962), 103.

[5] W. G. Hoskins (1950), 127–30.

The remainder of the sixteenth century saw an accentuation of this trend towards inequality and with it, the final obliteration of the medieval framework of landholding. Changes which had been taking place during the previous century and a half were hastened by inflation during the sixteenth century. The redistribution of land continued, and the dissolution of the monasteries added to the pool of land in the market from 1536 onwards, reducing further the direct influence of the Church upon the English landscape, and setting in motion the emergence of such ecclesiastical relict features as the ruins of Fountains abbey (Yorkshire).[1] One interesting cartographic by-product of the selling and exchanging of land and of the boundary disputes that often accompanied such transactions was the production of estate maps by improved techniques of land surveying.[2]

Changing land use and rural settlement

The *Nonarum Inquisitiones* of 1341 show that the abandonment of arable lands was under way before the Black Death.[3] Some lands, notably in the north (especially in Lancashire), were abandoned because of political disturbances; others were abandoned because of flooding by the sea (especially along the south coast); others because climatic hazards had reduced their supply of seed corn (especially in the Chilterns); others because soils were thought to be exhausted (Fig. 45). But probably the most common cause of this agricultural retrenchment was a shrinkage of village populations. Arable reverted to pasture and dwellings tumbled into ruins; many arable acres *solebant seminari et modo jacent ad pasturam* and many dwellings *sunt derelicta sine habitoribus*.[4] Such abandonment was not a linear retreat of settlement, it revealed itself 'not in a long thin high-water mark, the whole length of a shore, but in many scattered

[1] H. J. Habakkuk, 'The market for monastic property, 1539–1603', *Econ. Hist. Rev.*, 2nd ser., x (1957–8), 362–80; R. A. Donkin, 'The Cistercian Order in medieval England: some conclusions', *Trans. and Papers, Inst. Brit. Geog.*, XXXIII (1963), 181–98.

[2] H. C. Darby, 'The agrarian contribution to surveying in England', *Geog. Jour.*, LXXXII (1933), 529–35.

[3] G. Vanderzee (ed.), *Nonarum Inquisitiones in Curia Scaccarii* (London, 1807).

[4] A. R. H. Baker, 'Evidence in the *Nonarum Inquisitiones* of contracting arable lands in England during the early fourteenth century', *Econ. Hist. Rev.*, 2nd ser., XIX (1966), 518–32; A. R. H. Baker, 'Some evidence of a reduction in the acreage of cultivated lands in Sussex during the early fourteenth century', *Sussex Archaeol. Coll.*, CIV (1966), 1–5.

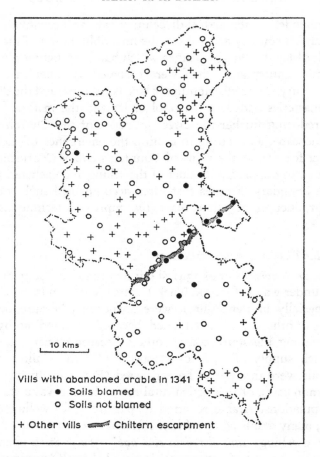

Fig. 45 Abandoned arable land in Buckinghamshire in 1341
Based on A. R. H. Baker, 'Evidence in the *Nonarum Inquisitiones* of con-
tracting arable lands in England during the early fourteenth century', *Econ.
Hist. Rev.*, 2nd ser., XIX (1966), 527–8.

rock-pools'.[1] Land reverted to pasture because it was not wanted as arable.
This process was particularly significant in the East Riding of Yorkshire,
in the Lincolnshire Wolds and in the light soil areas of Norfolk, but to
some extent it was characteristic over a much wider area and led to the
establishment of a new balance between grain lands and grass lands which
may have been reached sometime between 1420 and 1440. 'It was clearly

[1] M. W. Beresford (1954), 204.

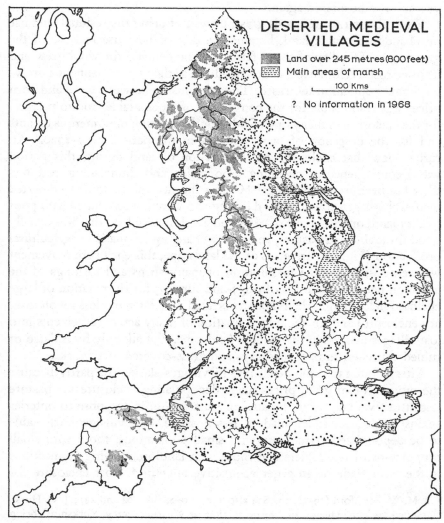

Fig. 46 Deserted medieval villages
  Based on M. W. Beresford and J. G. Hurst (eds.), *Deserted medieval villages* (London, 1971), 182–212.

a different balance from the balance of the late thirteenth century' and it was soon to be altered again.[1]

From about 1450 onwards grass, instead of being the residual land use on abandoned arable land, became the desired land use because of the growing profitability of sheep farming. The demand for wool grew and its price rose as cloth exports expanded and as the requirements of a growing home population increased. The relationship between wool and corn prices was responsible for the conversion of much arable land to pasture, for the enclosure of many open fields, for the eviction of numerous tenants and for the desertion of many villages in the period 1450–1520.[2] An earlier view that enclosure and conversion of land during this period was a consequence of declining crop yields and diminishing soil fertility has been found wanting.[3] Most of the two thousand or so deserted medieval villages now identified in England were not associated with poor quality land, or with a retreat of settlement, but with land that was equally good for arable and pasture, and with a changing emphasis in agricultural production (Fig. 46). In terms of the landscape, this enclosure movement was responsible for the fossilisation of many strips and furlongs of the common fields in the form of ridge-and-furrow, for the creation of large enclosed fields (larger than those associated with later and less revolutionary enclosure) and for the transformation of many active settlements into lost villages marked in the modern landscape, if at all, only by isolated or ruined churches, sunken roadways and grass-covered earthworks.

After about 1520, the death rate of villages slackened, partly because the movement of grain prices made depopulating enclosure for pasture less attractive, and partly because of state action, in response to outcries against enclosure, which made it less possible. The villages most liable to be depopulated were those with small populations, those with small proportions of freeholders and those with landlords having connections in the wool trade or an eager acquisitive appetite.[4] Most enclosure for

[1] M. W. Beresford (1954), 20; See also 150, 170–2, 198–204 and 241; J. G. Hurst, 'Deserted medieval villages and the excavations at Wharram Percy, Yorkshire', being ch. 10 of R. L. S. Bruce-Mitford, *Recent archaeological excavations in Britain* (London, 1956), 251–73.   [2] M. W. Beresford (1954), 27–77, 177–89 and 207–15.

[3] H. Bradley, *The enclosures in England: an economic reconstruction* (New York, 1918); R. Lennard, 'The alleged exhaustion of the soil in medieval England', *Econ. Jour.*, XXXII (1922), 12–27; M. K. Bennett, 'British wheat yields per acre for seven centuries', *Econ. Hist.*, III (1934–7), 12–29.

[4] M. W. Beresford (1954), 102–33, 212–14 and 228; M. W. Beresford and J. G. Hurst (eds.), *Deserted medieval villages* (London, 1971), 3–75.

livestock farming took place on the densely settled claylands of the Midlands, where mixed farming was practised and where crop rotations which included leys enabled farmers to keep more stock when market conditions favoured them. Enclosure here was often for cattle as well as for sheep, for beef and leather as well as for wool and mutton. In western Leicestershire, for example, in the neighbourhood of Charnwood Forest, enclosures were principally for dairying and cattle-rearing but in the eastern uplands enclosures were principally for sheep.[1] Other areas which underwent much enclosure in the Tudor period were the less fertile chalk and limestone uplands, the traditional sheep-rearing lands. A large farmer wanting to enlarge his sheep flock found it easier to enclose great tracts of common or to force enclosure of the open fields here, because his tenants were often few and their opposition weak. The economic incentive to enclose land in the uplands was the same as that which encouraged enclosure in the lowlands – the high prices of animal products, particularly wool and mutton. And these uplands produced a fine short wool which became more scarce and sought after as the sixteenth century wore on.[2]

Enclosure and its accompanying depopulation produced much controversy, many pamphlets, a number of government enquiries (in 1517–18, 1548–9, 1566 and 1607), a number of ineffective Acts of Parliament (e.g. in 1489, 1515, 1536 and 1563), and a revolt in the Midlands in 1607. The evidence is incomplete and difficult to interpret, but it would seem that between 1455 and 1607 about 8 to 9% of the area of some Midland counties were enclosed, and smaller percentages in a number of other counties (Fig. 47). The total area enclosed over 24 counties may have amounted to about half a million acres or less than 3% of the total area of England, a small amount when compared with the enclosures of the eighteenth century.[3] M. W. Beresford has suggested that the Midlands were particularly susceptible to depopulating enclosures for livestock farming because lands there were equally suitable for grain and grass; no such land use change was to be expected where the balance was set firmly towards the one or the other.[4]

Many parts of England did not see much depopulating enclosure.

---

[1] J. Thirsk, *Tudor enclosures* (Hist. Assoc., London, 1959), 20. [2] *Ibid.*, 18–19.
[3] E. Gay, 'The inquisitions of depopulation in 1517 and the Domesday of inclosures', *Trans. Roy. Hist. Soc.*, n.s., XIV (1900), 231–67, and 'Inclosures in England in the sixteenth century', *Quarterly Jour. Econ.*, XVII (1902–3), 576–97; A. H. Johnson, *The disappearance of the small landowner* (London, 1909), 47–59.
[4] M. W. Beresford (1954), 242.

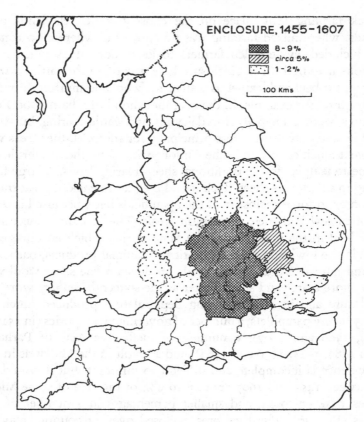

Fig. 47 Enclosure, 1455–1607
Recorded in the proceedings of the Courts of Exchequer, the Court of
Chancery, the Star Chamber and the Court of Requests. Based on: (1) A. H.
Johnson, *The disappearance of the small landowner* (Oxford, 1909), 48–9, and
map at end; (2) E. F. Gay, 'Inclosures in England in the sixteenth century',
Quart. Jour. Economics, XVII (1903), 585–6.

Little took place in the highland zone where pastoral pursuits had long
dominated the economy – none of the 113 villages in three upland hundreds
of the West Riding of Yorkshire was abandoned. In such places there was
little corn to be displaced, and plenty of room to expand grazing grounds
without destroying villages. Elsewhere, there were wooded areas whose
agriculture was biased away from arable husbandry, as in the Weald, the
Forest of Dean, Sherwood Forest and much of Essex and the Chilterns.
Within the Midlands, individual counties showed important contrasts

because of their far from uniform terrains; settlements in the Forest of Arden, to the north of the Avon, were relatively immune from depopulation, unlike villages to the south, in the classical common field country. Marshland and fenland settlements were also usually immune – not a single vill in the Holland division of Lincolnshire was deserted, although in Romney Marsh ruined churches testify to the swing from arable to pasture which occurred there during the later Middle Ages.[1]

There was, in addition to the dramatic enclosure for pasture during 1450–1520, a slow and piecemeal reclamation and enclosure for both pasture and arable throughout the later Middle Ages. Some new enclosures were made direct from the waste – the story, even in the first half of this period, was not entirely one of contraction. Some additional woodland was cleared, for example in the Weald and in the Forest of Arden, and some more marshland reclaimed, for example in Romney Marsh and in the fenlands of Lincolnshire.[2] In open-field districts, such as the west Midlands, the disintegration and regrouping of tenements in a fluid land market made possible a consolidation of scattered parcels so that eventually large enough blocks of land could be made into private enclosures. Certain districts which in 1334 lay within the region of the Midland open-field system had by 1600 passed silently into enclosed field systems. Such in particular were the counties of the west – Herefordshire, Shropshire, parts of Staffordshire, Worcestershire, Warwickshire, Gloucestershire and Somerset.[3] Enclosure by private agreements, often confirmed by the Court of Chancery or the Court of Exchequer, gathered pace during the sixteenth century, but such agreements required the unanimous consent of those involved.[4] Simultaneously, the pattern of old enclosed land was changing;

---

[1] M. W. Beresford (1954), 217–61.

[2] J. L. M. Gulley, 'The Wealden landscape in the early seventeenth century and its antecedents', unpublished Ph.D. thesis, University of London (1960), 259–67; P. F. Brandon, 'Medieval clearances in the East Sussex Weald', *Trans. and Papers, Inst. Brit. Geog.*, XLVIII (1969), 135–53; R. H. Hilton, 'Old enclosures in the west Midlands: a hypothesis about their late medieval development', *Geographie et histoire agraires* (Nancy, 1959), 272–83; R. A. L. Smith, 203; J. Thirsk, *Fenland farming in the sixteenth century* (Leicester, 1953).

[3] H. L. Gray, *English field systems* (Cambridge, Mass., 1915), 108.

[4] K. J. Allison, 'The sheep–corn husbandry of Norfolk in the sixteenth and seventeenth centuries', *Agric. Hist. Rev.*, V (1957), 12–30; M. W. Beresford, 'Glebe terriers and open-field Buckinghamshire. Part II', *Records of Bucks.*, XVI (1953–60), 5–28; J. Cornwall, 'Agricultural improvement, 1560–1640', *Sussex Archaeol. Coll.*, XCVIII (1960), 118–32; R. H. Hilton (1959), 274–80, and (1954), 189–94; E. Kerridge, 'Agriculture c. 1500 – c. 1793', in E. Crittall (ed.), *V.C.H. Wiltshire*, IV (1959), 43–64;

in the Chilterns and in parts of Kent, for example, the great demesne
fields of the early fourteenth century had been divided into smaller
enclosures by the end of the sixteenth century, and many small closes of
the peasantry and yeomen had been amalgamated into larger fields as
holdings became concentrated into fewer hands.[1]

Not all enclosure involved a change from arable to pasture. Landlords
could increase their flocks without recourse to converting arable land;
they could extend their 'fold-courses' or set up new ones; they could
enclose their own arable land to prevent livestock of tenants feeding there
after harvest; they could overstock commons; and they could enclose
commons for their own benefit. Enclosure in Norfolk in the sixteenth
century was almost entirely of this kind, and did not involve conversion
to pasture.[2] Enclosure here was a means of increasing productivity, for it
enabled land to be more efficiently cultivated and stock to be more carefully bred. Arable closes were better manured and sometimes carried crops
in the fallow year. There is some evidence from many different parts of
England that towards the end of the sixteenth century, under the stimulus
of high grain prices, there was both enclosure of existing open fields which
were kept as arable and also renewed reclamation to extend the cultivated
area. But the progress of enclosure of arable fields was limited in common
field areas by the system of common grazing. In lowland England,
especially in the densely settled Midlands, rights of common both in the
fields and on the wastes were of vital importance, for the supply of
grazing was very limited. What common grazings there were had to be
carefully stinted; and ultimately, when land shortage and food supplies
became critical, the wastes were enclosed and divided among individuals.
In highland England, where population pressure was generally less acute

M. R. Postgate, 'The field systems of Breckland', *Agric. Hist. Rev.*, x (1962), 80–101;
G. Youd, 'The common fields of Lancashire', *Trans. Hist. Soc. Lancs. and Cheshire*,
cxiii (1961), 1–41; R. H. Hilton, 'Social structure of rural Warwickshire in the Middle
Ages', *Dugdale Soc. Occasional Papers*, ix (1950); M. W. Beresford, 'Habitation versus
Improvement: the debate on enclosure by agreement', in F. J. Fisher (ed.), *Essays in
the economic and social history of Tudor and Stuart England* (Cambridge, 1961), 40–69;
S. A. Johnson, 'Some aspects of enclosure and changing agricultural landscapes in
Lindsey from the sixteenth to the nineteenth century', *Rep. and Papers, Lincs.
Archit. and Archaeol. Soc.*, ix (1962), 134–50.
  [1] D. Roden and A. R. H. Baker, 'Field systems of the Chiltern Hills and of parts
of Kent from the late thirteenth to the early seventeenth century', *Trans. and Papers,
Inst. Brit. Geog.*, xxxviii (1966), 73–88.
  [2] K. J. Allison (1957), 12–30.

than in the lowlands, where farming had a pastoral bias, and where there was abundant waste awaiting improvement, enclosures raised no great opposition. Here, enclosure of open fields proceeded piecemeal and by agreement, and much enclosure took the form of the division of the commons between intercommoning parishes, the introduction of stints on formerly unregulated waste and the taking in of waste in order to increase the arable acreage.[1]

Enclosure was one but not the only important technical change in agriculture during the later Middle Ages. Cereal production – in terms of yields per acre – certainly increased.[2] But probably the greatest single problem facing farmers was that of animal feed, and there were a number of attempts to increase fodder supplies. In open-field areas, the number of fields cultivated communally was often increased, thus reducing the fallow acreage; and rotations themselves, often based on furlongs rather than on fields, became more complicated than the number of open fields in any township might suggest. The fallow was further reduced by being sown, wholly or in part, often with legumes.[3] In some localities, field systems were rationalised during the later Middle Ages; thus the multiple open fields which existed at Bickenhill (Warwickshire) in 1350 had been simplified into three common arable fields by 1612.[4] In other localities, such as the dales of Cumberland, growing population pressure in the sixteenth century resulted in communal reclamation of waste lands, and in a multiplication of open fields.[5] Another important change was the more widespread adoption of convertible husbandry, single parcels and sometimes whole furlongs being laid down as temporary grass. This practice increased fodder supplies but not productivity per acre; ley

[1] J. Thirsk (1959), 4–8; R. H. Tawney, *The agrarian problem in the sixteenth century* (London, 1912), 237–53; W. G. Hoskins and L. D. Stamp, *The common lands of England and Wales* (London, 1963), 44–52.

[2] M. K. Bennett, 26–8.

[3] P. F. Brandon (1962), 65–66; H. L. Gray, 9, 73–81 and 109–56; F. G. Gurney, 'An agricultural agreement of the year 1345 at Mursley and Dunton – with a note upon Walter of "Henley"', *Records of Bucks.*, XIV (1941–6), 246–8; R. H. Hilton (1954), 159–61; (1947), 65 and 152–6; E. Kerridge (1959), 52; F. M. Page, 119; R. Scott, 15.

[4] V. H. T. Skipp and R. P. Hastings, *Discovering Bickenhill* (Birmingham, 1963), 15–29. See also J. Thirsk, 'The common fields', *Past and Present*, XXIX (1964), 3–25.

[5] G. G. Elliott, 'The system of cultivation and evidence of enclosure in the Cumberland open fields in the sixteenth century', *Geographie et histoires agraires* (Nancy, 1959), 118–36. See also A. J. Roderick, 'Open-field agriculture in Herefordshire in the later Middle Ages', *Trans. Woolhope Naturalists' Field Club*, XXXIII (1949–51), 55–67.

husbandry improved soil texture but not soil fertility, because on balance grazing beasts took out what they put in.[1]

Agriculture during the later Middle Ages was marked more by local diversity than by scientific experiment. Manuring of fields became more intensive and, as far as animal manure was concerned, more controlled and therefore more effective. Marling and liming were common practices by the end of the sixteenth century, and resulted in many pits and depressions over the surface of the ground.[2] But fertilising was still far from being scientific. The development of printing saw the appearance of agricultural writings such as John Fitzherbert's *Boke of Husbondrye* (1523) and Thomas Tusser's *Hundreth good pointes of Husbandrie* (1557), but new ideas spread only very slowly.[3] Turnips had long been grown as a garden vegetable, but Barnaby Googe, in his *Foure Bookes of Husbandrie* (1577), advised cultivation of turnips as a fodder crop, although it was not until the seventeenth century that field cultivation of turnips for livestock fodder began in England. After the middle of the sixteenth century experiments began in the floating of water-meadows to improve and increase fodder supplies, in the vales of Taunton and Hereford and elsewhere.[4] Experiments in fen drainage also became frequent, especially after 1560.[5] Agriculture in general was marked by increasing local specialisation, as the emergence of hop and fruit cultivation in Kent for the London market exemplifies. The commercialisation of agriculture made most advances in the neighbourhood of London and in the developing industrial areas, where there were large and growing populations employed partially at least in non-agricultural pursuits.[6]

[1] R. H. Hilton (1954), 197–8; W. G. Hoskins (1950), 140–4, and (1957), 67, 95, 152 and 164; E. Kerridge (1959), 52; E. L. Jones, 'Agriculture and economic growth in England, 1660–1750: agricultural change', *Jour. Econ. Hist.*, XXV (1965), 1–18.

[2] J. Cornwall, 121–4; H. C. Prince, 'The origin of pits and depressions in Norfolk', *Geography*, XLIX (1964), 15–32.

[3] G. E. Fussell, *The old English farming books from Fitzherbert to Tull, 1523 to 1730* (London, 1947), 1–20.

[4] E. Kerridge, *The agricultural revolution* (1967), 252–4   [5] *Ibid.*, 222–39.

[6] M. Campbell, *The English yeoman under Elizabeth and the early Tudors* (New Haven, 1942), 156–220; F. J. Fisher, 'The development of the London food market, 1540–1640', *Econ. Hist. Rev.*, V (1934–5), 46–64; D. C. D. Pocock, 'Some former hop-growing areas', *Agric. Hist. Rev.*, XIII (1965), 17–22; N. S. B. Gras, *The evolution of the English corn market* (Cambridge, Mass., 1915); 95–129 R. H. Hilton (1959), 276 and 281–3; E. Kneisel, 'The evolution of the English corn market', *Jour. Econ. Hist.*, XIV (1954), 46–52; C. W. Chalklin, *Seventeenth-century Kent: a social and economic history* (London, 1965), 88–95.

During the fourteenth and fifteenth centuries derelict buildings – products often of neglect and sometimes of destruction – were prominent relict features in the English landscape, not only in the countryside but also in the towns. Towards the end of the sixteenth century, a contrary process known as the Great Rebuilding gathered momentum.[1] Many of the edifices erected or altered in towns during this Great Rebuilding have been removed by subsequent developments; but in rural districts, particularly in areas already characterised in the sixteenth century by isolated farmsteads and small hamlets (such as the Weald), many structures were rebuilt in the sixteenth century and remain today.[2] Reconstruction usually took the form of inserting a ceiling in a medieval hall, formerly open to the rafters, so producing a living room and parlour on the ground floor and bedrooms above, reached by a staircase. Many farmsteads of the Cotswolds, built or rebuilt in local stone, date from 1570–1640, as do the characteristic 'black-and-white' timber-framed houses of the west Midlands and the Wealden houses of south-east England.[3] Indeed, J. L. M. Gulley has claimed that in the Weald many more houses were built between 1570 and 1640 than during any period of comparable length before or since, and that most of the older surviving buildings incorporate substantial structural alterations carried out between 1570 and 1640.[4] In addition, as population pressure increased during the late sixteenth century, numerous cottages were erected on newly reclaimed waste, for example on the moors and heaths of Devon, in the Lancashire forest of Rossendale, and on the commons of the Weald.[5] Most cottage building was by small farmers and labourers whereas the building and rebuilding of more substantial structures was the work of husbandmen, yeomen and lesser gentry with money accumulated from the gap between relatively fixed expenses (rents and wages largely) and rapidly rising prices of farm products.[6]

[1] W. G. Hoskins, 'The rebuilding of rural England, 1570–1640', *Past and Present*, IV (1953), 44–59; M. W. Barley, *The English farmhouse and cottage* (London, 1961), 55–125.

[2] J. L. M. Gulley, 'The Great Rebuilding in the Weald', *Gwerin*, III (1961), 1–16.

[3] W. G. Hoskins (1953), 46–7.

[4] J. L. M. Gulley (1961), 2.

[5] W. G. Hoskins (1952), 328–9; G. H. Tupling, *The economic history of Rossendale* (1927), 47, 62, 64 and 66–7; J. L. M. Gulley (1960), 105–14.

[6] W. G. Hoskins (1953), 50–5.

Forests, woodland and parks

The acreage of land subject to forest law declined during the later Middle Ages as the result both of wholesale disafforestations – the removal of their territories from forest jurisdiction and from control by royal officials – and of piecemeal enclosure. The Tudors increased their revenues at times by disafforesting or by selling areas of forest. Destruction of timber was heavy during the sixteenth century, and fears of a shortage were frequently voiced. In 1543 came the first important timber preservation act, one provision of which was that, wherever woods were cut down, at least twelve young trees must be left on every acre. Elizabeth's first parliament passed an act to preserve ship timber within fourteen miles of navigable water. The quantity of acts and the diversity of their provisions made it clear that during the sixteenth century real alarm over the timber situation was developing. Destruction of timber was going on apace, and local and regional shortages were attracting national attention.[1]

On the other hand, the area of parkland had greatly increased. Money from farming, industry and commerce was not only invested in improved housing but also in deer parks, particularly in the south-east, a reflection of the concentration of wealth in the London region (Fig. 58). By the end of the sixteenth century at least 69 parks had been established in the Chiltern Hills and not less than 10% of the surface area of the Weald was covered by parks. They were widely scattered, often remote from villages or public highways, and most seem to have been located on land of little value for agriculture. But deer parks were esteemed not only as social and recreational retreats; they were also valued for the income to be derived from their woods and trees at a time of growing timber shortage.[2]

INDUSTRY

The later Middle Ages were years of fundamental change in the industrial and commercial geography of England. They witnessed a radical transformation of the export trade, from the export of raw wool to the export of manufactured textiles. In addition, these years saw both a wider spread and a growing regionalisation of industry, and also increasing complexity in manufacturing processes and a broadening of domestic and overseas markets.

[1] R. G. Albion, *Forests and sea power* (Cambridge, Mass., 1926), 95–127.
[2] H. C. Prince, 'Parkland in the Chilterns', *Geog. Rev.*, XLIX (1959), 18–31; J. L. M. Gulley (1960), 61–72.

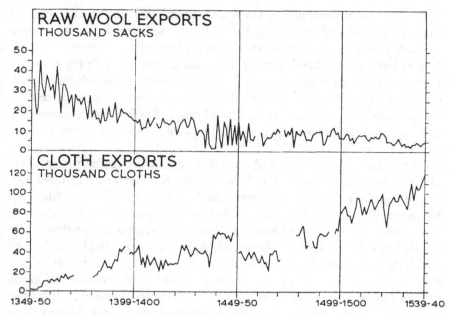

Fig. 48 Trends in the export of raw wool and cloth, 1349–1540
　　Based on E. M. Carus-Wilson and O. Coleman, *England's export trade,*
　　1275–1547 (Oxford, 1963), 122–3, 138–9.
　　Breaks in the graphs are the result of no data.

The textile industries

A remarkably precise picture of England's exports is provided by the
Enrolled Customs Accounts between 1347 and 1547, when there was
a sharp break in the continuity of both the customs and their records. The
cloth custom established in 1347 applied to cloth exported by natives as
well as by aliens, unlike the custom of 1303 which applied only to the
latter. The accounts show clearly the replacement of exports of raw wool
by exports of manufactured textiles (Fig. 48).[1] A turning point was reached
in the years around 1450, but the decline in wool and the increase in cloth
exports had begun a century earlier. *Exports of raw wool* suffered a serious
but temporary setback in the years immediately after the Black Death,
but in the 1350s and early 1360s they recovered to almost their previous
level of about 30,000 sacks annually. Renewed depression began in the

[1] E. M. Carus-Wilson and O. Coleman, *England's export trade, 1275–1547*
(Oxford, 1963), 1–33.

late 1360s, and exports of raw wool took a markedly downward turn; by 1440 the annual export averaged only some 18,000 sacks. *Cloth exports* showed a temporary setback in the years immediately after the Black Death; but recovered and showed an annual growth rate of 18% between 1353 and 1368. By 1392–5 exports averaged some 43,000 cloths, nearly tenfold the 1347–8 figure of 4,422. The increase in cloth export more than compensated for the decrease in that of wool. English cloth, moreover, also captured the home market and led to a decrease in imported foreign cloth.[1]

Total wool exports – raw and manufactured – were virtually unaffected by the considerable loss of population during the fourteenth century, the loss of population being counterbalanced by a higher level of productivity *per capita*. It was not until the first half of the fifteenth century that the total wool export declined substantially as a consequence of drastically reduced raw wool exports. The decade 1431–40 was probably the first in which the export of cloths exceeded in sack equivalents the export of raw wool.[2] Thereafter, with the exception of 1461–70, the amount of cloth exported exceeded the equivalent quantity of raw wool exported. The gap between them increased, as raw wool exports declined and cloth exports multiplied, until the mid-sixteenth century when the cloth export market contracted sharply, for reasons to be discussed shortly.

This radical transformation of the export trade had a number of related causes which sprang from a growing economic incentive to export cloth rather than wool. In the first place, of some negative importance were sporadic and temporary embargoes placed on the export of wool in war-time, but these could be evaded by special licence. Secondly, but more important, was the positive encouragement given to the domestic cloth industry when wars interrupted the import of cloth. In the period between the outbreak of the Hundred Years' War and the Black Death (i.e. between 1338 and 1348) war-time conditions provided a direct stimulus to the domestic industry; the government placed large orders for clothing the armed forces (2,000 pieces of cloth were bought for the navy alone

[1] H. L. Gray, 'The production and exportation of English woollens in the fourteenth century', *Eng. Hist. Rev.*, XXXIX (1924), 13–25; E. M. Carus-Wilson, 'The aulnage accounts: a criticism', *Econ. Hist. Rev.*, II (1929–30), 114–23, 'Trends in the export of English woollens in the fourteenth century', *Econ. Hist. Rev.*, 2nd ser., III (1950–1), 162–79, and 'The woollen industry', being ch. 6 of M. M. Postan and E. E. Rich (eds.), *The Cambridge economic history*, II (Cambridge, 1952), 355–428.

[2] Reckoning 4¼ cloths to one sack of wool: A. R. Bridbury (1962), 23–38.

in 1337). Moreover, the temporary closing of the English market to Flemish cloth, and the prohibition of the export of English wool to Flanders, gave protection to the English industry. When these extreme measures were withdrawn, heavily increased export duties on raw wool (amounting to some 33%) burdened foreign manufacturers with greatly increased costs and gave more permanent protection to the home industry; cloth exports, on the other hand, paid a duty of less than 2%. Cloth manufacturers in Flanders and Florence, to a great extent dependent upon supplies of English wool, thus had increased costs added to internal social problems, and they became decreasingly able to face the competition of English manufacturers.[1] A third factor, acting as a further disincentive to the export of English raw wool, was the establishment after 1337, and more especially after 1350, of a quasi-monopoly of the wool trade by the English Company of the Staple, replacing the relatively free export of wool by both aliens and natives and restricting the channels of export; cloth exports were not similarly restricted.[2] A fourth factor was the rising price of domestic wool during the fifteenth century which was an added brake on raw wool exports, as also in this and in the following century was a deterioration in the quality of many English wools which made them less sought after by foreign buyers.[3]

The boom period of English raw wool exports before 1334 had been characterised by large-scale production of wool based on demesne flocks: that period was one of wholesale contracts between large producers and exporters, of free trade, of low taxation and of the predominance of foreign merchants in the English trade. The period of declining raw wool exports after 1334, more especially after 1400, was characterised by smaller scale production based on peasant flocks; this period was one of the middleman dealer, of monopoly, of high taxation and of the English Company of the Staple.[4] The sixteenth century, however, saw a progressive decline in the relative importance of the small producer, as sheep ownership became concentrated in fewer hands and the economies of large-scale production came to be realised. By then, wool production was primarily to supply the domestic textile industry.[5]

[1] E. M. Carus-Wilson (1952), 413–15.
[2] E. Power, *The wool trade in English medieval history* (London, 1941), 36.
[3] P. J. Bowden, *The wool trade in Tudor and Stuart England* (London, 1962), 5, 26–7 and 109–10.
[4] E. Power, 37–40.
[5] P. J. Bowden (1962), 2.

During the later Middle Ages, the English cloth industry witnessed not only a prodigious increase in production but also important changes in location and in the organisation of manufacture and marketing. On a national scale, the industry became more widely distributed than at any time before or since. The urban exodus, initiated during the early Middle Ages, gained momentum with the more widespread adoption of the fulling-mill.[1] Some of the long-established urban centres of the industry, such as Salisbury, remained important, but the period was characterised by the growth of a rural industry, particularly during the fifteenth century.[2] Although cloth manufacture during the later Middle Ages was widespread, from Devon to Yorkshire and from Westmorland to Kent, it came to be concentrated in three main areas – the West Country, East Anglia and the West Riding (Fig. 49).[3] Within these areas it prospered principally along river valleys at sites providing water for cleansing the cloth and working the fulling-mills. Of secondary importance were local supplies of fine-quality wool and fuller's earth.

The West Country (Devon, Somerset, Gloucestershire and Wiltshire) emerged as England's prime cloth-producing region; here numerous new settlements developed in narrow valleys, and agricultural villages and hamlets acquired industrial proletariats. The Stroud valley provides a striking example of this industrial development, where expansion took place not in existing upland villages but in valley hamlets. Proximity to abundant supplies of high-quality Cotswold wool, to beds of fuller's earth, to Bristol as an outlet for exports, and, most important of all, to water of a quality suited to the finest dyes and of a quantity sufficient for driving large numbers of mills – all combined with a favourable manorial structure to encourage the growth of a prosperous cloth industry in the Stroud valley and its adjacent valleys. Growth was particularly noticeable in the mid-fifteenth century at such places as Stroudwater and Castle Combe. At Castle Combe, 50 new houses, most built by clothier tenants, were erected between 1409 and 1454.[4] Similar new centres arose in the kersey-

[1] E. M. Carus-Wilson, 'An industrial revolution of the thirteenth century', *Econ. Hist. Rev.*, XI (1941), 39–60.   [2] A. R. Bridbury (1962), 39–51.

[3] E. M. Carus-Wilson (1952), 412 and 417–22; E. Lipson, *A short history of wool and its manufacture (mainly in England)* (London, 1953); J. L. M. Gulley (1960), 204–10, 279–82 and 377–8; P. J. Bowden (1962), 46; G. D. Ramsay, 'The distribution of the cloth industry in 1561–2', *Eng. Hist. Rev.*, LVII (1942), 361–9; R. A. Pelham, 250.

[4] R. P. Beckinsale, 'Factors in the development of the Cotswold woollen industry', *Geog. Jour.*, XC (1937), 349–62; E. M. Carus-Wilson, 'Evidences of industrial growth on some fifteenth-century manors', *Econ. Hist. Rev.*, 2nd ser., XII (1959–60), 190–205;

producing region of Dorset, Devon and Cornwall and in the broadcloth- and kersey-producing areas of Berkshire, Hampshire and Kent.[1] In East Anglia, growth occurred along the Stour and its tributaries, in Essex and Suffolk, at such places as Sudbury, Long Melford, Clare and Lavenham.[2] In Norfolk, on the other hand, the manufacture of worsted – a light cloth of high quality made of long, not short wool, combed not carded, and requiring little milling – was, by the mid-fifteenth century, in serious decline. The manufacture of worsteds in Norfolk had probably reached its zenith in the late fourteenth century and thereafter declined as overseas markets in France, Spain and Portugal were slowly lost to the growing continental light cloth industry, especially to Dutch manufacturers.[3] But with the introduction of new types of cloth by Dutch and Walloon refugees after 1550, the East Anglian textile industry once more flourished. In the West Riding, industrial development was checked in the fourteenth century as much by Scottish devastations as by plague, but in the following century there was considerable growth, notably on the upper reaches of the Aire and Calder and this region emerged as an area of kersey pro- duction.[4] To the west of the Pennines, the coarse woollen cloths produced in Lancashire had gained outlets on the Continent while linens were, for the most part, consumed by the home market. Here, in the years around 1600, new branches of manufacture were introduced, cotton among them, which were in due course to become paramount.[5]

R. H. Kinvig, 'The historical geography of the West Country woollen industry', *Geog. Teacher*, VIII (1916), 243–54 and 290–306; R. Perry, 'The Gloucestershire woollen industry, 1100–1690', *Trans. Bristol and Gloucs. Archaeol. Soc.*, LXVI (1945), 49–137; K. G. Ponting, *A history of the west of England cloth trade* (London, 1957), 1–59; G. D. Ramsay, *The Wiltshire woollen industry in the sixteenth and seventeenth centuries* (London, 1943), 1–84; E. M. Carus-Wilson, 'The woollen industry before 1550' in E. Crittall (ed.), *V.C.H. Wiltshire*, IV (1959), 115–47.

[1] R. P. Beckinsale, 357–61; P. J. Bowden (1962), 50–1.

[2] B. McClenaghan, *The Springs of Lavenham and the Suffolk cloth trade in the XV and XVI centuries* (Ipswich, 1924), 1–28; J. E. Pilgrim, 'The rise of the "new draperies" in Essex', *Univ. Birmingham Hist. Jour.*, VII (1959–60), 36–59; P. J. Bowden (1962), 52–3; G. A. Thornton, *A history of Clare, Suffolk* (Cambridge, 1928), 141–211.

[3] K. J. Allison, 'The Norfolk worsted industry in the sixteenth and seventeenth centuries: 1. The traditional industry', *Yorks. Bull. Econ. and Soc. Research*, XII (1960), 73–83.

[4] H. Heaton, *The Yorkshire woollen and worsted industries* (Oxford, 1920), 1–88.

[5] A. P. Wadsworth and J. de L. Mann, *The cotton trade and industrial Lancashire* (Manchester, 1931), 3–23.

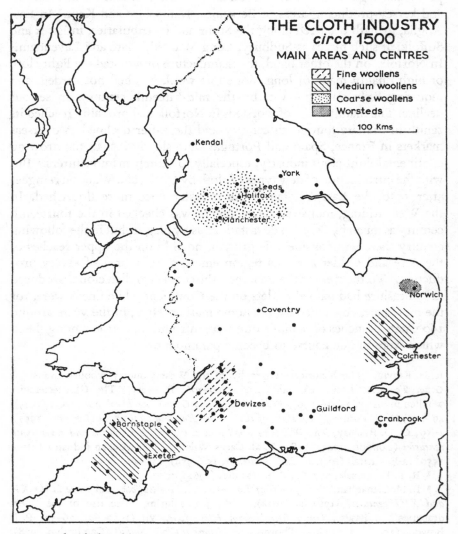

Fig. 49 The cloth industry *circa* 1500
    Based on P. J. Bowden, *The wool trade in Tudor and Stuart England* (London,
    1962), 46.

The location of the cloth industry was not related directly to wool supplies – the cloth-producing counties were not identical with the wool-producing ones, although the nature of local wool supplies strongly influenced the type of cloth produced. The highest quality wools in 1454 came from Shropshire and Herefordshire, which were not among the principal cloth-producing counties; the West Country broadcloth area relied more on these wools than on wools from the Cotswolds, while the serge industry of Devon and south-west Somerset was based on local long-woolled sheep. The industry does, on the other hand, seem to have been related directly to a supply of running water.[1] But not all areas with favourable sites for mills developed a cloth industry. Social factors may also have been important in influencing the emergence of Wiltshire, Suffolk and Yorkshire as the main cloth areas; J. Thirsk has suggested that rural industries may have developed where density of population was relatively high and farming essentially pastoral, so that there was both the need and the opportunity to supplement a meagre farming income by part-time industry.[2]

Industrial growth was associated with changes not only in location but also in the processes of manufacture and marketing. Fulling-mills were increasingly employed, as were other labour-saving devices such as the spinning wheel and the gig-mill for raising the nap on cloth; the first recorded instance of a gig-mill in England comes from Castle Combe in 1435.[3] Both fulling-mills and gig-mills aroused opposition from established cloth-workers, afraid of losing their livelihoods. But such protests, as well as legislation intended to limit the use of mills, were ineffectual. Indeed, an important feature of the cloth industry was its increasingly free operation, uncontrolled either by gild or state regulation. The only really effective regulation of the industry was that controlling the length and breadth of cloth put on the market, and even to this there were many permitted exceptions. Attempts to control the quality of cloth – for example, to prevent undue stretching of cloth and the use of waste or inferior wools – were never very effective, because they were not enforced by a regular system of inspection.[4]

[1] E. Lipson, 14; R. H. Kinvig, 249–54 and 296–300; R. A. Pelham (1936), 255.

[2] J. Thirsk, 'Industries in the countryside', in F. J. Fisher (ed.), 70–88.

[3] R. A. Pelham, 'The distribution of early fulling mills in England and Wales', *Geography*, XXIX (1944), 52–6; E. M. Carus-Wilson (1959–60), 201–2.

[4] K. J. Allison (1960), 80–1; H. Heaton, 124–40; G. D. Ramsay, 50–64; R. H. Tawney and E. Power (eds.), *Tudor economic documents*, III (London, 1924), 210–25 (Leake's treatise on the cloth industry, 1577).

The cloth manufacturer, the 'clothier', came to exercise great control over the rural industry. Clothiers were especially important in the West Country, where the industry was organised by capitalists who bought and owned the wool but who had all the work (except fulling) done in the workers' homes, usually with their own implements although some clothiers provided these as well. The clothiers then marketed the finished product.[1] These capitalist entrepreneurs, producing primarily undyed broadcloths for export, differed considerably from the 'meaner clothiers' of the West Riding, who made the cloth in their own homes, employing a few labourers, sometimes only their own family.[2] In the West Riding, cloth-workers were often also agricultural workers, thus differing from the West Country where both capitalist organisers and industrial proletariat came to be divorced from agriculture.[3] Some East Anglian clothiers, such as the Springs of Lavenham, became very prosperous, using part of their wealth to build or restore beautiful churches. A few clothiers, such as William Stumpe of Malmesbury (Wiltshire), operated on a grand scale, employing many workers in a single industrial establishment. But these were not typical.[4]

A widening of markets and of sources of wool supply was typical of the period. By the late fifteenth century 'Stroudwaters' and 'Castle Combes' were as well known on the Continent as in England.[5] In the sixteenth century, the internal market for wool began to develop on a wider-than-regional basis as the character of both English wool and cloth manufacture changed.[6] This widening of markets was more a consequence of the rise of the professional middleman (the economic catalyst of the later Middle Ages) and carrier than of improvements in the means of communication.[7]

The early Tudor period has been called 'the golden age of traditional

[1] E. M. Carus-Wilson (1959–60), 195 and 200–2; R. Perry, 110–19.
[2] H. Heaton, 89–123.
[3] E. M. Carus-Wilson (1959–60), 202; H. Heaton, 24–5 and 93.
[4] B. McClenaghan, 30–41 and 60–88; G. Unwin, 'Woollen cloth – the old draperies' in W. Page (ed.), *V.C.H. Suffolk*, II (1907), 254–66; R. Perry, 49–50; G. D. Ramsay, 31–49; P. H. Ditchfield, 'Cloth making' in W. Page (ed.), *V.C.H. Berkshire*, I (1906), 387–95; J. G. Oliver, 'Churches and wool: a study of the wool trade in 15th century England', *History Today*, I (1951).
[5] E. M. Carus-Wilson (1959–60), 190.
[6] P. J. Bowden (1962), 56–63 and 72–6.
[7] P. J. Bowden (1962), 77–106; T. C. Mendenhall, *The Shrewsbury drapers and the Welsh wool trade in the XVI and XVII centuries* (London, 1953), 1–119.

broadcloth manufacture'.[1] Cloth exports trebled in the first half of the sixteenth century, then fell during the second half. Earlier boom conditions never returned, although broadcloth manufacture did enjoy something of an 'Indian summer' at the end of the century.[2] During the sixteenth century wool from central England (which had produced much of the country's fine, short staple wool) became coarser and longer. P. J. Bowden has suggested that this was a direct consequence of the increased feed resulting from enclosure; as sheep pastures improved so fleeces deteriorated and fibres lengthened. M. L. Ryder, on the other hand, has argued that the change in the type of wool came about as a result of a change in the type of sheep; better nutrition allowed selective breeding of the primitive long-wool for increased length. It is clearly the case that the changing character of wool and of the textile industry in England were related: long and coarse wool was better suited to production of worsted than of woollen fabrics. The decline of the woollen and the revival of the worsted branches of the textile industry during the sixteenth century would seem to have been a cause rather than a consequence of the changing nature of the wool supply.[3] The Midlands became a source of supply for long-fibre wool for the 'new draperies' of East Anglia, introduced by immigrants from the Netherlands fleeing from Spanish persecution. In 1554 a few weavers were persuaded to settle in Norwich. In 1565 another 300 Dutchmen and Walloons came to the city. The harsh policy of the Duke of Alva in the Netherlands after 1567 increased the flow. By 1569 the number of refugees in Norwich had increased to 2,826; by 1571 to 3,900; and by 1582 to 4,678. The immigrants introduced two main types of fabric: (1) Walloon 'caugeantry' which differed from traditional worsteds only in size, colour, or in number of threads used in the warp or weft, and which were akin to worsteds in being fine, light-weight cloths; and (2) Dutch bays, which were heavier, more akin to woollens than to worsteds, and like woollens they were fulled. Only slowly did English workers in Norwich, or in Colchester (the second major concentration of Flemish immigrants), turn to producing these 'new draperies'. But by the

---

[1] P. J. Bowden (1962), 43; G. D. Ramsay, 65.

[2] P. J. Bowden (1962), 43–4; G. D. Ramsay, 65–7; F. J. Fisher, 'Commercial trends and policy in sixteenth-century England', *Econ. Hist. Rev.*, x (1940), 95–117.

[3] P. J. Bowden (1962), 25–7 and 44–5, and 'Wool supply and the woollen industry', *Econ. Hist. Rev.*, 2nd ser., ix (1956–7), 44–58; M. L. Ryder, 'The history of sheep breeds in Britain', *Agric. Hist. Rev.*, xii (1965), 65–82. See also correspondence between P. J. Bowden and M. L. Ryder in *Agric. Hist. Rev.*, xiii (1965), 125–6.

end of the century these new fabrics, in their many varieties, were being made both in the old worsted district of Norwich and in the traditional broadcloth district of Colchester.[1] Whereas the arrival of Flemish immigrants in England in the early and mid-fourteenth century was a symptom rather than a cause of the progress of English woollen manufacture, the arrival of immigrants in the mid-sixteenth century resulted in the introduction of new fabrics and in the increased importance of East Anglia in the English textile industry.

The iron industry

Unlike the cloth industry, which during the Middle Ages added a rural distribution to an existing urban distribution, the charcoal iron industry throughout this period maintained its rural location. But it became increasingly a regional industry; widely distributed itinerant bloomeries were superseded by more narrowly located furnaces and forges. The scale of iron-working continued to be limited throughout the fourteenth and fifteenth centuries. There is some evidence of a smaller production in the fourteenth than in the preceding century, but the fortunes of the industry between 1334 and 1500 are difficult to trace, because of its peripatetic nature.[2] *Forgiae errantes* had short lives and an irregular production of blooms. Accounts of the Tudeley works near Tonbridge, in the Weald, began in 1330 and show that they were rebuilt in 1343 but lay unused in 1346. By 1350 they were operating again, although production costs had risen because of plague, and another onset of plague closed the works finally in 1363. Output had varied greatly from year to year, being about 200 blooms per annum in 1330–4, 600 blooms in 1335, and 252 blooms in 1350–1.[3]

Many bloomeries have left no documentation, only cinder heaps; from sites of others, even their cinders have been removed for road-building.[4] In all of the main areas of production – the Forest of Dean, the Weald and the Cleveland Hills – it seems that working was limited more by the

[1] K. J. Allison, 'The Norfolk worsted industry in the sixteenth and seventeenth centuries: 2. The new draperies', *Yorks. Bull. Econ. and Soc. Research*, XIII (1961), 61–77; J. E. Pilgrim, *passim*; P. J. Bowden (1962), 52–4; D. C. Coleman, 'An innovation and its diffusion: the "new draperies"', *Econ. Hist. Rev.*, 2nd ser., XXII (1969), 417–29.

[2] H. R. Schubert, *History of the British iron and steel industry from c. 450 B.C. to A.D. 1775* (London, 1957), 112–13.

[3] J. L. M. Gulley (1960), 375.

[4] J. L. M. Gulley (1960), 278–9.

local exhaustion of supplies of wood fuel for charcoal-making than of supplies of iron ore.[1] There is increasing evidence of coppicing during the fifteenth century and of a growing use of young and small trees in preference to dead wood and the branches and roots of trees felled in woodland clearance. Coppices were planted even in remote places where iron production was on a small scale and there was no lack of wood; such were the coppices near Barnard Castle (Co. Durham) in 1437. Timber shortages were to become of more consequence during the sixteenth century when production expanded significantly.[2]

Increased iron production was made possible by technological advances, and was stimulated by the demands of war. The application of water power to bellows, providing an artificial draught for smelting, was more widely adopted after about 1350, and may, in some instances, have been encouraged by labour shortages after the Black Death.[3] The use of water power to work bellows was not new in the later Middle Ages, but its application to hammers was an innovation. The first unequivocal reference to water-powered hammers comes from the Weald in the 1490s, where the works at Newbridge, in the parish of Hartfield in Ashdown Forest, included a 'great water hammer', but it was probably in use some decades before this.[4]

The most important change in the iron industry during the later Middle Ages, however, was the introduction of the blast furnace. Instead of obtaining malleable iron with a very low carbon content directly, as did bloomeries, the blast furnaces produced a highly carbonised or pig-iron too brittle for the smith's hammer; this pig-iron had then to be freed from the surplus of carbon and other impurities by smelting in a hearth or 'finery' before the iron could be drawn out and shaped by the forge hammer. Refining introduced an intermediate stage between the two stages of smelting the ore into a bloom and hammering the reheated bloom – hence the term 'indirect process' of iron-working. This new process needed larger and more complex plant, and greater capital investment, than did the old direct process; and the use of water power made necessary a local separation of furnace and forge, because separate water

[1] F. T. Baber, 'The historical geography of the iron industry in the Forest of Dean', *Geography*, XXVII (1942), 54–62; B. Waites, 'Medieval iron working in northeast Yorkshire', *Geography*, XLIX (1964), 33–43; J. L. M. Gulley (1960), 189–200, 275–9 and 375; E. Melling, 93–104; E. Straker, *Wealden iron* (London, 1931), 101–40.
[2] H. R. Schubert, 122–3 and 145.   [3] H. R. Schubert, 133–4.
[4] H. R. Schubert, 134–40 and 147; J. L. M. Gulley (1960), 277–8.

wheels were needed for blast production, one for the furnace, and one each for the 'finery' and power hammer. The sites of iron production consequently became more permanent than hitherto, and furnaces were often distinct from forges.[1]

The first definite reference to a blast furnace comes from Newbridge, in the Weald, in 1496, an ironworks commissioned by the Crown to manufacture iron for armaments to be used in the Scottish war. An account for 1496-7 tells of axle-trees, wheel-rims, cast-iron bullets and shot to be carried to the Tower of London. Here at Newbridge, too, in 1509, guns of cast-iron were successfully manufactured for the first time in England. By 1542 two groups of blast furnaces had appeared in the Weald, one in south-east Sussex and the other around Ashdown Forest.[2]

At the time of the introduction of the blast furnace, bloomeries using the direct process existed in sufficient numbers to satisfy local demands for iron in rural districts of England, and native production was supplemented by imports of better quality iron for use in tool-making crafts and armaments. Furthermore, blast furnaces and forges were comparatively expensive structures requiring large ponds, dams and equipment. The indirect process spread only slowly, and new bloomeries using the direct process were still being erected well into the sixteenth century. The fact that a blast furnace could produce seven times as much iron per day as a bloomery was one factor encouraging the diffusion of the innovation. Another was that throughout the early sixteenth century the chief market for the new industry was the demand for ordnance, and for carts and other military equipment. Adoption of the indirect process was, however, slow, because of difficulties encountered in producing a pig-iron completely suitable for casting and conversion into malleable iron, and because, whereas Henry VII had stimulated a domestic iron industry with the skill of French founders, Henry VIII placed large orders for arms abroad, especially in the Low Countries. There were in 1542 only nine English blast furnaces, all in Sussex. Proximity to London encouraged the development of the iron industry in the Weald.[3]

During the 1540s the productive capacity of the industry grew enormously, and eleven new furnaces were built in the Weald in 1543-8. Threat of war with France greatly enlarged the demand for ordnance. By 1548 there were some 20 furnaces and 28 forges in the Sussex Weald, and in the

[1] H. R. Schubert, 157-8.
[2] H. R. Schubert, 162-6; J. L. M. Gulley (1960), 277-8; E. Straker, 38-52.
[3] H. R. Schubert, 158-61 and 166-70; J. L. M. Gulley (1960), 278.

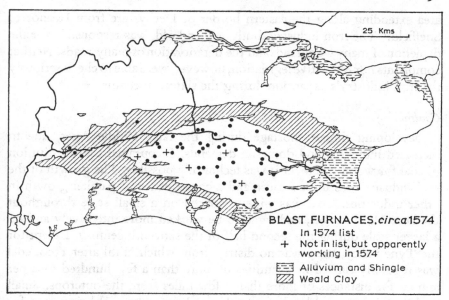

Fig. 50 Charcoal blast furnaces in Sussex, Kent and Surrey *circa* 1574
Based on: (1) E. Straker, 'Wealden ironworks in 1574', *Sussex Notes and Queries*, VII (1938), 97–103; (2) H. R. Schubert, *History of the British iron and steel industry from c. 450 B.C. to A.D. 1775* (London, 1957), 354–92.

following decade the new process spread into Kent and Surrey. Until about 1560 the Weald had a monopoly of iron production by blast furnace. In 1574, when the industry was at a peak, there were at least 51 blast furnaces and 58 forges operating in the Weald (Fig. 50); but there were only seven other blast furnaces throughout the country, three in the Midlands and four in South Wales and Monmouthshire. The traditional iron area of the Forest of Dean was slow in adopting the blast furnace in place of the bloomery, and the first blast furnace did not appear there until the 1590s. In the meantime, much Forest of Dean pig-iron was made for sale upstream along the Severn to the metalcraft centres of the west Midlands. When Leland visited Birmingham about 1540, he reported 'many smithes in the town', but it was not until later in the century that there is evidence not only of smithies, processing bar iron into tools and weapons in the town, but also of furnaces and forges, producing pig and bar iron. The first recorded blast furnaces in northern England were those at Heanor (Derbyshire), about 1576, and at Rievaulx (North Riding), about 1582. Gradually other blast furnaces and forges were built in the area extending along the eastern border of Derbyshire from Heanor to Sheffield.[1] The iron industry, both new and old, was responsible for the depletion of many woodlands and the deterioration of many roads. Neither complaints nor restrictive legislation, however, were able to check seriously the iron industry's expansion during the sixteenth century.[2]

Coalmining

A developing timber shortage during the later Middle Ages, due to increased industrial and domestic demands, encouraged the substitution of coal for wood wherever it was technically possible. The growth of the coal industry was both a cause and a consequence of parallel growth in other industries. Coal was widely worked on a small scale throughout the later Middle Ages, and came to be worked more intensively and on a larger scale during the second half of the sixteenth century. Except for the Tyne valley, there was no district from which, until after 1500, coal was regularly carried in quantities of more than a few hundred tons per annum for distances of more than a few miles from the outcrops. Small coal-workings added local pits to the landscape, the coal being used for lime-burning for agricultural and building purposes, for smith's work (although charcoal was preferred if obtainable) and for baking, brewing and salt-making.[3] During the fifteenth century, pits were sunk deeper than previously and precautions taken against flooding. A 'colepytte' at Kilmersdon (Somerset) was deep enough in 1437 to have an adit or drainage channel, and a pit at the same place was said in 1489 to be deep and dangerous: the 'wark' or spoil from these pits in the outcrop areas remains today in mounds of considerable size.[4] It was not until the late

---

[1] H. R. Schubert, 170–84; J. L. M. Gulley (1960), 275–7; L. T. Smith (ed.), *The Itinerary of John Leland*, III (London, 1910), 97; J. A. Langford, 'Birmingham at the time of Leland's visit', *Trans. Birmingham Archaeol. Soc.* (1882–3), 32–42; R. A. Pelham, 'The migration of the iron industry towards Birmingham during the six-teenth century', *Trans. Birmingham Archaeol. Soc.*, LXVI (1945–6), 142–9, and 'The establishment of the Willoughby ironworks in north Warwickshire in the sixteenth century', *Univ. Birmingham Hist. Jour.*, IV (1953), 18–29; W. H. B. Court, *The rise of the Midland industries, 1600–1838* (London, 1938), 33–44.

[2] State Papers Domestic, Elizabeth: Book 117, No. 39 – transcribed in D. and G. Mathew, 'Iron furnaces in south-eastern England and English ports and landing places, 1578', *Eng. Hist. Rev.*, XLVIII (1933), 91–9; B. Waites, 37; J. L. M. Gulley (1960), 198–200.

[3] J. U. Nef, *The rise of the British coal industry*, I (London, 1932), 8–9.

[4] J. A. Bulley, '"To Mendip for coal" – a study of the Somerset coalfield before 1830', *Proc. Som. Archaeol. and Nat. Hist. Soc.*, XCVII (1952), 46–78.

sixteenth century that deeper mines and mechanical pumping became common (although a pump worked by horse power had been used at Moorhouse, Co. Durham, in 1486).[1]

Coal mined in the Tyne valley was marketed beyond its immediate vicinity, and was exported from Newcastle by sea, in particular to London but also to other English ports, and to Flemish, Dutch and occasionally to French and German ports. Newcastle's coal trade declined during the fourteenth century, but became increasingly important when the port's trade in wool and hides collapsed at the end of the century. Before 1500, however, total annual shipments from the Tyne probably rarely exceeded 15,000 tons; by 1563-4 they reached 33,000 tons, and by 1597-8 as many as 163,000 tons. Much of this considerably increased production and export was a response to the rapidly growing demands of London. Mining from pits sunk within the manors of Whickham and Gateshead, to the south of the Tyne, was intensified; and, farther west, mining began at Winlaton, Stella and Ryton. Winlaton colliery alone produced more than 20,000 tons in 1581-2.[2]

During the late sixteenth century many new and deeper pits were sunk on most English coalfields. Production expanded as the market, both domestic and industrial, grew, and as supplies of wood fuel declined. Development was greater where there was access to water transport; the manor of Wollaton, for example, because of its proximity to the Trent and to abundant and accessible seams, was a favoured site in Nottinghamshire.[3] Many pits were still operated on a small scale and with primitive techniques, serving only local markets; such was the Earl of Shrewsbury's mine at Sheffield (Yorkshire) which between June 1579 and December 1582 had an annual output of about 1,200 tons.[4]

Other industries

The lead-smelting industry of the Peak District of Derbyshire saw an eastward migration away from its location on the limestone upland on to the gritstone edges to the east. This migration was largely the consequence

[1] 'The charters of endowment, inventories, and account rolls of the Priory of Finchale, in the County of Durham', *Publ. Surtees Soc.*, VI (1837), cccxci: cited in L. F. Salzman, *English industries of the Middle Ages* (Oxford, 1923), 10.

[2] J. U. Nef, 9–26; C. M. Fraser, 'The north-east coal trade until 1421', *Trans. Archit. and Archaeol. Soc. Durham and Northumberland*, 11 (1962), 209–20.

[3] J. U. Nef, 57–109.

[4] L. Stone, 'An Elizabethan coalmine', *Econ. Hist. Rev.*, 2nd ser., III (1950–1), 97–106.

of a search for higher and draughtier sites for the hearths, and it resulted in a fundamental structural change in the industry towards the end of the fourteenth century. The smelting of lead had become separated from the other branches of the industry, creating a class of middlemen who bought dressed ore from the miners, smelted it, and resold it. An advantage of the new location of the hearths was their proximity to the main lead-marketing routes which led not only to Chesterfield, an important exchange centre for the industry, but also to the Continent through the ports of Boston and, later, Hull. The new siting of the hearths had a considerable advantage over the old from the point of view of fuel supply. Resources of suitable timber on the upland had been virtually exhausted but there had been as yet little exploitation of the wood on the slopes of the gritstone edges. These new local sources of fuel were apparently sufficient to last until the latter half of the sixteenth century, when supplies of wood were imported into the region at a time when production of lead increased to meet the demand resulting from the wave of new building at home and abroad.[1] Lead mining in the Mendips also increased from the middle of the sixteenth century, and lead was exported from Bristol. Here, too, calamine, the ore of zinc, was discovered in 1566.[2]

The second half of the sixteenth century saw a similar intensification of activity in the exploitation of other minerals, often encouraged by royal patronage. Two great companies were chartered in 1568, the Mines Royal and the Mineral and Battery Works; these were the first companies to be formed in England for the manufacture of a product (copper and brass respectively) as distinct from companies formed for trading purposes.[3] German capitalists and workmen were invited to develop mineral industries, notably the copper industry of the Lake District, centred on Keswick.[4] Much mineral exploitation towards the end of the sixteenth century was being checked by difficulties, especially that of flooding, but the period was also marked by technical experiments: shafts were sunk deeper; adits were pushed almost horizontally into hillsides; pumping and other machinery was introduced; and ores were smelted in furnaces

[1] J. P. Carr, 'The rise and fall of Peak District lead mining' in J. B. Whittow and P. D. Wood (eds.), *Essays in geography for Austin Miller* (Reading, 1965), 212–13. See also L. F. Salzman (1923), 41–68.

[2] J. W. Gough, *The mines of Mendip* (Oxford, 1930), 65–6, 82, 112.

[3] M. B. Donald, *Elizabethan copper* (London, 1955) and *Elizabethan monopolies* (London, 1961).

[4] F. J. Monkhouse, 'Some features of the historical geography of the German mining enterprise in Elizabethan Lakeland', *Geography*, xxviii (1943), 107–13.

instead of open hearths.[1] Mineral production was extremely erratic, as that of tin in Devon and Cornwall illustrates. The introduction of shaft-mining in the late fifteenth century must have contributed to the sudden increase in the output of Devon tin round about 1500, while production fell at the end of the sixteenth century when production costs rose considerably and when drainage techniques were unable to cope with progressive flooding as the level of the mining sank deeper.[2]

The salt industry of Worcestershire and Cheshire continued to consume great quantities of turves and wood in the boiling of brine from salt springs.[3] Along the east coast, the last recorded evidence of salt-making at Fleet in Lincolnshire comes from 1455; here and elsewhere along the east coast, decayed salterns remain even today as relict features – clustered masses of irregularly shaped mounds, some of them 16 to 20 feet high.[4] In the north-east, along the coasts of Durham and Northumberland, coal began to be used for boiling brine in the sixteenth century, and resulted in considerable development, especially at South Shields at the mouth of the Tyne.[5]

The English salt industry underwent a 'commercial revolution' during this period. In 1334 England exported salt, but as the cloth industry developed the coastal salter had every incentive to leave salt-making for employment in the cloth industry. The rise in wages during the fourteenth century, in the absence of a general rise in the market price of salt, destroyed any cost advantages that English salt enjoyed in continental markets over the salt of northern Germany and the Low Countries. Furthermore, competition during the fourteenth century from cheaply produced salt in the Bay of Bourgneuf, to the south of the Loire estuary in France, helped to transform England from an exporter to an importer of coarse

---

[1] L. Stone, 98; F. J. Monkhouse, 'Pre-Elizabethan mining law, with special reference to Alston Moor', *Trans. Cumberland and Westmorland Antiq. and Archaeol. Soc.,* XLII (1942), 43–55; A. Raistrick, *Mines and miners of Swaledale* (Clapham, Yorks., 1955), 21–31.

[2] G. R. Lewis, *The Stannaries. A study of the English tin miner* (Cambridge, Mass., 1924), 39–64 and 252–65; A. R. Bridbury, 24–6; W. G. Hoskins, *Devon* (London, 1954), 133.

[3] H. J. Hewitt, *Mediaeval Cheshire* (Manchester, 1929), 110–14.

[4] E. H. Rudkin and D. M. Owen, 'The medieval salt industry in the Lindsey marshland', *Rep. and Papers, Lincs. Archit. and Archaeol. Soc.,* n.s., VIII (1960), 76–84; H. E. Hallam, 'Salt-making in the Lincolnshire Fenland during the Middle Ages', *Rep. and Papers, Lincs. Archit. and Archaeol. Soc.,* n.s., VIII (1960), 85–112.

[5] P. Pilbin, 'A geographical analysis of the sea-salt industry of north-east England', *Scot. Geog. Mag.,* LI (1935), 22–8; A. E. Smailes, *North England* (London and Edinburgh, 1960), 132–6.

salt. By 1364 more was imported than was exported, and exports almost entirely ceased by 1500. Bristol, Yarmouth (because of its connections with the herring industry) and London dominated the import of salt. With the dislocation of the salt trade by wars against France during the middle of the sixteenth century, and with increasing use of coal, the salt industry again became a large-scale enterprise, particularly at the mouths of the Wear and Tyne, where cheap coal was obtainable. Capital costs in the industry were high (one salt-works on the Wear in 1589 involved an investment of £4,000). The use of coal thus concentrated the industry and changed its character from domestic to capitalistic production.[1]

Another major industrial change was seen in the paper industry. Paper came increasingly to be used instead of parchment. When Caxton set up his printing office at Westminster in 1476, he used imported paper, and the overwhelming majority of books printed here in the sixteenth century continued to use imported paper. But the needs of printing, as well as of wrapping and writing, produced a rising demand for paper in the sixteenth century. A paper mill was at work near Hertford in 1495 but it had probably failed by 1507. In the 1540s there may have been no paper at all produced in England. Some of the mills established in the 1550s had short lives, and it was not until the latter decades of the century that the industry began to flourish. In 1558, a mill was established at Dartford (Kent) by a German, John Spilman, employing German workmen; and in 1589 Spilman was granted a monopoly for the making of white paper. The industry grew up along river valleys (water being needed as a raw material as well as for power in the pulping process) and in proximity to towns which provided supplies of rags (the principal raw material) as well as markets for the finished product. The growth of paper mills in the south-east owed much to the proximity of London: here was the largest English market for paper, and here was the largest English supply of linen rags (linen rags were much more important than woollen rags, and England imported most of her linens from France, Holland and Germany through London). The use of old cordage and sails as raw materials probably helped to attract brown (wrapping) paper mills to the proximity of such ports as Dover, Exeter and Southampton.[2]

[1] A. R. Bridbury, *England and the salt trade in the later Middle Ages* (Oxford, 1955), *passim*; E. Hughes, 'The English monopoly of salt in the years 1563–71', *Eng. Hist. Rev.*, XL (1925), 334–50.

[2] D. C. Coleman, *The British paper industry 1495–1860: a study in industrial growth* (Oxford, 1958), 1–88; A. H. Shorter, *Paper mills and paper makers in England,*

TRANSPORT AND TRADE

Roads and rivers

There were minor, but no fundamental, changes in the network of internal communications between 1334 and 1600. Improvement of roads was highly localised, and concentrated especially on bridges. Maintenance of road conditions was very haphazard, and was the responsibility of local communities whose interests seldom transcended their own parish boundaries. The Highways Act of 1555 – significant as the first legislation ever passed applying to roads in general in England – achieved little more than the transference of responsibility from manorial officials to parochial surveyors of highways. It did little to improve the conditions of travel.[1]

River navigation remained important for the transport of bulky articles such as grain and timber, and a number of statutes were passed, more especially after 1500, to facilitate the removal of obstructions from rivers; the channel of the Lea, for example, so important for London, was improved between 1571 and 1581. After various attempts to improve the Exe below Exeter, the Exeter ship canal was dug in 1564–6 between Exeter and its outport of Topsham. It incorporated the earliest pound-locks in England, and has been called 'the first true canal of modern times in the British Isles'.[2] But the beginning of the 'canal age' was yet nearly two centuries away.

Although there was no fundamental improvement in the physical condition of the internal communications in England, there was, however, a concentration of inland trade into fewer centres; everywhere agricultural traffic, for example, tended to be drawn away from the smaller markets, ports, and fairs, and to be centred in the larger provincial towns such as Canterbury and Reading. There were far fewer market towns and villages in 1600 than in 1334, probably less than a third. In Norfolk, for example, where there had been 130 markets, there were only 31 by the sixteenth century; and the 53 in Gloucestershire had become 34. Moreover, the larger market towns began to specialise in particular types of products such as grain, cheese and butter, poultry, horses, sheep and leather products.[3]

1495–1800 (Hilversum, 1957), 22–50, and Fig. 1, p. 92; A. H. Shorter, *Paper making in the British Isles* (Newton Abbot, 1971), 13–19.

[1] H. J. Dyos and D. H. Aldcroft, *British transport* (London, 1969), 30–1.

[2] *Ibid.*, 36–8. See E. A. G. Clarke, *The ports of the Exe estuary, 1660–1860* (Exeter, 1960).

[3] A. Everitt, 'The marketing of agricultural produce', in J. Thirsk (ed.), *The agrarian history of England and Wales*, IV, 1500–1640 (Cambridge, 1967), 467–9 and 490–6.

Maritime trade: coastal and overseas

Coastwise traffic, so often overlooked, was an essential element in the economy of the country, and a very substantial amount of inter-regional trade was carried on by this means between the many harbours around the irregular coasts of the British Isles. One indication of this traffic was the growth of London, which drew its supplies from a very wide area, much of them by sea. It is difficult to see how London could have grown as it did during this period without the advantages of cheap seaborne traffic. It was in the sixteenth century that the Corporation of Trinity House received its first charter (in 1514) from Henry VIII. Later in the century it was granted authority to erect beacons and other marks for the guidance of mariners around the coasts of England.

There was also foreign traffic – in an ever-increasing quantity. In this connection, there were two important technical developments in ship-building: firstly, the transition from the one-masted to the three-masted ship, thus giving scope for a variety of sails with particular functions; secondly, the lengthening of a ship in relation to its beam, the transition from the 'round ship' to the 'long ship'. The sixteenth century in particular saw considerable expansion of the shipping industry, in response to the navy's growing requirements and to the expansion of trade (especially the east coast coal and the Atlantic fishing trades).[1]

Some idea of the scope of European commerce can be obtained for the fifteenth century from *The Libelle of Englyshe Polycye*, by an anonymous author, which appeared in 1436,[2] and from Sir John Fortescue's *Comodytes of England*, which was written at about the same time.[3] Trade in the spheres of the Baltic and North Sea was essentially in necessities – fish, salt, timber and forest products such as pitch, tar and potash. Export of wool to the Low Countries and northern France was from east coast ports of fundamental – but declining – significance (Fig. 51). By the end of the fifteenth century, on the other hand, the export of English cloth had

---

[1] H. D. Burwash, *English merchant shipping, 1460–1540* (Toronto, 1947); G. V. Scammell, 'English merchant shipping at the end of the Middle Ages', *Econ. Hist. Rev.*, 2nd ser., XIII (1960–1), 327–42, and 'Ship-owning in England *circa* 1450–1550', *Trans. Roy. Hist. Soc.*, 5th ser., XII (1962), 105–22; R. Davis, *The rise of the English shipping industry in the seventeenth and eighteenth centuries* (London, 1962), 1–8 and 44–6; G. J. Marcus, *A naval history of England. I. The formative centuries* (London, 1961), 1–67.

[2] G. Warner (ed.), *The Libelle of Englyshe Polycye* (Oxford, 1926).

[3] Printed in *The works of Sir John Fortescue*, 2 vols. (London, 1869).

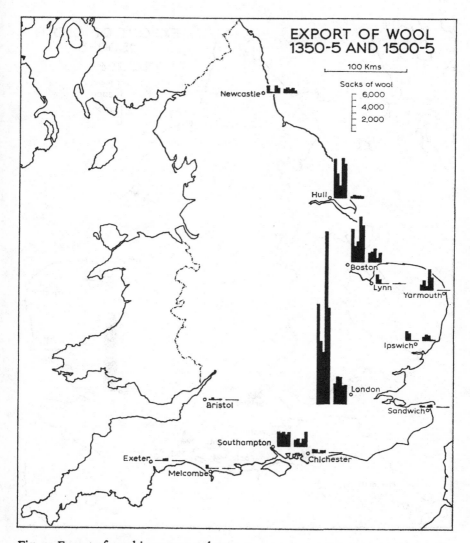

Fig. 51 Export of wool in 1350–5 and 1500–5
    Based on E. M. Carus-Wilson and O. Coleman, *England's export trade,
    1275–1547* (Oxford, 1963), 47–8, 70.

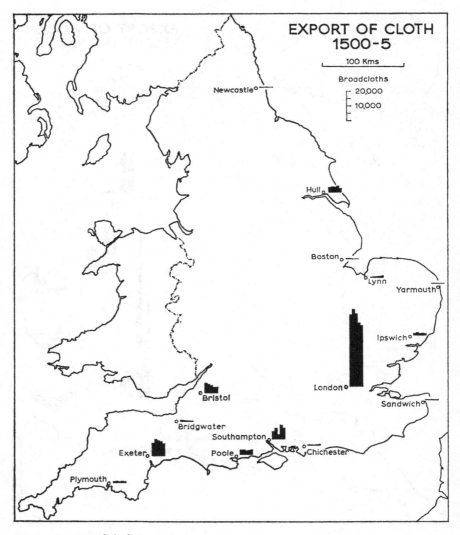

Fig. 52 Export of cloth in 1500–5
Based on E. M. Carus-Wilson and O. Coleman, *England's export trade, 1275–1547* (Oxford, 1963), 112.

become a very prominent element in the international trade of Europe. The export was primarily from London – as had been that of raw wool – and secondly from the ports of the south and south-west, in contrast to the earlier concentration on exporting of raw wool from the east coast ports (Fig. 52). From the west coast of France were imported salt and wine, the latter coming from south-west France through the ports of London, Southampton, Bristol and Hull.[1] Trade with Mediterranean countries and beyond was essentially in imported luxuries, such as spices, drugs, perfumes, sugar, precious stones, dyestuffs (indigo, madder, saffron), alum and carpets. The outward commodities were mainly woollens and linens together with some raw wool, metals and hides.

During the sixteenth century new trading companies extended the range of commercial activity, diversifying exports and more particularly imports. To the traditional northern and Mediterranean trades was added emergent Atlantic and Far Eastern trades. Thus, the main source of imported sugar seems already by about 1590 to have been shifting to Brazil and the West Indies. Furthermore, by this time refineries had been set up in England, and the finished product was being exported to continental markets.[2] From about 1566 onwards tobacco began to be imported from North America, and pipe smoking had become general by the end of the century. The formation of the East India Company of London in 1600 was another indication of the changing pattern of England's overseas trade.

TOWNS AND CITIES

The fortunes of English towns during this period were very varied and were characterised by great diversity both in time and place. Even so, some general changes can be discerned, such as the relative growth of many new industrial villages and small towns, the relative decline and later recovery of many of the old corporate towns, the close association of the condition of towns with the state of trade, and the increasing economic supremacy of towns in general and of London in particular. A comparison of the subsidies of 1334 and 1524 shows that urban wealth constituted a far larger proportion of total lay wealth at the end of this period than at the beginning. The fortunes of individual towns differed but collectively they strengthened their grip upon the national economy.[3]

[1] M. K. James, *Studies in the medieval wine trade* (Oxford, 1971).
[2] L. Stone, 'Elizabethan overseas trade', *Econ. Hist. Rev.*, 2nd ser., II (1949–50), 30–58.   [3] A. R. Bridbury (1962), 77–82 and 111–13.

DHG

Industrial villages and towns

A striking feature of the later Middle Ages was the growth of industrial villages, some of which developed (or were later to develop) into towns. The pattern of this new industrial settlement was markedly different from that of the old. Not confined within the walls or even the suburbs of the old towns, it was often a straggling growth integrated only by its market place and by the parish church, itself often wholly or partly rebuilt. Kersey and Long Melford, two cloth-making centres in Suffolk, still exhibit a linear pattern of unregulated growth along a main street.[1] The expansion of industries in the countryside is still exemplified at Castle Combe, in Wiltshire, where, during the first half of the fifteenth century, there was an impressive industrial growth alongside the stream, with the building of new and rebuilding of old houses, many of them in local stone. Among these new buildings at Castle Combe was a fifteenth-century church tower with its decorations based on cloth-working implements. Here, a growing class of craftsmen with no agricultural holdings lived in the valley at 'Nethercombe' while yeoman cultivators lived on the heights above the wooded valley at 'Overcombe'.[2] Industrial settlements emerged in the West Riding, such as Halifax, Leeds and Wakefield; and in the Midlands many small settlements developed important tanning industries.[3] In the mid-sixteenth century, Birmingham with 'its open and countrified aspect' and a population of about 1,500, was more important for its tanning and clothing than for its metal industries, although the latter were beginning to assume increasing importance.[4]

But it was the cloth industry which, more than any other during the later Middle Ages, promoted small settlements into prosperous towns. For example, in Devon, quiet Totnes blossomed suddenly during the fifteenth century into such a flourishing woollen manufacturing town that by 1523–7 it was among the twenty most prosperous provincial towns in England.[5] Involvement in cloth-making was enough to raise a

[1] M. W. Beresford and J. K. S. St Joseph, *Medieval England. An aerial survey* (Cambridge, 1958), 242–3 and 245–7.

[2] M. W. Beresford and J. K. St Joseph, 247–9; E. M. Carus-Wilson (1959–60), 197–204.

[3] H. Heaton, 54–79.   [4] W. H. B. Court, 33–43.

[5] E. M. Carus-Wilson, *The expansion of Exeter at the close of the Middle Ages* (Exeter, 1963), 17; J. Cornwall, 'English country towns in the fifteen-twenties', *Econ. Hist. Rev.*, 2nd ser., XV (1962–3), 54–69; W. G. Hoskins (1959), 177.

Table 5.1 *The ranking of towns, 1334–1525 (excluding the City of London, Westminster and Southwark)*

	1334	1524–5		1334	1524–5
Bristol	1	2	Newbury	21	21
York	2	11	Plymouth	22	
Newcastle upon Tyne	3	a	Newark on Trent	23	
Boston	4	22	Peterborough	24	
Great Yarmouth	5	20	(*cum membris*)		
Lincoln	6	15	Nottingham	25	
Norwich	7	1	Exeter	26	5
Oxford	8	29	Bury St Edmunds	27	13
Shrewsbury	9	26	Stamford	28	30
Lynn (King's and South)	10	8	Ely (*cum membris*)	29	
Salisbury	11	4	Luton	30	
Coventry	12	3	Reading		9
Ipswich	13	6	Colchester		10
Hereford	14	19	Lavenham		12
Canterbury	15	7	Worcester		14
Gloucester	16	17	Totnes		16
Winchester	17		Hull		18
Southampton	18	27	Hadleigh		23
Beverley	19		St Albans		24
Cambridge	20	28	Leicester		25

a Not taxed in 1524–5 but may have had second place.

place like Lavenham, in Suffolk, with scarcely a thousand inhabitants, also to a position among the first twenty towns. A comparison of the thirty most prosperous provincial towns in 1334 with those in 1524–5 shows clearly the rise of new towns as Lavenham, Hadleigh and Totnes and the relative decline of such old towns as York, Lincoln and Oxford.

The fate of the corporate towns

Many of the old-established towns had their populations drastically reduced by plague during the fourteenth century, and their economies checked during the malaise of the early fifteenth century. The physical spread of some seems to have been temporarily halted and there was even structural decay. During the later Middle Ages, plague appears to have become mainly an urban phenomenon, because of crowded and insanitary conditions.[1] In York, for example, there occurred a serious outbreak of bubonic plague accompanied by sweating sickness in 1550 and the summer of 1551.[2] Reduced populations resulted in derelict buildings. Numerous buildings fell into disuse and disrepair in Lincoln, for example, and at least 12 of its 46 parish churches seem to have decayed during the four-teenth and fifteenth centuries. In 1428, seventeen parishes in the city were returned as not having more than ten inhabitants each. Lincoln's suburbs were shrinking and so also were the back-streets of the city itself.[3] By the mid-sixteenth century many town castles were also in decay and town walls in ruins.[4] The age of private local wars had ended, and the decline of the town garrison had an impact upon its associated dependants. After the removal of the castle from Leicester, for example, a commission of enquiry into the state of house-property belonging to the Crown in the town disclosed, in the autumn of 1587, that there were no fewer than 235 tenements 'in great decay'.[5]

The prospects of both old and new towns, moreover, were closely associated with their involvement in industry and trade. The fortunes of Norwich, for example, were equally those of its worsted industry. Trade fluctuations checked any notable increase of population between the fourteenth and the early sixteenth century, by which time some of the city's buildings were in a bad state of repair. Many craftsmen avoided paying the city's charges, which included payments for street-paving, for the replacement of thatch with tiles and slates after disastrous fires, for the rebuilding of burnt houses and for the cleansing of rivers and streets. Both the worsted trade and the face of the city were given a lift during the late sixteenth century under the influence of Dutch and Walloon refugees who, by the 1580s, probably numbered one-third of its population.[6] In York, the growth of the cloth industry, trade, population and prosperity during the fourteenth century resulted in much building and rebuilding. On the other hand, the decline of the city's economy and numbers during the following century threw many houses into decay

[1] J. M. W. Bean (1962–3), 430.

[2] A. G. Dickens, 'Tudor York' in P. M. Tillott (ed.), *V.C.H. Yorkshire: The city of York* (1961), 120.

[3] J. W. F. Hill, *Medieval Lincoln* (Cambridge, 1948), 286–8.

[4] J. D. Mackie, *The earlier Tudors 1485–1558* (Oxford, 1952), 37–8.

[5] W. G. Hoskins, 'An Elizabethan provincial town: Leicester', in J. H. Plumb (ed.), *Studies in social history: a tribute to G. M. Trevelyan* (London, 1955), 33–67.

[6] K. J. Allison (1960), 73–4 and 79, and (1961), 61–2.

and dereliction.[1] Southampton also suffered a similar contrast of fortunes during the fifteenth and sixteenth centuries.[2] Towns intimately engaged in cloth production and marketing on a considerable scale suffered less than those without this economic backbone. Thus Winchester, lacking a staple industry, fell into serious decline, while Exeter rose as a centre for the manufacture and export of Devon kerseys and tin, particularly during the late fifteenth century. In addition to those industries common to every provincial town, such as milling and baking, Exeter acquired highly developed tanning, bell-making, cloth-making and cloth-finishing industries.[3]

Within towns, industrial quarters became more apparent. Much of Exeter's industrial activity was carried on beyond the city walls; thus below the west wall on Exe Island a multitude of leats, drawn from the Exe, were established to drive corn mills, fulling-mills and probably tanning mills.[4] There developed an increasing separation of the poor in the extra-mural suburbs and in the back-lanes and side-streets within the walled areas. Certain parishes in the old cities like Exeter, York and Leicester were almost entirely populated by labouring classes in the 1520s and other parishes were equally reserved for the wealthier.[5] Social and economic inequality in towns was aggravated during the later Middle Ages. W. G. Hoskins has calculated from the 1524 Lay Subsidy that in the larger English towns approximately half of the taxable population belonged to the wage-earning class.[6] In sixteen towns in Sussex, Buckinghamshire and Rutland the wage-earning class amounted to nearly 40% and yet it rarely owned as much as 10% of the wealth.[7] In the country as a whole, wealth was increasingly becoming concentrated in the capital.

London

Whereas in 1334 London had been not quite five times as wealthy as the richest provincial town (Bristol), in the 1520s the City of London alone was almost ten times as wealthy as Norwich, then the leading provincial

[1] J. N. Bartlett, 'The expansion and decline of York in the later Middle Ages', *Econ. Hist. Rev.*, 2nd ser., XII (1959–60), 17–33; E. Miller, 84–6.

[2] A. A. Ruddock, *Italian merchants and shipping in Southampton, 1270–1600* (Southampton, 1951).

[3] T. Atkinson, *Elizabethan Winchester* (London, 1963), 29–33; E. M. Carus-Wilson (1963), 5–16 and 22–4.

[4] E. M. Carus-Wilson (1963), 22.

[5] W. G. Hoskins (1955), 43–4, and 'English provincial towns in the early sixteenth century', *Trans. Roy. Hist. Soc.*, 5th ser., VI (1956), 19.

[6] W. G. Hoskins (1956), 17–19.   [7] J. Cornwall (1962–3), 63.

city, and more than fifteen times as wealthy as Bristol. In the subsidy of 1543–4, London paid thirty times as much tax as Norwich, and well over forty times as much as Bristol. Even the suburb of Southwark, across the river, paid more tax than Bristol. London contributed as much in 1543–4 as all the other English towns put together, from Norwich down to the smallest place that functioned as a local market centre.[1] The population of London increased threefold during the sixteenth century, to a large extent because of considerable immigration.[2] Its population in 1600 was probably about four to five times as great as it had been in 1334. Its prosperity was based upon an increasing share of England's trade, particularly of the cloth trade. A weekly cloth market was established in 1396 at Blackwell Hall, near the Guildhall, and the addition of several annexes in the course of the fifteenth century, and finally its destruction and rebuilding in 1588 as a 'new, strong and beautiful store-house', bore witness to the expanding textile trade and to London's control of the trade.[3]

Gradually London usurped the trading functions of many provincial towns. During the fifteenth century the distribution of cloth in York, for example, was taken out of the hands of the city's traders first by West Riding merchants and later by London merchants able to supply, in return for textiles, a large variety of imported goods hitherto unobtainable from local traders.[4] By the early sixteenth century, much cloth manufactured in Devon was being sent to London and thence exported to the Low Countries, rather than being exported to France from diverse West Country ports as it had been during the fifteenth century.[5] The ever-increasing influence of London's merchant class is clearly demonstrated in Southampton, whose prosperity during the fifteenth century was fostered by Italian merchants, nurtured by local merchants and ultimately killed by London merchants. From the middle of the fifteenth century, Londoners had taken a leading part in Southampton's commerce, using the town as an outport for trade between England and the Mediterranean. From trading with the Mediterranean, London merchants spread into every branch of commerce in Southampton, gradually swamping local merchants by their superior capital resources and large-scale ventures.

[1] W. G. Hoskins (1955), 35.
[2] P. Ramsey, *Tudor economic problems* (London, 1963), 110.
[3] E. M. Carus-Wilson (1952), 420–1, and G. D. Ramsay, 25.
[4] J. N. Bartlett, 30.
[5] E. M. Carus-Wilson (1952), 28–9.

Southampton's role as London's outport was undermined when, during the first half of the sixteenth century, improvements in shipbuilding (especially in rigging) made the large square-rigger less cumbersome and less difficult to manoeuvre in the restricted channels amid the sandbanks of the Thames' estuary. And when, in 1514, the control of pilotage in the Thames was placed in the charge of a body of experienced lodesmen, the Brethren of Trinity House at Deptford Strand, navigational safety was considerably furthered. From 1540 onwards especially, London merchants forsook Southampton and increasing numbers of ships journeyed to the Thames rather than to Southampton Water. Southampton became moribund, with considerable structural decay.[1] Overland trade between Southampton and London was drastically reduced and intermediate towns like Salisbury were absorbed within the capital's economic hinterland as their cloth merchants exported more to the Low Countries via the Thames and less to France and Spain via Southampton Water.[2] Increasingly, London was dominating overseas trade as it came to dominate the English economy in general.

[1] A. A. Ruddock, 255–72.   [2] G. D. Ramsay, 21–2.

Chapter 6

ENGLAND *circa* 1600

F. V. EMERY

One aspect of the flowering of Elizabethan England, drawing its strength
from a variety of sources, was the beginning of a tradition of topographical
writing.[1] John Leland's notes for a proposed 'Description of the realm of
England' had not been published when he died in 1552 and, though used
by many people, they were not printed until 1712. In the meantime, in
1577, there appeared William Harrison's *Historical description of the island
of Britain* as an introduction to Holinshed's chronicles, and this was
revised for a second edition in 1588, part of which was called *The description of England*. It was during these years that William Camden began
the work that appeared in 1586 as the *Britannia*. Written in Latin, it was
organised upon a county basis, and soon ran into a number of editions.
The sixth edition in 1607 was illustrated by a series of county maps en-
graved from the surveys of Saxton, Norden and others. It was translated
into English in 1610 by Philemon Holland, who called it a 'chorographical
description'. Then in the following year came John Speed's *The theatre of
the empire of Great Britaine* which aimed at 'presenting an exact geography'
of the realm. It, too, was organised on a county basis; its text was abridged
from Camden's *Britannia*, but there was a new set of maps, based likewise
on those of Saxton, Norden and others.

The county maps that Camden and Speed were using constituted an
innovation made possible by developments in surveying instruments and
the art of surveying. Laurence Nowell in the 1560s had proposed to make
maps of the English counties;[2] but what Nowell proposed, Christopher
Saxton accomplished. He worked during the 1570s under the patronage
of Thomas Seckford, lawyer and courtier, who obtained official support

[1] A. L. Rowse, 'The Elizabethan discovery of England', being ch. 2 (pp. 31–65) of
The England of Eliȝabeth (London, 1951).
[2] R. Flower, 'Laurence Nowell and the discovery of England in Tudor times',
Proc. British Academy, XXI (1935), 54.

for the enterprise. Saxton's atlas of England and Wales on a county basis appeared in 1579; it has been described as 'the first national atlas',[1] and it inaugurated a new era in cartography. Another surveyor of note was John Norden who aimed at producing a *Speculum Britanniae*, a series of county chorographies illustrated by maps. He succeeded in publishing only those for Middlesex (1593) and Hertfordshire (1598), but a number of his county maps appeared alone. These maps of Saxton and Norden provide much geographical information – not only about villages and market towns but also about woods, parks and hills; Norden's county maps were the first to show roads. John Speed's *Theatre* included on each county map an inset showing a plan or a bird's-eye view of the chief town of the shire, and so provides a unique collection of town plans.

The text of Camden and Speed's works is heavily weighted with references to antiquities. There are, it is true, many references to economic circumstances – to the ironworks of Sussex, to the grazing on Romney Marsh, to sheep on the Cotswolds, to the rich vales of Aylesbury and Belvoir, to the orchards of Kent and the saffron of Essex. But it is difficult to piece such fragments of information into a balanced picture of the national use of land. Camden, moreover, too often simply refers in a conventional way to the main kinds of land use as, for instance, in the Thames valley below Oxford, 'chequered with corn fields and green meadows, clothed on each side with groves'.

Parallel with the general accounts by Harrison, Camden and Speed, there were more detailed descriptions of particular counties. The county had for long been an important element in the administration of the realm, but now a new 'county self-consciousness', or provincial patriotism, was emerging. William Lambarde's *A perambulation of Kent* (1576) was the first, and was full of pride in the orchards, the deer parks and the countryside of Kent in general. He expressed the wish that 'some one able man in each shire' would 'describe his own country'. Richard Carew's *Survey of Cornwall* appeared in 1602, although he had been at work on it since the 1580s; it tells us much about tin-mining and fishing and about cattle and crops. Tristram Risdon's *The chorographicall description or survey of the county of Devon* was written in 1630, although it was not printed until 1714. It describes the six various combinations of soil and agriculture, ranging in quality from the South Hams, the very 'Garden of Devonshire',

---

[1] E. Lynam, *The mapmaker's art* (London, 1955), 63.

to Dartmoor 'barren and full of brakes and briars'.[1] These are only three
of a number of county descriptions, some of which, like Risdon's *Devonshire*, did not see the light of print for many years. Some writers devoted
themselves to towns and cities, and we are fortunate that among these was
John Stow whose detailed *Survey of London* appeared in 1598.

Others wrote about their 'countries' on a more circumscribed scale
than the county, as John Smith did in his profile of the hundred of
Berkeley. 'In the body of this hundred', he said, 'are observed three steps
or degrees', beginning with the vale alongside the Severn, 'which hath
wealth without health'; farther upslope was the best land of all, and above it
the Cotswold, 'which affordeth health in that sharpe aire, but lesse wealth'.[2]
As we read these descriptions, and also those of foreign travellers, the
emphasis varies. Sometimes it is upon tillage and the diligence of the
husbandman, as with Camden, or upon the rich pastures and stores of
cattle and sheep that so impressed the duke of Württemberg when he
visited England in 1592.[3] All contributed to the geographical variations
in what Michael Drayton was calling 'Albion's glorious isle'.[4]

POPULATION

Estimates of the total population of England about 1600 vary greatly.
They can be based only upon such indirect evidence as that of parish
registers, subsidy rolls, lists of communicants, and muster rolls of able-
bodied men. There is also a variety of opinion about the multipliers that
should be used to produce total populations from such figures. Most
acceptable estimates vary around the figure of 4 million. William Harrison, about 1580, gave the number of 'able men for service' in the musters
of 1574–5 as 1,172,674.[5] E. E. Rich, using the muster returns, with a
multiplier of 4.0, suggested a total of 'something over four millions'.[6]

---

[1] F. V. Emery, 'English regional studies from Aubrey to Defoe', *Geog. Jour.*,
CXXIV (1958), 315.

[2] *A description of the hundred of Berkeley. . .and of its inhabitants, by John Smith*,
being vol. 3 of J. Maclean (ed.) *The Berkeley Manuscripts* (Gloucester, 1885).

[3] W. B. Rye, *England as seen by foreigners in the days of Elizabeth and James I*
(London, 1865), 30–1.

[4] M. Drayton, *Polyolbion, or a chorographicall description of Great Britaine*
(London, 1612).

[5] W. Harrison, *Description of England (1577–1587)*, ed. G. Edelen (Cornell,
1968), 235.

[6] E. E. Rich, 'The population of Elizabethan England', *Econ. Hist. Rev.*, 2nd ser.,
II (1949), 247–65.

Whatever multiplier is used for 1574–5, the population some twenty-five years later was appreciably higher, but by 1600 there was a steadily decreasing rate of growth of population in most parts of England, making it a suitable point in time for a cross-sectional view.[1] The 'Liber Cleri' returns of 1603 put the number of communicants, recusants and non-conformists for the county as a whole at about 2 million, and using conventional multipliers this would yield a figure of about 4 million.[2] John Rickman, in the introduction to the 1841 Census, attempted to calculate the population of each county for a number of pre-Census years, i.e. before 1801. By applying certain assumptions to the figures for births, deaths and marriages, supplied by the parish clergy, he produced a total of 4.3 million for England in 1600 (Wales, including Monmouthshire, accounted for another 380,000). Subsequent research has shown how uncertain such estimates based upon parish registers must be, but, even so, Rickman's figure may be not far from the truth.

In view of the uncertainties, Fig. 53, based upon Rickman's tables, shows not densities per square mile, but variations, county by county, from the national density of 87.6 per square mile, and it may provide a basis for comparing different areas one with another. The most densely peopled part of the country comprised the metropolitan counties, a reflection of the intensive agriculture and varied industry in and around London. The high figures for Devon and Somerset also reflect a varied economy based not only on agriculture but on cloth manufacture, mining and commercial activities. Industrial Lancashire was beginning to emerge, and Worcestershire, with its vale of Evesham, its orchards, its salt industry and its metal-working in the north-east, also stood high. The figures for the counties of Northampton, Nottingham and Oxford may be thought surprisingly high, and that for Gloucestershire surprisingly low. The low figures for most northern counties are to be expected.

Fig. 53 is unsatisfactory not only because of the nature of the evidence upon which it is based but also because county averages are misleading. Thus southern Cambridgeshire was very much more populous than the northern half of the county with its undrained fen. Within each county,

---

[1] E. A. Wrigley (ed.), *An introduction to English historical demography* (London, 1966), 266.

[2] J. C. Russell, *British medieval population* (Albuquerque, New Mexico, 1948), 270–81; H. J. Habakkuk, 'The economic history of modern Britain', *Jour. Econ. Hist.*, XVIII (1958), 486–501; G. S. L. Tucker, 'English pre-industrial population trends', *Econ. Hist. Rev.*, 2nd ser., XVI (1963), 205–18.

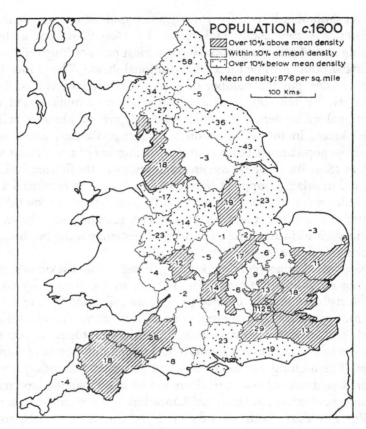

Fig. 53 Population *circa* 1600
    Based on John Rickman's estimates in *Census of 1841: Enumeration Abstract*,
    36 (P.P. 1843, xxii).

the ability of the land to support a farming population, and the intensity
of various agrarian systems, were reflected in densities both higher and
lower than the average for the county as a whole. Nowhere was this more
true than in the uplands of the north. The justices of the peace for
Westmorland claimed that their county was smaller, more barren, and
more densely populated than any other county in England. Their view
is substantiated by what is known about the concentration and com-
pression of settlement in the dale or valley tracts within each fell parish.[1]

[1] J. Thirsk, 'Industries in the countryside', in F. J. Fisher (ed.) *Essays in the
economic and social history of Tudor England* (Cambridge, 1961), 82.

We may take as a standard of measurement the average parish of southern England and the Midlands, say in the champion country of Leicestershire, with not more than 40 or 50 households accommodated on about 1,000 acres of land. Placed alongside this, the populations of 100, 200 and 300 families in the fell parishes of Westmorland, even though their acreages were ten times larger, must be regarded as dense.

Furthermore, the county totals include the populations of towns which held a significant and growing position in the realm. Any attempt to calculate the proportion of the total population of England that lived in towns is full of uncertainty, but a figure of about one-fifth may not be far wide of the mark.[1] If, however, London with some 250,000 people be excluded, the percentage drops to about 13 or so. It must be remembered, of course, that the line between country and town was at least as difficult to draw then as it is now. Many towns had open fields within their limits, and generally contained unmistakable reminders of the country in the form of barns, straw-ricks and hay-stacks, stalls, pens and penfolds for farm animals. On the other hand, in the countryside, farming pursuits were often combined with part-time occupations in rural industries, in cloth-making, weaving and spinning and in coalmining or iron-working.

An important feature of the population in general was its mobility. S. A. Peyton has shown from an examination of the subsidy rolls for Nottinghamshire that there was continuous movement.[2] When successive rolls are compared, 'names continually disappear, while new names occur, themselves in turn vanishing'. In a number of hundreds the majority of freeholders in 1612 appear to have held land for less than two generations. Some families whose names disappeared may have continued to reside in their parishes, although in poorer circumstances, but the evidence as a whole points to continuous and widespread migration. 'The population of the county of Nottingham between the years 1558 and 1641 was in a highly mobile condition.' The subsidy rolls for other counties point to similar movement – those for Bedfordshire,[3] Lincolnshire[4] and elsewhere.

[1] G. M. Trevelyan, *English social history* (London, 1942), 141.
[2] S. A. Peyton, 'The village population in Tudor Lay Subsidy rolls', *Eng. Hist. Rev.*, XXX (1915), 234–50.
[3] L. M. Marshall, 'The rural population of Bedfordshire, 1671–1921', *Beds. Hist. Rec. Soc.*, XVI (1934), 54 *et seq.*
[4] M. Campbell, *The English yeoman under Elizabeth and the early Stuarts* (London, 1942), 37–8.

254 F. V. EMERY

Other evidence for Kent[1] and for Sussex[2] shows movement within the county and also from outside, sometimes from as far away as the north and west of England. E. E. Rich found that the Muster Rolls indicate the same mobility, and they sometimes include complaints about the continual 'remove' of men. 'Most of the Elizabethans who emerge from their backgrounds are members of families who have moved at least once in two generations. The Shakespear family moved from Snettisham to Warwick before William moved to London.'[3]

Some migrants went merely to neighbouring parishes; certainly this was so in Devonshire where it was unusual to leave the county unless one lived near the Somerset, Dorset or Cornish border. Dorset experienced a remarkable immigration of small farmers and gentry from Devon at this time.[4] Other migrants moved farther afield and frequently to the growing towns near and far. As at earlier periods, London was a great centre of attraction, but the numbers now involved must have been very considerable to account for its great growth during the sixteenth century to some 250,000 in spite of its high mortality rate.[5] Wherever they lived, Englishmen were still at risk when the harvest failed, and were likely to find themselves in the grip of starvation and sickness. The year 1600 itself was poised between a harvest pattern in the 1590s with more deficient, bad or dearth years ('the Great Famine') than in the following decade, which had six abundant or good years.[6]

THE COUNTRYSIDE

The man-made landscape of England in 1600, as at all other times, rested upon the basic physical division of the country into an upland zone and a lowland zone. The former, with its damp climate and with much of its

[1] P. Clark, 'The migrant in Kentish towns', being ch. 4 of P. Clark and P. Slack (eds.), *Crisis and order in English towns, 1500–1700* (London, 1971).

[2] J. Cornwall, 'Evidence of population mobility in the seventeenth century', *Bull. Inst. Hist. Research*, XL (1967), 143–52.

[3] E. E. Rich, 260; E. J. Buckatzsch, 'The constancy of local populations and migration in England before 1800', *Population Studies*, V (1951), 62–9.

[4] W. G. Hoskins, *Devon* (London, 1954), 173; C. Taylor, *Dorset* (London, 1970), 136.

[5] D. Cressey, 'Occupation, migration and literacy in East London, 1580–1640', *Local Population Studies*, V (1970), 53–60.

[6] W. G. Hoskins, 'Harvest fluctuations and English economic history, 1480–1619', *Agric. Hist. Rev.*, XII (1964), 28–46; C. J. Harrison, 'Grain price analysis and harvest qualities', *Agric. Hist. Rev.*, XIX (1971), 135–55.

land over 800 ft. above sea-level was largely a pastoral area. It included, it is true, lowland enclaves with tillage, but this was largely subordinate to pastoral pursuits – along the coastlands of Durham and Cumberland, in the Lancashire–Cheshire plain, and elsewhere. Most of this lowland was an enclosed countryside. Some of it had always been so, and had been taken in directly from the waste. Other parts of it had been enclosed from open fields, but open fields quite different in their functioning from those of the two- or three-field 'Midland' system; and the open fields that still remained were soon to disappear.

In the lowland zone, the main distinction was between enclosed country, wooded or hedged land, on the one hand, and open-field countryside, 'champaign' or 'champion' land on the other (Fig. 54). Leland had constantly referred to the presence of woodland in the enclosed districts and its absence in the open-field areas;[1] and Arthur Standish, in 1613, drew a contrast between enclosed country with timber, and the bare fields of 'champaign countries' where the countryside was open and fuel was scarce.[2] It was a distinction that could split a single county; so was it in Warwickshire between wooded Arden to the north and open Feldon to the south, or in Buckinghamshire between the open vale of Aylesbury and the wooded Chilterns. The open-field area was itself far from uniform. The Midland two- or three-field system was characterised, amongst other things, by the common pasturing of the village animals on the fallow land and on the stubble of the arable after harvest, hence the term 'common fields'. But there were many areas of open field where such common pasturing did not take place, and where there were also other differences. Lowland England included a variety of irregular open fields, and also areas where enclosures and open fields existed side by side.

One element in the English landscape to be found in upland and lowland zones alike comprised forests, woodland and the deer parks which were marked on the county maps of the time. The forests were used primarily for pleasure and for hunting, but, as well as producing venison, they sometimes provided grazing for animals and horses, and parts of some were even cultivated. They were not as important as they once had been in the geography of the realm. There were indeed still many deer parks, but some of them had started to disappear. Over the countryside in general much woodland remained, but complaints about its disappearance had begun to grow numerous.

[1] E. C. K. Gonner, *Common land and inclosure* (London, 1912), 322–3.
[2] A. Standish, *New directions of experience to the commons complaint* (London, 1613), 6.

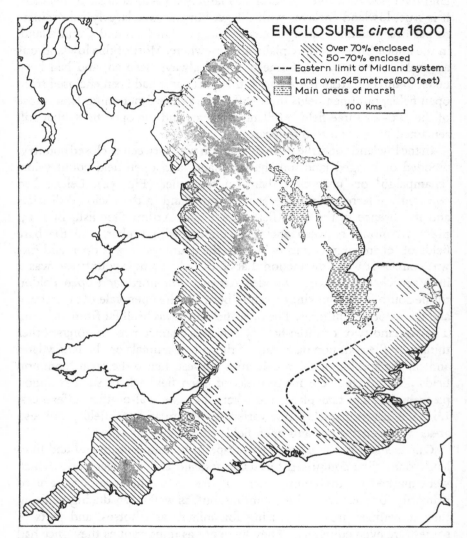

Fig. 54 Enclosure *circa* 1600
   Based on: (1) E. C. K. Gonner, *Common land and inclosure* (London, 1912),
   map D; (2) H. L. Gray, *English field systems* (Cambridge, Mass., 1915).

While all this variety of arable and pasture and woodland cannot be portrayed accurately on maps, it is at any rate possible to make some broad generalisations; likewise about the distribution of types of rural settlement. Such generalisations must inevitably ignore the special features of this or that locality.

Rural settlement

The evidence does not permit us to discuss in any detail the relative distribution of the different types of settlement in which the population was disposed. But the basic contrast between nucleated and dispersed settlement was well appreciated by the men of the time. William Harrison about 1580 made the point clearly. In 'champaign' county, the houses stood 'all together by streets and joining one to another'; whereas 'in the woodland countries', they were 'dispersed here and there, each one upon the several grounds of their owners'.[1] It was essentially a contrast between open-field and enclosed districts, as could be seen in the historic division of Warwickshire into open-field Feldon with nucleated villages and wooded Arden with dispersed settlements. Even as Harrison wrote, dispersed houses were occasionally appearing in newly enclosed districts within open-field country. Francis Trigge, who strongly protested against enclosers, pointed out one of the results of their work in 1604: 'We may see many of their houses built alone like raven's nests, no birds building neere them.'[2] We also know from a variety of sources that the upland districts of England were characterised by small hamlets and scattered homesteads, although villages and small towns were to be found in the more fertile valleys and coastal plains.

The buildings that made up these settlements were still mostly of timber, and Harrison noted that 'as yet few of the houses of the common-alty' were 'made of stone'. He drew a contrast between the 'strong and well-timbered' houses of the 'woody soils' and those of 'the open country and champaign countries', having only a few upright and cross-posts with clay-covered panels between them.[3] But changes were already afoot, and 'the rebuilding of rural England' was well advanced.[4] What has been

[1] W. Harrison, 199 and 217.

[2] F. Trigge, *The humble petition of two sisters; the church and the commonwealth* (London, 1604).

[3] W. Harrison, 195.

[4] W. G. Hoskins, 'The rebuilding of rural England, 1570–1640', *Past and Present*, IV (1953), 44–59.

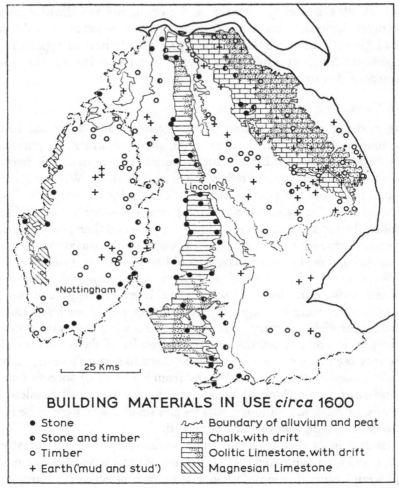

Fig. 55 Lincolnshire and Nottinghamshire: building materials in use *circa* 1600
Based on M. W. Barley, *The English farmhouse and cottage* (London, 1961), 82.

called the first phase of the housing revolution (1570–1615) saw the modernisation of medieval houses in the south-eastern counties and rather more new building farther away from London – among, for example, the prosperous yeomen farmers of Devonshire and the clothiers of the West Riding as well as among cottagers generally.

Stone was beginning to replace timber as the most usual building material, but only for the larger houses; it did not appear in the majority of farmhouses, even in stone country like the Cotswolds, until much later in the seventeenth century.[1] Brick came into general use later than stone. Even 'the ancient manors and houses' of gentlemen, so Harrison tells us, were for the most part of timber, except that those 'lately builded' were 'commonly either of brick or hard stone or both'. Newly built houses were to be seen on all sides, not only the grandiose piles like Holdenby in Northamptonshire and Wollaton in Nottinghamshire but in many manor-houses and farmsteads. That they still reflected regional characteristics may be seen from the details given in parsonage terriers of the sixteenth and seventeenth centuries (Fig. 55). One of the most striking was the 'magpie' half-timbered style of Lancashire, Cheshire and the west Midlands. Here, in a damper environment than that of counties like Essex, oak timbers were given a protective coating of tar or pitch, and plaster panels or brick nogging were whitewashed, making the contrast of black and white.[2]

Old men, wrote William Harrison about 1580, were noting a number of things 'to be marvelously altered in England within their sound remembrance'.[3] Many of the newly built mansions, such as Montacute House in Somerset, were set in gardens designed with the 'elaborate symmetry that delighted the Elizabethan eye'.[4] Chimneys and glazed windows were becoming more common; minor comforts, too, were increasing, such as feather beds instead of straw pallets, and pillows instead of wooden head rests, and platters and spoons made of metal instead of wood. In all such changes the south of England was in advance of the north.

The upland zone

Northern England. The Pennine uplands were largely lands of heath, moor and bog, and of rough grazing and poor pasture farming. The Border adjoining Scotland was moreover subject to raids from the Scots at least until 1603 and the union of the two Crowns. The main business of upland farmers was the breeding of sheep for wool and of young cattle for sale to lowland graziers for fattening. Transhumance was a feature of upland life, and stock was sent in early summer to hill pastures or 'grasshouses' on the better parts of the moorlands. Camden wrote of the men

---

[1] M. W. Barley, *The English farmhouse and cottage* (London, 1961), 101–3 and 123–5.

[2] A. Clifton-Taylor, *The pattern of English building* (London, 2nd ed., 1965), 32–3 and 46–51.

[3] W. Harrison, 200 *et seq.*

[4] E. J. M. Buxton, *Elizabethan taste* (London, 1963), 43–4.

of the northern Pennines 'who from the month of April lie out scattering and summering (as they term it) with their cattle in little cottages here and there, which they call sheals or sheilings'. Such outposts might also provide suitable temporary tillage plots, after which the land was allowed to revert to moor. Temporary intakes for cultivation were also to be found in the southern Pennines – in Rossendale, for example, and the Peak District. Fertile arable land was confined to the narrow valleys and dales that pierced the mountains and that provided a basis for a few small open fields and for meadow and permanent grass. Similar conditions characterised the Lake District and the North York Moors; here, too, cattle and sheep were the mainstays of the economy, together with some shifting cultivation.[1]

Larger expanses of more fertile land bordered the Pennine upland to the east and west. Along the eastern lowland much of the arable was both open and common, and lay in large fields. Such arrangements frequently resembled those of the Midlands in that a holder's strips were divided approximately between three fields; but they differed in having nearby large areas of pasture and waste that provided opportunities for a temporary expansion of the arable. In places there were arrangements similar to those of the infield–outfield system. This was a district transitional in field arrangements as well as in location between the Midlands and Scotland. By 1600, in many places, the arable open fields were in decay, and enclosure was spreading. The demands of the coalmining district around Newcastle, and those of the port itself, stimulated meat production supplemented by dairying, the rearing of oxen for draught, and horse breeding.[2]

Along the western lowlands, similar features could be encountered in Cumberland – some open fields, infield–outfield arrangements, and emphasis on cattle and sheep, with a tendency towards enclosure. Farther south, in Lancashire, much of the plain seems never to have been in open

[1] T. S. Willan and E. W. Crossley, *Three seventeenth century Yorkshire surveys* (Richmond, 1605; Middleham, 1605; Wensleydale, 1614), Yorks. Archaeol. Soc. Ser., 104 (1941), xx–xxiv; C. M. L. Bouch and G. P. Jones, *A short economic history of the Lake counties, 1500–1830* (Manchester, 1961), 94–101; G. Elliott, 'The decline of the woollen trade in Cumberland, Westmorland and Northumberland in the late sixteenth century', *Trans. Cumb. and West. Antiq. and Archaeol. Soc.*, LXI (1961), 112–19.

[2] J. Thirsk (ed.), *The agrarian history of England and Wales*, vol. IV, 1500–1640 (Cambridge, 1967), 27. For further details of agrarian practice here and in the other districts surveyed below, see Mrs Thirsk's chapter 'The farming regions of England', *ibid.*, 1–112.

field but had been enclosed directly from wood-pasture and rough grazing. The small open fields, where they existed, were rapidly disappearing. In 1600, this was predominantly a pastoral country characterised by the rearing and fattening of cattle and sheep together with some dairying.[1]

The Welsh border counties. Enclosure in the counties bordering Wales had gone far by 1600. In many localities the land had been cleared from the waste as enclosed fields, held in severalty, with hedge or fence from the start. Elsewhere the small and irregular open fields were being enclosed, either in a piecemeal fashion or by more general agreement. Yet other areas remained in heath and waste and common. Some parts of these were ploughed up for a few years and then allowed to revert to rough grazing. It was predominantly a pasture-farming countryside with relatively little grain and with much cattle-rearing, dairying and pig-keeping.

There were variations dependent upon soil and other circumstances. The most important grain-growing area was in Herefordshire, but it was also a dairying county, and Camden's description of the county as 'fruitful for corn and cattle feeding' was echoed by other writers. Here, in the 'Golden Valley' of the Dore, Rowland Vaughan's experiments in watering meadows were taking place, and he was not the only man in the county interested in dairying. Herefordshire was also known for its orchards, so was Worcestershire and the adjoining parts of Gloucestershire; from the apples and pears cider and perry were made. Cheshire, on the other hand, had long been famous for its cheese, 'the best in Europe' according to Speed. Its small irregular open fields were quietly disappearing by agreement, many into pasture closes.[2]

The south-west. Much of the south-western peninsula consists of infertile upland country. Expanses of rough grazing on Dartmoor, Exmoor, Bodmin Moor and other hills were used as summer pastures for cattle and

[1] G. Elliott, 'The system of cultivation and evidence for enclosure in Cumberland open fields in the sixteenth century', *Trans. Cumb. and West. Antiq. and Archaeol. Soc.*, LIX (1959), 84–104; G. Youd, 'The common fields of Lancashire', *Trans. Hist. Soc. Lancs. and Cheshire*, CXIII (1961), 1–41; F. J. Singleton, 'The influence of geographical factors on the development of the common fields of Lancashire', *ibid.*, CXV (1963), 31–40; R. A. Butlin, 'Northumberland field systems', *Agric. Hist. Rev.*, XII (1964), 99–120.
[2] D. Sylvester, *The rural landscape of the Welsh borderland* (London, 1969), 230–1, 253–4, 262–70. See also T. Rowley, *The Shropshire landscape* (London, 1972).

sheep. As in the north, transhumance was practised. There were likewise various forms of shifting cultivation and of infield–outfield cultivation. Plots selected for tillage were fertilised by paring off and burning the turf and mixing the ash with the soil ('denshiring').

Around the moors were the more fertile areas of the coastal plains and the vales. There were some open fields, but 'nothing to show that the tenants' holdings were normally distributed in equal parcels, one or more in each field'.[1] Moreover, in such a region, where most parishes had their own tracts of rough grazing, 'there was no absolute necessity to leave a half or a third of the ploughland under grass each year'.[2] In some areas, the open fields had undergone silent enclosure by agreement; other areas may have been enclosed directly from rough pasture.[3] Writers of the sixteenth century, in speaking of Cornwall and Devon, refer characteristically to enclosed fields surrounded by great hedges. Enclosed fields were also to be found over much of Somerset, west Dorset and south Gloucestershire. The south-west as a whole was predominantly a land of pastoral husbandry with patches of mixed farming and grain growing on the coastlands and in the broader vales. John Norden in 1607 could describe the vale of Taunton Deane as the paradise of England, with its fields of wheat, barley and oats and its dairies.[4] One corollary of enclosure was increasing local specialisation and production for the expanding markets. In Devonshire, men were enlarging their orchards and Richard Hooker, about 1600, referred to the careful management of orchards and apple gardens; cider was sold for the provisioning of ships.[5] East Somerset and west Dorset was a dairying countryside, and Yeovil a great cheese market. Domestic handicrafts (such as the making of cloth, gloves and bone-lace) supplemented the dairying occupations. To the north, the vales of Berkeley and of Gloucester were areas of mixed farming with dairying as an important element.

One of the most distinctive countrysides in the south-west was that of the Somerset Levels. Contemporary descriptions are similar to those of

---

[1] W. G. Hoskins and H. P. R. Finberg, *Devonshire studies* (London, 1952), 283.
[2] *Ibid.*, 287.
[3] P. J. Fowler and A. C. Thomas, 'Arable fields of the pre-Norman period at Gwithian', *Cornish Archaeology*, I (1962), 61–84; P. D. Wood, 'Open field strips at Forrabury Common, near Boscastle', *ibid.*, II (1963), 26–33; E. M. Yates, 'Dark Age and medieval settlement on the edge of wastes and forests', *Field Studies*, II (1965), 133–53.
[4] J. Norden, *The surveyor's dialogue* (London, 1607).
[5] W. G. Hoskins (1954), 94.

the Fenland in eastern England – a watery spectacle with fowling, fishing and turf cutting, and with much grazing of cattle on pastures made fertile by winter flooding. Here, too, dairying was important, with an emphasis on butter and on Cheddar cheese. Some localities had been drained by 1600, but it is difficult to say how much of this was for arable and how much for pasture.

The lowland zone

The Midland area. The main physical characteristics of the Midland open-field system were the large open arable fields, generally two or three in number, each covering up to 400 acres. Under the three-field system, one was in winter-sown grain (wheat or rye), one in spring-grown grain (barley or oats) or pulse, and the third in fallow. Under the two-field system, one was in fallow and the other was divided between winter and spring grains. The fields were composed of 'furlongs' or blocks of land of varying size, and each furlong in turn was divided into strips, selions or ridges, usually of between one-quarter and one-third of an acre apiece. The arable land of a village may thus have comprised as many as two thousand separate strips or more. The holding attached to a farmstead consisted of a number of these strips, widely scattered and intermixed with the strips of other holdings. There were also common meadows allocated to each holding in small strips or 'doles'. Essential features in the working of such a system were: (1) the communal regulation of cultivation, with the strips in each field growing the same crop in any particular year; (2) the more or less equal division of the scattered strips of a holding between the two or three fields; and (3) the throwing open of the arable and meadow for common grazing in the fallow season and after the crops had been harvested. This was an invaluable right enjoyed by all cultivators in a village, and was the essential feature of the Midland system as opposed to other varieties of open field.[1]

By 1600 these arrangements still survived in their entirety on many Midland manors and partially on most. But it is clear that the system was

---

[1] H. L. Gray, *English field systems* (Cambridge, Mass., 1915), 39–49; C. S. and C. S. Orwin, *The open fields* (Oxford, 3rd ed., 1967), 59–66; M. W. Beresford and J. K. S. St Joseph, *Medieval England: an aerial survey* (Cambridge, 1958), 21–45; A. R. H. Baker, 'Howard Levi Gray and English field systems', *Agric. Hist.*, XXXIX (1964), 1–6. For comment on the terms 'open field' and 'common field', A. R. H. Baker, 'Some terminological problems in studies of British field systems', *Agric. Hist. Rev.*, XVII (1969), 138–40.

much more flexible than it was once thought to be. The manorial maps that become available in the latter part of the sixteenth century often reveal quite complex arrangements. They show that the unit of rotation was as likely to have been the furlong as the field. Moreover, individual furlongs could be put under grass although the rest of a field was in arable. Even individual strips could be left in grass by communal agreement; a farmer would then tether or hurdle his beasts during the growing season, and, after harvest, his strips would be thrown open, together with the stubble of his neighbours, for general grazing by the village animals. From such practices within the common fields sprang the incentive to consolidate strips, to enclose open land and to embark upon the convertible husbandry of alternate crops and grass. This was so especially in the densely settled claylands where mixed farming prevailed and where the possibilities of heavier crops and more livestock were realised. Innovations and variations became increasingly frequent; thus, at Harwell in Berkshire the rotation in two common fields was extended by grazing a fodder crop of vetches on the fallow field. Or again, in some villages there were blocks of enclosed land in the midst of fields which were still open. In Oxfordshire 19% of the townships were enclosed, although with variations in different parts of the county; on the Redlands around Banbury enclosure had touched only 13% of the townships, but 35% of those on the Chilterns were already enclosed.[1]

It was under such circumstances that the enclosures of the fifteenth and sixteenth centuries had taken place and had frequently involved the conversion of arable to pasture. In nine Midland counties, and especially in those of Leicester, Northampton, Rutland and south-east Warwick, over 8% of the total area had been enclosed between 1455 and 1607 (Fig. 47). Here, the new fields within which the flocks and herds were so carefully managed by the graziers were often as large as 50 to 100 acres, their strong fences and hedges sometimes following the rectilinear outlines of former furlongs, at other times transgressing them and cutting across the pattern of ridge and furrow.[2]

---

[1] R. H. Hilton, 'Medieval agrarian history', *V.C.H. Leicestershire*, II (1954), 145–98, 161, 197; M. A. Havinden, 'Agricultural progress in open-field Oxfordshire', *Agric. Hist. Rev.*, IX (1961), 73–83; W. G. Hoskins, *The Midland peasant* (London, 1957), 67–70, 95; G. E. Fussell, *Robert Loder's farm accounts, 1610–1620*, Camden Society, 3rd ser., 53 (London, 1936).

[2] M. W. Beresford and J. K. S. St Joseph, 114–20; K. J. Allison *et al.*, *The deserted villages of Oxfordshire* (Leicestershire, 1965), 8.

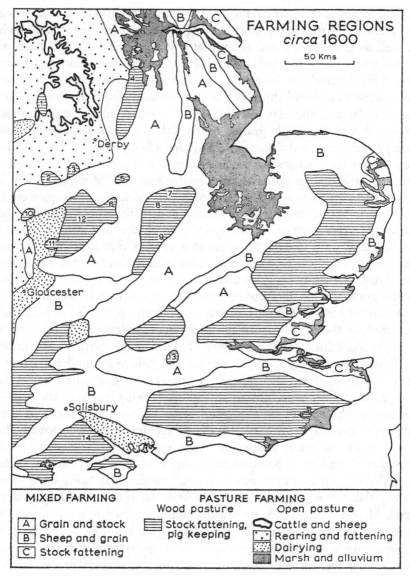

Fig. 56 Farming regions in lowland England *circa* 1600

Based on J. Thirsk, *The agrarian history of England and Wales*, vol. IV, 1500–1640 (Cambridge, 1967), 4.

Forests are numbered as follows: 1, Macclesfield; 2, Cannock; 3, Needwood; 4, Sherwood; 5, Charnwood; 6, Leicester; 7, Leighfield; 8, Rockingham; 9, Whittlewood and Salcey; 10, Kinver; 11, Feckenham; 12, Arden; 13, Windsor; 14, New Forest.

In the mixed farming of the Midland area, grain and livestock were complementary to one another. Crops (including peas, beans and vetches as well as grain) were fed to the fattening animals; plough oxen and horses were needed to work the arable land; sheep were folded and cattle grazed on the fields to manure them. Various combinations of crops and stock were widespread through the clay lowlands from the vale of York southwards. On the clay plains around Oxford, for example, farmers grew and sold the staple grains, nearly half the sown land being under barley for malting and a quarter of it under wheat; much of their produce went down the Thames to London. Cattle were reared on most farms, fattened or added to dairy herds; smaller numbers of sheep and a few horses and pigs were also kept. In north Oxfordshire, the crops were fed to fat cattle which could walk to market and, with larger flocks of sheep, the farmer made most of his money from wool, beef and mutton. Graziers on the grand scale, like the Spencers at Wormleighton in south Warwickshire, kept as many as 14,000 sheep, and they also sold fat cattle to London; their 'stock ranching on great enclosed pastures' was made possible both by the grass and by the feeding-stuffs grown on their arable land.[1]

A second kind of mixed farming was the sheep and grain husbandry practised on the light soils of the chalk and Oolitic limestone outcrops – on the Yorkshire and Lincolnshire Wolds, the Lincolnshire Heath belt, the Cotswolds and on the chalk downs that radiate in all directions from Salisbury Plain. Here the farmer's mainstay was grain, first barley and then wheat. They were grown in the open fields, the fertility and soil texture of which were kept in good order by the large flocks of sheep folded on them. The primary purpose of keeping sheep was not for their wool but for the dung of the fold. They were fed in the daytime on the sheepdown, and then brought in at night to be folded on the arable. In Wiltshire, for example, about one half of the chalklands were sheepdown, about three-eighths in arable and the remainder in permanent grass. Meadow grass and the smaller arable crops (oats, peas, vetches) were used to supplement the downland grazing. The sheep of some villages numbered many thousands, even up to 10,000 and more. On thinner soils the inferior sheepdown merged into rabbit warrens. Parts of

---

[1] H. Thorpe, 'The lord and the landscape', *Jour. Birm. Archaeol. Soc.*, LXXX (1962), 38–77; J. A. Yelling, 'Common land and enclosure in east Worcestershire, 1540–1870', *Trans. and Papers, Inst. Brit. Geog.*, XLV (1968), 157–68; J. A. Yelling, 'The combination and rotation of crops in east Worcestershire, 1540–1660', *Agric. Hist. Rev.*, XVII (1969), 24–43.

the downland were occasionally broken up, tilled for one or more years and laid to sheep pasture. On the wolds of Yorkshire and Lincolnshire arrangements akin to the infield–outfield system were sometimes to be found.[1]

There were other variants in the Midland area (Fig. 56). The ancient royal forests were often characterised by pasture farming and by irregular field systems that had passed into enclosure by 1600. With large commons, it was not necessary to keep so much arable in fallow each year; moreover, assarts from the waste had not necessarily been incorporated into the open-field system. Thus the Northamptonshire forests of Rockingham, Salcey and Whittlewood, on heavy Boulder Clay, had some large commons and many old enclosures. Forested districts on light soils were characterised by temporary cultivation, and arrangements reminiscent of the infield–outfield system were also to be found – on the Bunter Sands of Sherwood Forest where 'breaks or temporary enclosures' were kept in tillage for five or six years and then allowed to revert to pasture; on the Bunter Sands of the Cannock Chase area and also in the Forest of Arden. The fertility of the arable on the light soils in and around the New Forest was maintained by the folding of sheep that grazed on the heaths and commons by day. This, too, was a pasture-farming area with pigs, cattle, and horses that also fed on the heaths; much of the area had passed into enclosure by 1600. Yet other variants were to be found in villages with substantial amounts of meadow along such rivers as the Thames, the Warwickshire Avon and the Wiltshire Avon; here again there were often irregular field arrangements and enclosure by 1600.[2] To sum up: the area of the traditional two- and three-field Midland system had many variants that seem to have been associated with differences in soil and geographical circumstance.

The Fenland. This was a distinctive area covering some 1,300 square miles and including two distinct types of country – siltlands towards the sea and peatlands inland.[3] The silt area, bordering the Wash, was quite as wealthy and prosperous as the uplands around, in places even more so. If open fields had ever existed here, they had certainly long disappeared. In any case, the amount of arable was small, and the main feature of the economy was its emphasis upon grazing, and the rearing and fattening of animals

---

[1] E. Kerridge, *The agricultural revolution* (London, 1967), 61–2, 105–7.

[2] H. L. Gray, 83, 107.

[3] H. C. Darby, *The draining of the Fens* (Cambridge, 2nd ed., 1956); J. Thirsk, *English peasant farming* (London, 1957).

for meat and hides. Sheep were also important, and Camden referred to the great pasture of Tilney Smeeth, in Norfolk, which supported 30,000 sheep. Seawards, towards the Wash, successive areas of marsh had been reclaimed, and beyond the outermost bank there was salt marsh liable to be flooded at spring tides.

On the landward side of the silt belt lay the peat area marked, as Camden wrote, by 'foule and flabby quavemires', and varied by stretches of water – the meres of Whittlesey, Ramsey, Soham and the like. Projecting above the general level of the peat surface were 'islands' where settlement was possible and where the cultivation of the arable followed a three-field system; the largest of the islands was that of Ely with a dozen or so villages. The characteristic occupations of the area were fishing and fowling, turf-cutting and reed-gathering, and, not least, the pasturing of animals, cattle, sheep, horses. Winter floods produced excellent summer pastures and a whole array of complicated regulations controlled these economic activities and also the waterways that flowed through the district. During the closing decades of the sixteenth century there had been attempts to drain various localities, and the idea of a large-scale project of draining took shape. At last, in 1600, came 'An Act for the recovering of many hundred thousand acres of marshes.' Of the main stretches of marsh in the kingdom that of the Fenland itself promised the most spectacular transformation, but it was not to come for thirty years or so.

To the north of the Fenland there was a smaller tract of marsh that straddled the Lincolnshire–Yorkshire–Nottinghamshire boundary, and that included the Isle of Axholme and the Hatfield Levels. Here, conditions resembled those of the Fenland, with fishing, fowling and pasturing, and with settlements sited on the islands.

East Anglia. The open-field system extended on to the light and medium soils of East Anglia, but with differences. One difference was that the strips constituting a holding did not lie widely scattered over two or three large open fields, but were concentrated in one particular part of the arable, maybe in adjacent furlongs. Thus a 1583 survey of Castle Acre in north-west Norfolk shows that it had three fields – East, Middle and West; but there the resemblance to a Midland village ceased, for 80% of one holding lay in Middle Field, about 75% of each of two holdings lay in West Field, and nearly 70% of a fourth in East Field.[1] Clearly the term 'field' was merely a topographical designation and had no functional

[1] H. L. Gray, 314–15.

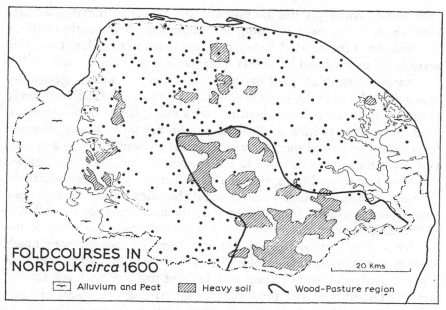

FOLDCOURSES IN
NORFOLK *circa* 1600

| ⌐~⌐ Alluvium and Peat | ▨ Heavy soil | ⌐ Wood-Pasture region |

20 Kms

Fig. 57 Townships with fold-courses in Norfolk *circa* 1600
     Based on: (1) K. J. Allison, 'The sheep–corn husbandry of Norfolk in the
     sixteenth and seventeenth centuries', *Agric. Hist. Rev.*, v (1957), 14; (2) *An
     economic survey of agriculture in the eastern counties of England* (Heffer, Cam-
     bridge, 1932), viii (for soil map).

significance. There was also considerable flexibility; adjoining strips in the
same field were not necessarily under the same crop, and there were even
strips of grass amidst growing grain.

    Another difference lay in the fact that East Anglian villages usually
contained two or three manors, and the flock of each manorial land had
a 'fold-course' extending over arable and heath within clearly defined
limits (Fig. 57).[1] Within the arable of each fold-course some areas must
have been under winter crops, some under spring crops, while others lay
fallow. During the autumn and winter both sheep and cattle grazed
generally over the stubble of the unsown fields and over the heath; but for
the rest of the year (from, say, March to October) access to the fallow
arable was strictly reserved for the manorial flocks, each within its

    [1] K. J. Allison, 'The sheep–corn husbandry of Norfolk in the sixteenth and seven-
teenth centuries', *Agric. Hist. Rev.*, v (1957), 12–30; A. Simpson, 'The East Anglian
fold-course: some queries', *ibid.*, vi (1958), 87–96.

fold-course. Although this necessitated the labour of moving hurdles, it did secure an intensive manuring of the light soil. 'Some of the thriftless convenience of the midland system may have been sacrificed, but superior agricultural method and profitable sheep-raising were compensations.'[1]

A variant of this East Anglian open-field system was an arrangement by which the arable of a village was divided into two parts – a small intensively cropped infield near the settlement, and a much larger outfield on the heath or waste; a portion of the latter was broken up each year, cropped for a season or so and then allowed to revert to heath.[2] It was a system that was later to give its name to the Breckland, the area of the lightest and most hungry soil in East Anglia. One feature of the heathlands was the large number of rabbit warrens, and rabbit skins formed an important element in the exports from Blakeney on the north coast of Norfolk.[3]

The sheep–corn husbandry did not characteristically extend on to the heavier soils of central Norfolk. Here, pasture and meadow were much more important than arable; fold-courses were rare (Fig. 57); more land was enclosed; and cattle-rearing and dairying together with pig-rearing were important activities producing butter, cheese and bacon. It was a district that had much in common with the 'wood pasture district' of Suffolk.

South-east England. A large part of south-eastern England was almost entirely enclosed in 1600. Essex and Kent had been described in 1549 as counties 'which be most enclosed'. To them could be added much of Suffolk, Hertfordshire, Surrey and Sussex (Fig. 54). Some of this enclosed land may have been taken directly from the waste and wood; others parts had been enclosed, sometimes recently, from irregular open fields by agreement in a piecemeal fashion. Unenclosed parcels of strips were still to be found lying within enclosed fields. They were not the relics of a two- or three-field system; they were not subject to common grazing by the livestock of all the tenants; and they were tilled on a variety of rotations.[4]

---

[1] H. L. Gray, 329.

[2] J. Saltmarsh and H. C. Darby, 'The infield–outfield system on a Norfolk manor', *Econ. Hist.*, III (1935), 30–44; M. R. Postgate, 'The field system of Breckland', *Agric. Hist. Rev.*, X (1962), 80–101.

[3] B. Cozens-Hardy, 'The maritime trade of the port of Blakeney, Norfolk, 1587–90', *Norfolk Rec. Soc.*, VIII (1936), 19.

[4] A. R. H. Baker, 'The field systems of an East Kent parish (Deal)', *Archaeologia Cantiana*, LXXVIII (1963), 96–117; 'Open fields and partible inheritance on a Kent manor (Gillingham)', *Econ. Hist. Rev.*, 2nd ser., XVII (1964), 1–23; 'Field patterns in

Between this area of almost complete enclosure and the Midland com-mon-field area there was a belt of country with irregular field arrange-ments and some enclosure – in the Chiltern parts of Hertfordshire, Buckinghamshire and Oxfordshire, and in Middlesex, east Berkshire and Surrey. In some parishes there were no traces of open field; in others, as well as closes, there were up to a dozen or so small open fields with strips lying in only a few of these, sometimes in only one field. As in the more fully enclosed area, such arrangements were very different from those of the two- or three-field systems, although in the Chilterns there were rights of common grazing over the fallow.[1]

An important element in the agriculture of the south-east was the demand of the London food market, and the large amount of enclosed land facilitated the development of individual practices and an intensified commercial agriculture. Access was provided by the many roads that converged upon the city and also by river and coastwise routes. Local farmers dealt with drovers and merchants who frequented the central London markets. Grain was a considerable crop in the Chilterns and in Essex and Kent; the soil was kept fertile by London's refuse and dung, and by sheep grazed on downs and waste and folded at night on the arable. A convertible husbandry was frequently practised. On most farms of the enclosed district of central Suffolk sheep were outnumbered by cattle, and some of the largest dairy herds in the country were to be found here. Cheese and butter were marketed at places like Woodbridge and Ipswich, and were dispatched by sea to London.

There was, in addition, a variety of land use associated with differing geographical circumstances. On the numerous heaths and commons there were extensive sheep-walks – on Blackheath and Hounslow Heath, for example, and on the Suffolk Sandlings. Parts of the heathland were en-closed for temporary tillage, and parts were occupied by numerous rabbit warrens and cattle commons. The heathy upland of the High Weald also carried many sheep, but was especially noted as a nursery of cattle. Along the coast the marshes were mostly under grass for cattle and sheep brought from elsewhere – on Romney Marsh, on the Pevensey Levels and on the marshes along the Thames estuary. In Kent, there were

seventeenth-century Kent', *Geography*, L (1965), 18–30; 'Field systems in the vale of Holmesdale', *Agric. Hist. Rev.*, XIV (1966), 1–24.

[1] D. Roden and A. R. H. Baker, 'Field systems of the Chiltern Hills and of parts of Kent from the late thirteenth to the early seventeenth century', *Trans. and Papers, Inst. Brit. Geog.*, XXXVIII (1966), 76 and 85.

numerous cherry and apple orchards, especially in the Maidstone district and to the north between Rainham and Blean. Hopfields, too, were frequent here and elsewhere – in Surrey, Essex and Suffolk. Among other specialised crops were saffron which gave its name to Saffron Walden in Essex; weld, 'the dyer's weed' used for making yellow dye, which flourished on the chalklands near Canterbury and Wye; and carrots which were characteristic of the Suffolk Sandlings. In Middlesex there were dairy farms producing milk, cheese and butter, and from many places nearby large quantities of hay were brought to the Haymarket for the many horses in the city.

One of the most visible signs of the presence of London was the spread of market gardening, maintained to a great extent by immigrants from the Low Countries. Closes were being transformed into market gardens at nearby places such as Lambeth, Fulham, Putney, Whitechapel, Stepney and Greenwich, all within easy access of the city's manure. Among the produce were cabbages, carrots, parsnips, turnips and cauliflowers, some for the table and some to support the stall-fed cows that helped the milk supply of the city. That the London Gardeners Company received its charter in 1605 is indicative of this new element in the scene around the city.

Forests, woodland and parks

The acreage once subject to forest law had greatly diminished as the result of successive disafforestations. Some forests had completely disappeared in the sense that their territories had been relieved of forest jurisdiction and of control by an army of royal officials, the verderers, the regarders, the bow-bearers and others. In other forests a piecemeal nibbling had taken place; large or small tracts had been enclosed in return for payment, and were now in tillage or pasture. Even so, the amount of land under forest law was still considerable (Fig. 56), and Camden was prompted to reflect upon the waste involved. 'It is incredible', he wrote, 'how much ground the kings of England have suffered every-where to lie waste and have set apart and enclosed for deer.'[1] The red deer, the fallow deer and the roe deer were plentiful and they provided venison for the royal household and for the king's friends. Other beasts of the chase were the fox and the marten. The wild boar was still to be found in Lancashire, Durham and Staffordshire, but the wolf had disappeared from England, though not from Scotland.

[1] W. Camden, *Britain, or a chorographicall description* (London, 1610), 293.

In spite of Camden's strictures, some good had resulted from the severity of the game laws in that much timber had been saved from the destruction involved in clearing land for agriculture and for the needs of industry. This was now available to meet the growing demand for timber for shipbuilding as a result of Tudor maritime expansion. The navy drew its timber from a number of royal forests, but its three main sources were the Forest of Dean, the New Forest and Alice Holt Forest in north-east Hampshire. The royal forests in general were far from being effective sources of supply, and contemporary surveys of the royal forests, such as that of 1608,[1] show large numbers of decaying trees as well as those fit for naval timber. Much of the supply came from private estates and from the woodlands of the nobility and landed gentry. An oak tree was at its best for shipbuilding when it was about one hundred years old, and there was much complaint about the premature cutting of half-grown oaks to meet unusual financial demands. The sale of timber provided the easiest way of obtaining ready money in the days of entailed estates.

During the sixteenth century, a number of Acts of Parliament had tried to restrict the cutting of timber for industrial purposes, and by 1600 there were many complaints about the loss of wood as a result of such activities as iron-making, glass-making and the provision of bark for tanning. William Harrison about 1580 had pointed to the fact that a man could often ride ten or twenty miles and find very little wood, or even none at all, 'except where the inhabitants have planted a few elms, oaks, hazels or ashes about their dwellings' as a protection 'from the rough winds'.[2] Arthur Standish, in 1611, pointed to the destruction that had taken place during the preceding twenty or thirty years, and urged that trees should be replanted.[3] Many of these complaints may well have been exaggerated. There were still great woods of beech and hazel as in the Chilterns and eastern Berkshire, and of oak as in the Weald. There were also many other districts characterised by a mixture of pasture and variegated forest, groves and glades, with stands of oak, ash, hazel, hawthorn, elm, birch, beech and holly. The fact was that not only had the woodland been reduced by clearance, but that the demand for timber had grown greater in pace with an increasing population and a general quickening of economic life. The real shortage was yet to come.

[1] Reprinted in H. C. Darby (ed.), *An historical geography of England before A.D. 1800* (Cambridge, 1936), 398–9.
[2] W. Harrison, 275.
[3] A. Standish, *The commons complaint* (London, 1611).

10

DHG

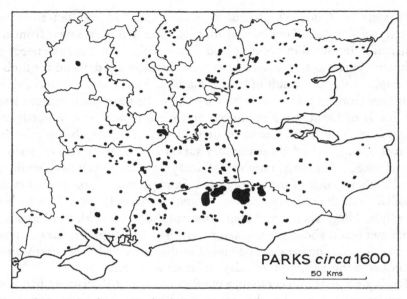

Fig. 58 Parks in south-east England *circa* 1600
   Based on the county maps in John Speed, *The theatre of the empire of Great
   Britaine* (London, 1611).

There were, moreover, some landowners who refused to sell their
trees and who preferred to keep them for ornamental purposes. This was
so in the many deer parks that had come into existence around, or near to,
the country houses of the realm. There were many royal deer parks;
and, moreover, established or aspiring landowners, rich merchants and
wealthy wool-staplers among them, made parks to signify their status.
The parks were generally enclosed, wrote William Harrison, 'with
strong pale made of oak' to prevent the deer from escaping.[1] The fences
were sometimes raised on earth banks, often with a circumference 'of four
or five miles, and sometimes more or less'. They are marked on the
county maps of the time (Fig. 58); and they number nearly 800 in England,
with another 30 or so in Wales.[2] But already many were disappearing.
Richard Carew in his *Survey of Cornwall* (1602) pointed to a number that
had disappeared within living memory and said that some owners had
abolished the deer 'to give the bullockes place'. It was more profitable to

[1] W. Harrison, 254–5.
[2] H. C. Prince, *Parks in England* (Shalfleet, I.O.W., 1967). 1–2; E. P. Shirley,
Some account of English deer parks (London, 1867).

breed bullocks than to graze deer. Even when parks were retained or created, they were reduced in area to provide land for grazing or tillage. Although deer parks were still numerous in 1600, they had begun to decline.

INDUSTRY

Manufacturing industry did not figure very prominently in the life of the English towns, apart from London and such clothing centres as Norwich and Colchester, together with a few other centres of growing industry. The occupational structure of most towns was varied, with no special emphasis on manufactures, and reflected the regional character of farming. Thus at Leicester, according to probate inventories made between 1560 and 1599, the principal occupations were as follows:[1]

	Per cent
Leather crafts (tanners, glovers, etc.)	19
Textiles (weavers, etc.)	19
Husbandry (graziers, etc.)	19
Victualling crafts (butchers, etc.)	14
Housing (carpenters, joiners, glaziers, etc.)	9
Retail trades (mercers, grocers)	6
Miscellaneous	14

A similar pattern was found in such provincial towns as Gloucester, Cirencester and Tewkesbury; they were not manufacturing centres, but handled the finishing processes and served as markets and sources of supply for the countryside.[2]

It follows that industries were to be found in rural areas, where they were very widely dispersed. The reasons for this are not difficult to see. In the first place, mechanical processes were generally carried out by means of water power, with which the country was well and generally supplied. Streams turned water-wheels for smelting tin in Cornwall, for draining collieries in Nottinghamshire, for forging iron in Sussex, and for fulling cloth in Lancashire. Likewise an important source of fuel was still charcoal, especially abundant in the wooded areas of late settlement in the Midlands or the Weald. Secondly, not only were the major raw materials like wool or iron well distributed, but so too was the population

[1] E. W. Kerridge in R. A. McKinley (ed.), *V.C.H. Leicestershire*, IV (1959), 178.

[2] A. J. and R. H. Tawney, 'An occupational census of the seventeenth century', *Econ. Hist. Rev.*, V (1934), 37–8; L. A. Clarkson, 'The leather crafts in Tudor and Stuart England', *Agric. Hist. Rev.*, XIV (1966), 25–39.

and the domestic market. London, as a great centre of population and trade, was exceptional in attracting to itself a variety of industry. One example must suffice: the copper and brass works sited on the Thames between Isleworth and Worton in Middlesex and shown on Norden's map as 'Coppermills'.

Some weight should also be given to the social reasons suggested by Joan Thirsk in her study of the handicraft industries that developed in communities already engaged in farming.[1] There was a coincidence of these industries with a populous society of small farmers, often mainly free-holders pursuing a pastoral economy such as dairying. This was typical of places as varied as north Wiltshire, central Suffolk, the Kentish Weald, Westmorland, the Fens, and the Yorkshire dales. The population of these areas was sometimes increased by immigrants attracted by their ample commons and weak manorial organisation. It was often maintained by the partible inheritance of land, which kept people in their native places and gave rise to large and immobile populations. There were thus many men whose farming left them time to engage in the subsidiary occupation of a handicraft industry; dairying, for instance, did not require as much hand labour as arable farming.

It is possible to show fairly exactly the importance of rural industry in Gloucestershire, a county in which manufacturing was important. Although agriculture was preponderant, farming in fact occupied only some 50% of the adult male population; the proportion varied so much in different parts of Gloucestershire that it might be safer to acknowledge that 'agriculture and industry were inextricably intertwined. Not only corn and cattle, but corn, wool, cloth, and, in some districts, even corn, coal, and iron, were almost joint products.'[2]

With these general considerations in mind, the long-standing industries of cloth-making and iron-working must be examined, together with the more recently developed coal industry and also a variety of other industrial activities.

The textile industries

Woollen cloth was England's main export, and its manufacture the largest industry in the realm. The cloth was of widely different character, depending upon the nature of the wool used and the method of manufacture. In a general way it was of three kinds; broadcloths usually

[1] J. Thirsk in F. J. Fisher (1961), 70–88.   [2] A. J. and R. H. Tawney, 42.

of fine quality; kerseys which were lighter, cheaper and maybe coarser; and worsteds which, unlike the other two types, did not need fulling. A bewildering variety of names were used for various local types and products – arras, callimancoes, carrells, frisadoes, minikins, pomettes, stamelles and very many others. The general distribution of the industry was still largely that of the latter part of the Middle Ages; cloth-making, although widespread, was associated mainly with three areas, the West Country, East Anglia and Yorkshire and Lancashire (Fig. 49).

The West Country included two cloth-making districts. Broadcloth was made over an area comprising the Cotswolds and extending into Wiltshire, Oxfordshire and Somerset. Kerseys, on the other hand, were made in west Somerset and Devon. In the broadcloth area Gloucestershire held a foremost place. Of all the spring-fed streams flowing across the Cotswold scarp, those of the Frome, the Cam and the Little Avon were the most constant and furnished the most reliable supply of power for water-wheels. The water was also at its softest here, best suited to the cleansing, scouring and fulling processes. Another natural advantage lay in the deeply incised valleys with their floors on Upper Liassic Clay; they were well suited to the making of reservoirs for large fulling mills, and, later, for the dye-houses for coloured cloth that supplanted the unfinished broadcloth. We are fortunate in being able to plot the distribution of the Gloucestershire industry for the year 1608 from John Smith's census of occupations (Fig. 59).[1] In the villages around Stroud in the Frome valley, around Dursley in the Cam valley and around Wotton-under-Edge, as many as 40 to 50% and over of the able-bodied men were employed solely in cloth-making. Such densities of clothiers, weavers, fullers and dyers had emerged primarily because of local advantages in water for power and washing.[2]

The location of cloth-making was not usually influenced by special resources of raw materials, because wool, fuller's earth and dyestuffs were generally available and few areas were decisively well off in them. From Gloucestershire the industry extended into the adjoining parts of Oxfordshire, where there were fulling mills at Witney and other places along the Windrush, and into Wiltshire where the Avon and its tributaries provided

[1] *The names and surnames of all the able and sufficient men in body...in 1608...* compiled by John Smith (London, 1902).

[2] R. Perry, 'The Gloucestershire woollen industry, 1100–1690', *Trans. Bristol and Gloucs. Archaeol. Soc.*, LXVI (1945), 49–137; J. Tann, 'Some problems of water power: a study of mill siting in Gloucestershire', *ibid.*, LXXXIV (1965), 53–77.

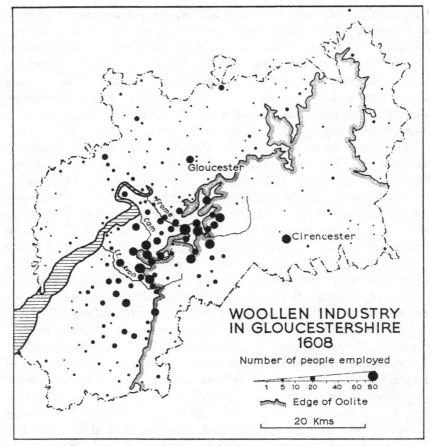

Fig. 59 The woollen industry in Gloucestershire, 1608
Based on *The names and surnames of all the able and sufficient men in body . . . in 1608 . . . compiled by John Smith* (London, 1902).

power for places extending from Malmesbury to Bradford on Avon and so to Bath. Across the border the industry continued into Somerset, where places such as Frome, Wells and Shepton Mallet developed specialities.[1]

Westwards in Somerset, at Taunton and Wellington, kerseys were made, and they were also a characteristic product of Devonshire with its

[1] E. M. Carus-Wilson and J. de L. Mann in E. Crittall (ed.), *V.C.H. Wiltshire*, IV (1959), 133, 139, 141, 147, 150–1; E. M. Hewitt in W. Page, *V.C.H. Somerset*, II (1911), 408, 411.

numerous streams suitable for fulling mills. Exeter, Crediton, Tiverton, Totnes and a large number of other places developed a degree of specialisation as between spinning and weaving. By 1600, the 'new draperies' of East Anglia had scarcely appeared in Devonshire except in and around Barnstaple in the north, but soon kerseys were to be supplemented and supplanted by serges or 'perpetuanos', so called for their long-lasting quality.[1]

In East Anglia, two separate districts were outstanding for cloth-making. In the north, Norwich was the centre of a group of towns and villages engaged in producing worsted for which the fulling mill and water power were not required; a few weavers of other kinds of textiles worked in Norwich itself (Fig. 60). In the south, along the Suffolk–Essex border, a number of villages made coarse cloths such as baize and kerseys. They included places such as Clare, Long Melford, Lavenham, Sudbury, Hadleigh and others situated along the south-eastward flowing rivers; this area also supplied yarn for the Norfolk weavers, and the manufacture of linen was gaining in importance. At the middle of the sixteenth century the industries of both East Anglian areas were in a depressed condition until all was changed as a result of the admission of Dutch and Walloon weavers fleeing from religious persecution. A few came to Norwich in 1554, and more in 1565. Others came to Colchester about 1570 and to other places. Their numbers increased in the years that followed. The manufactures they introduced included a varied range of cloth under a variety of names and spellings–bays, sayes, barracans, rashes, shalloons, bombazines and 'other outlandish commodities'. These were some of the lighter and finer fabrics of the so-called 'new draperies' and they did not involve fulling. By 1600 the East Anglian cloth areas were once more in a flourishing condition.[2]

The third main area for cloth-making was in the north. Here, the older 'clothing' towns, such as York, Beverley and Selby, had lost much of their industry; but other centres were rising, favoured by the abundance of water power for fulling mills. In the West Riding, a vigorous manufacture of coarse kersey, some coloured, was carried on at Halifax, Wakefield, Leeds and other places. The flourishing condition of the inhabitants, wrote Camden, 'confirms the truth of that old observation

[1] W. G. Hoskins (1954), 127; P. J. Bowden, *The wool trade in Tudor and Stuart England* (London, 1962), 50.

[2] G. Unwin in W. Page (ed.), *V.C.H. Suffolk*, II (1907), 249, 255, 257–8, 262, 267; D. C. Coleman, 'An innovation and its diffusion: the new draperies', *Econ. Hist. Rev.*, 2nd ser., XXII (1969), 417–29.

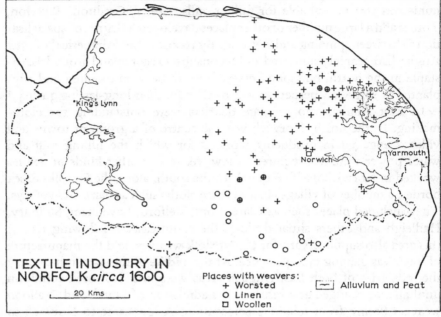

Fig. 60 The textile industry in Norfolk *circa* 1600
        Based on K. J. Allison, 'The Norfolk worsted industry in the sixteenth and
        seventeenth centuries: I. The traditional industry', *Yorks. Bull. Econ. and
        Soc. Research*, XII (1960), 75, and on additional information provided by
        Dr Allison.

that a barren country is a great whet to the industry of the natives'.[1]
On the other side of the Pennines, the manufacture of coarse cloth was
also growing around Manchester, Bolton and Rochdale, towns with few
gilds and few special restrictions on trade and therefore well suited to
be centres for the finishing and distributive sides of the industry; most of
the weavers lived in the countryside around. By 1600 the making of
kerseys had been supplemented by that of bays, one of the new draperies.
A more important innovation was the making of fustian with a linen
warp and a cotton weft; this seems to have been one of the new draperies
that never flourished in East Anglia. This early introduction of cotton
was to be of the greatest consequence for Lancashire, although all-cotton
goods were not much made there for some years after 1600.[2]

[1] Quoted in H. C. Darby (1936), 374; M. Sellers in W. Page (ed.), *V.C.H.
Yorkshire*, II (1912), 413–16.
[2] W. H. Chaloner and A. E. Musson, *Industry and technology* (London, 1963), 23;

The iron industry

Coal was not yet used in the iron-smelting industry, the chief centre of which was the Weald in south-eastern England (Fig. 61). There was plenty of ironstone in the Weald Clay and the Hastings Beds (principally in the Wadhurst Clay and the Ashdown Sands); the upper courses of the rivers, such as the Rother, Cuckmere and Ouse in Sussex, had deeply incised valleys and were therefore easily dammed, giving a steady head of water power for the bellows and hammers; finally, fuel was abundant locally in the form of charcoal burned from the brushwood of this wooded region. The Weald was also reasonably well placed to supply the London market. The blast furnaces were usually built of stone with buttresses and a bracing framework of wooden beams, about 18 ft. high. They were worked by two powerful pairs of bellows, in turn operated by a water-wheel; and they produced sows and pigs of cast-iron. These were re-heated and hammered at the finery, forge, or ironmill, thus giving blooms and bars of malleable wrought-iron which were suitable for general use by the smiths. Castings could also be made direct from the furnaces, including cannon and shot (for which the Wealden ironworks were famous) and domestic articles like iron firebacks.[1]

The supremacy of the Wealden area was still unchallenged in 1600, when it had 49 furnaces out of a total of 73 in England. There were also two furnaces not far away in Hampshire. An annual output of between 100 and 200 tons per furnace was usual. By then there were eleven in the west Midlands with its local Coal Measure ironstone and with charcoal from Wyre Forest. The 'free miners' of the Forest of Dean were slow in adopting the blast furnace, and there were only three in this area by 1620.[2] They nevertheless made much pig-iron for sale upstream along the Severn to the metalcraft centres of the west Midlands. South Staffordshire and north-east Worcestershire constituted the largest nail-producing area in England. Other products included such things as stirrups and spurs for horses, locks, fire-arms, cutlery and edged tools and a variety of ironmongery. We must envisage this as 'a countryside in

A. P. Wadsworth and J. de L. Mann, *The cotton trade and industrial Lancashire, 1600–1780* (Manchester, 1931), 14–16.

[1] M. C. Delany, *The historical geography of the Wealden iron industry* (London, 1921), 8–9, 19–21, 27–9, 32–3, 36–42; E. Straker, *Wealden iron* (London, 1931), 101–40; D. W. Crossley, 'The management of a sixteenth-century iron works', *Econ. Hist. Rev.*, 2nd ser., xix (1966), 273–88.

[2] C. Hart, *The industrial history of Dean* (Newton Abbot, 1971), 8.

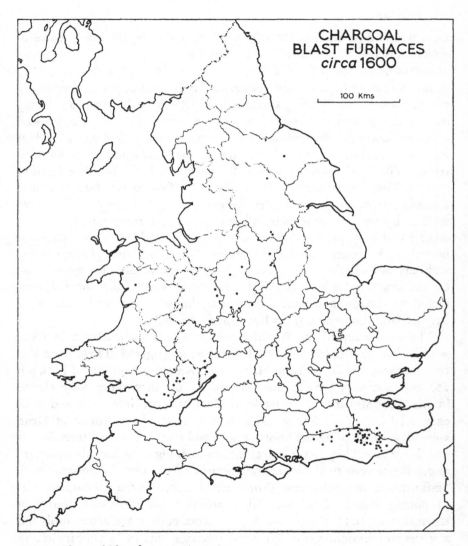

Fig. 61 Charcoal blast furnaces *circa* 1600
        Based on H. R. Schubert, *History of the British iron and steel industry from
        c. 450 B.C. to A.D. 1775* (London, 1957), 354–92.

course of becoming industrialised; more and more a strung-out web of iron-working villages, market towns next door to collieries, heaths and wastes gradually and very slowly being covered by the cottages of the nailers and other persons carrying on industrial occupations in rural surroundings'.[1] But populous centres were beginning to emerge – places like Stourbridge, Dudley and Walsall, and especially Birmingham. Camden found Birmingham 'swarming with inhabitants, and echoing with the noise of anvils, for here are great numbers of smiths', and its tanneries and fulling mills were being displaced by forges and blade-mills. It may then have had a population of not more than about 2,000.[2]

There was another group of blast furnaces in an area stretching from east Derbyshire into south Yorkshire, also using Coal Measure ironstone. Associated with iron locally produced there was a variety of metal trades – nails, guns, ironmongery and, in particular, cutlery. The cutlers of Sheffield already had an established reputation throughout the country.

Coalmining

Coalmining had grown phenomenally during the sixteenth century, and by 1600 all the fields, apart from those of Kent, were being worked to some extent. Many of the monastic lands in northern and Midland counties which had passed into lay hands by 1540 were rich in coal, and their new owners had a thoroughly enterprising attitude to mining the seams. Long leases encouraged the heavy investment of capital in ambitious works; the Willoughbys of Wollaton in Nottinghamshire, for example, were putting nearly £20,000 a year into collieries and ironworks by 1600, installing rag-and-chain force-pumps to help draw the water from the deeper pits.[3] But it would be a mistake to assume that large-scale workings of this kind were typical of the industry as a whole. Far more characteristic was the colliery worked for the earl of Shrewsbury in Sheffield Park. Five men were generally employed in pits no deeper than 90 feet, producing between 1,200 and 1,400 tons a year; it was sold chiefly to the metal trades at Sheffield at a net profit of no more than £50 a year. Water was drained

[1] W. H. B. Court, *The rise of the midland industries, 1600–1838* (Oxford, 1938), 22; H. R. Schubert, *History of the British iron and steel industry from circa 450 B.C. to A.D. 1775* (London, 1957), 174–5 and 179–82.

[2] C. R. Elrington, P. M. Tillott and D. E. C. Eversley in W. B. Stephens (ed.), *V.C.H. Warwickshire*, VII (1964), 6–7, 81–3.

[3] W. H. Chaloner and A. E. Musson, 23; J. U. Nef, *The rise of the British coal industry*, I (London, 1932), 324.

away by adit or sough, but there was little machinery at pit-head or underground; much timber was used, being obtained freely from the earl's park. Output fluctuated with the high degree of absenteeism, for not only did the colliers observe a large number of feasts and fairs as holidays, they also followed other employment, probably on the land. 'The unmechanized, cheap, unenterprising little concern such as that at Sheffield', bringing in a regular income, was very common in the English coalfields.[1]

Coal was used not only in metal work at the forge but in a variety of other industries – in salt-making, soap-boiling, sugar-refining, glass-making, dyeing and brewing. It was also used in place of wood as domestic fuel, and a radical change in the method of heating houses had taken place over a short period. It is true that even late in Queen Elizabeth's reign some people objected to the smell of coal fires, but improved fireplaces and chimneys were helping to make coal the general fuel of the country. Moreover, not only was the consumption per head rising but the total population was increasing.

The cost of land carriage prohibited the haulage of coal for long distances, but much was transported along waterways such as the Trent and the Severn, and also around the coast by sea. In this respect the Northumberland–Durham field was well placed and its development had been particularly outstanding. Some of its mines were so situated that coal could be loaded direct into ships and the export from Newcastle alone reached about 160,000 tons by 1600. About a fifth of this went to foreign countries, to France, the Netherlands and Germany. The remainder went to ports along the south and south-east coasts, and especially to London, which, with about 250,000 inhabitants, provided a vast market in the context of the time. The average size of a cargo of coal increased from 56 tons in 1592 to 73 tons in 1606. A prodigious number of vessels bearing coal crowded the river below London Bridge. The number of lightermen and seamen employed was considerable, while the demand for coal-carrying ships was revolutionising the shipbuilding industry.[2]

Other industries

Other 'furnace industries', and the mines that kept them going, were well represented in the upland counties of northern and western England. In

[1] L. Stone, 'An Elizabethan coal mine', *Econ. Hist. Rev.*, 2nd ser., III (1950–1), 105.
[2] J. U. Nef, 390; J. Simmons, *Transport: visual history of modern Britain* (London, 1962), 17.

Derbyshire, the local justices of the peace felt that farmers were in danger of being outnumbered by the industrial workers: 'Many thousands live in work at lead mines, coal mines, stone pits, and iron works', becoming increasingly dependent for their food on Danzig corn brought from Hull along the Trent.[1] Derbyshire lead was transported downstream to Hull, the Peak being a main producer. The demand for lead had greatly increased; at home, the wave of new building meant that more lead was needed for roofing and plumbing, while (with tin) it was used by the pewterers; abroad, lead was one of England's traditional exports.[2] Shallow mines were thickly aligned along the veins of lead ore in the Carboniferous Limestone, often on moorlands up to 1,000 ft. above sea-level, between the rivers Dove and Derwent. The returns of the local justices showed that there were few parishes without lead mining, 'whereof our Hundred of High Peak hath much employment'. Around Wirksworth there was 'so great a multitude of poor miners' that the justices were compelled to tolerate a large number of ale-houses.[3] Lead was also mined in the northern Pennines, in the district around Alston, and again to some extent in the Lake District. Another lead-producing district lay in the Mendips, which, said Camden, were rich in lead mines.[4]

The West Country was also renowned for its tin mines, some in Devon but mostly in western Cornwall. Mine-shafts driven into the lodes (some of them worked by German miners) were replacing the streaming of placer deposits as traditionally practised in the eastern parts of Cornwall. The lodes of metallic ores were most abundant and at their richest in the narrow zone of contact between the slates and each of the granite masses, especially the western blocks of St Just, Carnmenellis, and Hensbarrow. Thus the western coinage towns of Helston and Truro, where the blocks of white tin received the Duchy seal as a guarantee of quality, were handling eight times as much tin as Liskeard and Lostwithiel in the east. The Cornish miners formed a distinct community, regarded somewhat unsympathetically at the time as 'ten thousand or twelve thousand of the roughest and most mutinous men in England'.[5] Carew described how they were 'let down and taken up in a stirrup by two men who wind

[1] J. Thirsk in F. J. Fisher (1961), 73.

[2] L. Stone, 'Elizabethan overseas trade', *Econ. Hist. Rev.*, 2nd ser., II (1949), 45.

[3] J. H. Lander in W. Page (ed.), *V.C.H. Derbyshire*, II (1907), 177–8, 331.

[4] W. Camden, 230. See J. W. Gough, *The mines of Mendip* (Oxford, 1930), 112.

[5] A. L. Rowse, *Tudor Cornwall* (London, 1941), 54–62; W. S. Lewis, *The West of England tin mining* (Exeter, 1923), 5–6, 10, 39.

the rope', even where some of the workings were 300 feet deep. They toiled by candle light with pick-axe and wedges, draining their mines by means of 'sundry devices, as adits, pumps, and wheels driven by a stream'. Water power was as essential here as in the clothing districts: it worked the stamping-mills and crazing-mills that pulverised the ore; it washed the black tin sufficiently clean for the blowing-houses, and this was 'melted with charcoal fire, blown by a great pair of bellows moved with a water-wheel'.[1]

There was also a deliberate policy of trying to reduce England's dependence on imported copper, which brought new industries into being. Helped by German capital and hundreds of skilled workmen from Saxony, copper ore was mined and smelted at Brigham, near Keswick in Cumberland, under the auspices of the Society of Mines Royal, incorporated in 1568.[2] At the same time calamine, the ore of zinc, was found and worked in the Mendips, thus furnishing the means of making brass (from copper and zinc). Brass foundries and batteries were set up by the Society of Mineral and Battery Works.

The manufacture of salt was marked by the increasing use of coal for boiling. From early times there had been salt-pans at a large number of places around the coast, but in 1600 salt-making was especially flourishing at the mouths of the Tyne and Wear where coal was easily accessible and cheap. There were also important sources inland, some were in Cheshire where 'great store of white salt' was made at Nantwich, Northwich, and Middlewich; the brine was taken from the pits in wooden troughs to the wich-houses, where it was 'seethed in lead cauldrons'.[3] Charcoal was still the main fuel here and also at the Droitwich saltworks in Worcestershire; 'what a prodigious quantity of wood these salt works consume', complained Camden, 'though men be silent, yet Feckenham-forest, once very thick with trees...by their thinness declare'. William Harrison devoted a whole chapter to salt-making.[4]

A variety of other industries were also prospering in 1600, and were able to do so largely because of the substitution of coal for wood as a fuel. Glass-making, which had benefited from the French methods

[1] F. E. Halliday (ed.), *Richard Carew of Antony: The survey of Cornwall* (London, 1953), 92–4.

[2] C. M. L. Bouch and G. P. Jones, 118–27.

[3] W. Smith, *The particular description of England, 1588*, ed. by H. B. Wheatley and E. W. Ashbee (London, 1879), 9.

[4] W. Harrison, 375–8.

introduced in the 1550s, was shortly to be transformed by the cheaper process using an improved crucible and coal as its fuel. Glassworks were sited widely, from London to the New Forest, from the Weald to Staffordshire. Alum, much in demand as a fixing agent in dyeing and in curing skins, was mined and boiled at Whitby and in the Isle of Wight, again in an attempt to foster home production. Coal was also increasingly used in such activities as sugar-refining, soap-boiling and candle-making. Moreover, the transport of coal by sea 'probably accounts more than any other single factor for the growth of the English shipping industry in the late sixteenth and seventeenth centuries'.[1] The building of ships was already 'very much in practice' at Ipswich.[2] Some miscellaneous in-dustries had been established with the aid of foreign immigrants. Many of them brought new skills which, in Kent alone, resulted in silk-weaving in Canterbury, a paper mill at Dartford, an iron-plate mill at Crayford, and in Sheppey 'a certain Brabanter' used pyrites to make copperas (iron sulphate), used as a mordant or fixing agent in dyeing cloth. The pull of the metropolitan market may also be seen in the siting of many paper mills along the Wye valley near High Wycombe.[3]

TRANSPORT AND TRADE

Roads and rivers

Tudor England inherited a well-developed road system based upon London, and incorporating many roads of Roman origin such as Watling Street. There were also cross-country highways such as that from Bristol to Gloucester and so by the Severn valley to Chester. These were supple-mented by a network of many minor roads, most frequent where the market towns were most numerous. The extent to which all these various roads were used is only now being realised. Most of Southampton's imports were distributed by carriers' services to London, Bristol, and even as far afield as Manchester or Kendal. It has been said that the transport of goods and passengers before 1500 'could be easily and efficiently under-taken by road throughout southern England and the Midlands';[4] but it

[1] J. U. Nef, 325.

[2] F. Hervey (ed.), *Suffolk in the XVIIth century. The breviary of Robert Reyce, 1618* (London, 1902), 97–8.

[3] D. C. Coleman, *The British paper industry, 1495–1860* (Oxford, 1958), 49–50; A. H. Shorter, *Paper mills and paper makers in England, 1495–1800* (Hilversum, 1957), 22–50, 92; F. W. Jessup, *A history of Kent* (London, 1958), 98, 105.

[4] J. Simmons, 6–8 and 11.

must be remembered that the roads were unmetalled and unfit for heavy traffic.

The pattern of main roads in England by 1600 is outlined in Fig. 62, from which the focal position of London can be appreciated. Major post-roads went from it to the extremities of the kingdom, the longest making its way via Stamford, York, and Newcastle to the Scottish border at Berwick. Another ran to Coventry and Chester, with extensions to Carlisle and to North Wales. Other roads led to Bristol or to Gloucester and South Wales. The great road to Land's End made its way through Salisbury and Exeter. A post-road went through Canterbury to Dover, by no means the only significant Channel port, as Rye was 'the chiefest for passage betwixt England and France'.[1] Most English regions were within easy access of a main road. The chief exception was in the north, where the network of intersecting roads stopped short at the trans-Pennine connection from York to Chester. North of this cross-country line only the western and eastern post-roads were recorded. The pattern seems to be sparse beyond Lincoln, but there was a much-used ferry into the East Riding at Barton-upon-Humber, and Camden noted a road from Doncaster crossing the Trent below Gainsborough: 'it is a great road for pack-horses, which travel from the west of Yorkshire, to Lincoln, Lynn, and Norwich'.

London's primacy in the national pattern was repeated on a lesser scale in those cities which acted as provincial capitals: William Smith in 1588 entitled his summary of the English roads 'The highways from any notable town in England to the city of London, and likewise from one notable town to another'. The influence of York or Coventry may be deduced from their road connections not only with London, but with half-a-dozen major towns besides. Places such as Exeter, Bristol, Salisbury and Gloucester each had four or five main roads converging on it. For natural reasons, however, stretches of these roads could become impassable in the winter: the Exeter–Bristol road had a short-cut from Bridgwater to Axbridge through the marshes, 'but no man can travel it well except it be in summertime, or else when it is a great frost'. River-crossings do not seem to have been a deterrent, for by using no fewer than six ferries it was possible to follow a road from Southampton to Helford in Cornwall 'all along the sea coast'.[2]

Another feature of the road pattern was the advantage shared by counties immediately to the north and west of London, in the angle

[1] W. Smith (1588, ed. 1879), 9.   [2] Ibid., 69–72.

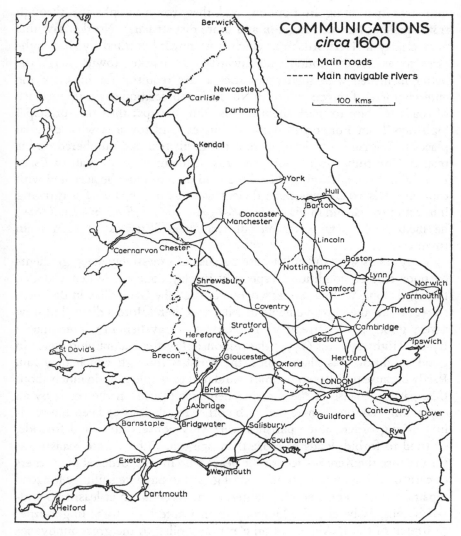

Fig. 62 Communications *circa* 1600

    Based on: (1) *Map of XVII century England* (Ordnance Survey, 1930);
    (2) T. S. Willan, *River navigation in England, 1600–1750* (Oxford, 1936),
    map 1.

formed by the roads to York and Bristol. These counties had alternative
means of marketing in London, and they could supply the through
traffic passing between London and more distant parts. Norden saw this
very clearly in Hertfordshire, which was 'much benefited by thorough-
fares to and from London northwards': its market towns were 'the
better furnished with such necessaries as are requisite for inns, for the
entertainment of travellers'. John Norden's appreciation of the importance
of roads led him to mark them on his county maps; thus the 'principal
highways from London through Middlesex, and towards what especial
places in England they lead' were listed in his text and numbered on his
maps.[1] Unfortunately the four variants of his manuscript map of Essex
reveal that his depiction of the whole road pattern must be accepted with
caution.[2] His road-maps nevertheless were symptomatic of a growing
interest in roads, and in his *An intended guyde for English travailers* (1625)
he included 'a new invention' of triangular distances between main
towns.

Many roads were difficult to use because of excessively steep gradients
or poor surface conditions, especially in clay country. When Camden
travelled along Watling Street where it crossed the Gault Clay in Bedfordshire, he saw the improvements in what was 'heretofore a dirty (but now
a very good) road, extremely troublesome to travellers in winter-time'.[3]
By the Highways Act (1555) the responsibility for maintaining roads in
good condition was placed upon the parishes through which they ran.
Partly because parishes sometimes could not, or would not, do much about
this, partly because of the pressing demands placed on the roads by an
expanding economy, there had to be other regulations to keep the roads
in working order. Some parishes were very conservative, and forbade,
or tried to forbid, loads of more than one ton. Others were realistic, as
for instance the Kent justices of the peace who in 1604 ordered the owners
of carts carrying more than one ton of goods to pay five shillings for road
repairs. Loads were becoming larger, some reaching to at least two-and-
a-half tons, 'whereby the highway from Canterbury to Sittingbourne'
(a stretch of the Dover Road on clay) 'is spoiled to the great annoyance
of all travellers'. Some coal was being moved on the Midland roads, taken

[1] J. Norden, *Speculum Britanniae: An historicall and chorographicall description of
Middlesex* (London, 1593).
[2] F. G. Emmison and R. A. Skelton, 'The description of Essex, by John Norden,
1594', *Geog. Jour.*, CXXIII (1957), 39.
[3] H. C. Darby (1936), 361.

by pack-horse trains along Watling Street from the Shropshire and Warwickshire coalfields, and (in 1581) carried 40 miles from Coleorton in Leicestershire. Wagons took heavy goods of all description: Ipswich, for example, had a thrice-weekly carriers' service to London. By the time John Taylor's *Carriers Cosmography* appeared in 1637, 'an organised system of goods transport' was in being.[1]

Increasing use was also made of navigable rivers. The only artificial canal (built with pound locks in 1564–6) bypassed obstructions in the estuary of the Exe, and joined Exeter with the sea at Topsham.[2] There were, however, a number of attempts to improve river courses themselves. Eight Acts of Parliament were passed for this purpose in the sixteenth century, such as that for improving the river Lea by a new cut (1571); the Lea supplied London with cheap corn and malt, but brought the bargemen into sharp conflict with millers, fishermen, and road-carriers who were content to see the river remain as it was.[3] For bulky loads such as grain, wool, coal, and building materials, river transport was far cheaper than road transport, especially on long journeys in the big trows and barges sailing the Severn or the Thames. Coal moved down the Severn and the Trent from the coalfields, and upstream from east-coast ports to areas served by the Thames, the Great Ouse, and the Welland.[4] The movement of grain, cheese, butter, and timber was contrary to this, going down the Ouse to King's Lynn, for example, or upstream to Nottingham on the Trent. There were at least 685 miles of navigable rivers, giving the country 'the foundation of its future system of internal water communications' (Fig. 62), but there were still 'great tracts of the country which lay more than 15 miles or one day's carriage by land from the sea or from a navigable river'.[5]

Maritime trade: coastal and overseas

Rivers either collectively or separately were not a self-contained transport system, because they were inseparable from the coasting trade (and, for that matter, the foreign trade) that passed through the English seaports.

---

[1] J. Simmons, 16–18.

[2] W. B. Stephens, 'The Exeter lighter canal, 1566–1698', *Jour. Transport History*, III (1957), 1.

[3] W. T. Jackman, *The development of modern transportation in England*, I (Cambridge, 1916), 165–8; G. B. G. Bull, 'Elizabethan maps of the lower Lea valley', *Geog. Jour.*, CXXIV (1958), 357–8.

[4] T. S. Willan, 'Yorkshire river navigation, 1600–1750', *Geography*, XXII (1937), 189.

[5] T. S. Willan, *River navigation in England, 1600–1750* (Oxford, 1936), 133.

The evaluation of inland navigation 'must take into account both the rivers and the sea'. The sea, so to speak, was 'merely a river round England, a river with peculiar dangers, peculiar conditions and peculiar advantages'.[1] A score of head ports and their many members were situated fairly regularly around the coastline, and few regions were unable to enjoy the very cheap transport provided by coasters.[2] Above all, the eastern ports were influenced by the London market, with its gigantic appetite for grain and coal. Newcastle sent each year, salt, glass and over 70,000 tons of coal. Hull shipped lead, iron and much else primarily to London; from King's Lynn went a constant flow of grain, malt, and butter. London itself shipped all kinds of goods to, and received cargoes from, almost every other port of any size in England and Wales.[3]

In foreign trade, furthermore, London's share 'was greater than that of all the provincial ports put together', and comprised between two-thirds and three-quarters of the nation's trade abroad.[4] Its superiority overshadowed Newcastle's shipments of 27,000 tons of coal a year to Emden, Hamburg, and other north European ports, or Bristol's trade with Ireland, France, and Spain (yielding customs dues of £2,112 in 1598–9). Of the other significant provincial ports, the ranking in 1594–5 (based on customs revenue for both imports and exports, but excluding prize cargoes) was Exeter (with its members Dartmouth and Barnstaple), Sandwich (and Dover), Hull, Ipswich, and Southampton. The London and provincial trades alike rested on the export of a few staple products, especially cloth, and the import of a wide range of raw materials, manufactured goods, foodstuffs and wines.[5] By 1600 the most striking development was the growing share of eastern ports in the export of grain, to the order of about 75% of the national whole. Another feature was the concentration of trade in fewer ports, as in King's Lynn and Yarmouth with their handling of 80% of the malt and barley, rye and wheat shipped from East Anglia.[6]

[1] T. S. Willan (1936), 5.
[2] W. Smith (1588, ed. 1879), 18, 50.
[3] T. S. Willan, *The English coasting trade, 1600–1750* (1938), 111, 120, 126, 167, 192.
[4] T. S. Willan, *Studies in Elizabethan foreign trade* (Manchester, 1959), 65; F. J. Fisher, 'London's export trade in the early seventeenth century', *Econ. Hist. Rev.*, 2nd ser., III (1950), 151–61; L. Stone, 'Elizabethan overseas trade', *Econ. Hist. Rev.*, 2nd ser., II (1949), 30–58.
[5] T. S. Willan (1959), 90.
[6] A. Everitt in J. Thirsk (1967), 525–6.

TOWNS AND CITIES

John Speed in *The theatre of the empire of Great Britaine* (1611), besides bringing together fifty town plans, marked on his county maps the market towns 'fit for buying and selling and other affairs of commerce'. They numbered 605, but a recent estimate has put the total at 760 for England with another 50 for Wales.[1] They varied greatly in size downwards from London, which may have had a population of about 250,000. Chief among the provincial centres was Norwich with about 15,000 people, and there were four other regional capitals each with a population of about 10,000 or over – York, Bristol, Newcastle upon Tyne and Exeter. Below these came another ten or so substantial towns each with a population of 5,000 or more. They were followed by about thirty towns with over 3,000 inhabitants apiece, and these were succeeded in turn by a variety of smaller towns, often with under 2,000, and sometimes with only 1,000 people or less.

A contemporary estimate placed the number of walled towns at about 100.[2] The walls had, for the most part, ceased to be defensible, but on the Scottish Border and along the coastline, walls had played a part in the anti-invasion plans of Elizabeth. Thus at Berwick 'the utmost town in England and the strongest hold', the earlier walls had been replaced by elaborate defences complete with Italianate fortifications and gun-emplacements.[3] Some towns were still confined within their ancient walls as at Nottingham and Northampton, but Speed's maps show that quite a number had expanded, frequently in the form of linear suburbs along main highways leading from the town gates. Such spread was also to be seen leading from unwalled towns such as Reading, where a sprawling built-up area was of long standing, and where the increasing population of Elizabethan times was accommodated by the subdivision of houses into tenements.[4]

In varying degrees, the towns were centres of trade and craftsmanship, being intimately bound to their hinterlands in the countryside and

[1] *Ibid.*, 467.   [2] W. Smith (1588, ed. 1879), 2.
[3] J. Speed, *The theatre of the empire of Great Britaine* (London, 1611), 89; M. W. Beresford and J. K. S. St Joseph, 177–9; I. MacIvor, 'The Elizabethan fortifications at Berwick-upon-Tweed', *Antiq. Jour.*, 45 (1965), 64–96.
[4] C. F. Slade, 'Reading', in M. D. Lobel (ed.), *Historic towns*, I (London and Oxford, 1969), 5–6; Banbury, Gloucester, Hereford, Nottingham and Salisbury are also included in this volume.

dependent upon good road and waterborne communications. But most of them had not yet lost their rural character. In Maitland's phrase, those who would study the history of towns 'have fields and pastures on their hands'.[1] Leading tradesmen were often farmers or graziers. At Leicester, 'pigs and cows went their way about the town, though ringed and herded, and as late as 1610 it was necessary to forbid winnowing in the streets'.[2]

Major provincial towns

Norwich seems to have had a population of about 15,000 or so. Virtually the whole of the town was contained within its medieval walls which enclosed an area of nearly one square mile, although as much as a quarter of this was not continuously built over. The city had suffered grievously in the sixteenth century from repeated fires and epidemics of plagues. In spite of these reverses it had begun to grow again when, in 1565, Dutch and Walloon refugees from religious persecution were invited to settle in the city and there weave their 'outlandish commodities', the so-called new draperies. At the end of the century they numbered about a quarter of the total population, and were reviving the depressed textile industry. 'By their means', so we read in a document of 1575, 'the city is well inhabited, and decayed houses re-edified and repaired that were in ruins', although initially the refugees had accentuated the poverty of overcrowded parishes such as St Paul's and All Saints.[3] Moreover, the central core of the city around the castle was improving its amenities and changing its appearance. River water was piped from New Mills to the Market Hall and Cross (1582); all new-built roofs were required to be of tile, slate or lead, and not of thatch, so as to reduce the risk of fire (1583); and two of the principal bridges across the Wensum were rebuilt in stone in place of wood (1591). These changes were consonant with Norwich's many functions as a regional capital, concerned with distributive and service industries as well as with the textile trades.[4]

York, with a population of about 12,000, was the most prominent city

---

[1] F. W. Maitland, *Township and borough* (Cambridge, 1898), 9.

[2] E. W. J. Kerridge in R. A. McKinley (ed.), *V.C.H. Leicestershire*, IV (1958), 99.

[3] Quoted in E. Lipson, *The history of the woollen and worsted industries* (1921), 23; J. F. Pound, 'An Elizabethan census of the poor', *Univ. Birm. Hist. Jour.*, VIII (1962), 140–4 and 150.

[4] J. F. Pound, 'The social and trade structure of Norwich, 1525–1575', *Past and Present*, XXXIV (1966), 49–69; Anon., *The history of the city and county of Norwich* (Norwich, 1768), 225–8, 242–4, 246.

in the north. It was the seat of an archbishopric and an important administrative centre; it was well served by inland waterways; and it had escaped
the worst ravages of plague during the sixteenth century. Its walls enclosed
some 143 acres, and linear suburbs extended from it on all sides. But
much of its medieval glory was departing. It had lost a great deal of its
clothing industry to other Yorkshire towns and villages; and it was losing
its position as a port to places nearer the sea where larger sea-going ships
could be accommodated. Hull gained much of what York lost.[1]

Bristol also may have contained about 12,000 people, and it had long
grown beyond the circle of its medieval walls. 'For trade of merchandize',
it was 'a second London, and for beauty and account next unto York',
wrote Speed.[2] Set between the Frome and the Avon with their sheltered
tidal harbours, it was well placed for trade. It was the port for one of the
most vigorous cloth-making areas in England. It had connections with
the Midlands and South Wales by means of the Severn and the Bristol
Channel. Beyond, it traded with Ireland, Gascony and the Iberian
peninsula, and its mariners and merchants were among the first Englishmen to take advantage of wider opportunities beyond Europe. Around
the ancient centre had grown up a large industrial suburb and a fashionable
residential area. It was far more salubrious than most of the towns of the
realm. As William Smith could say in 1588, 'there is no dunghill in all
the city', and Speed could point to its underground sewers removing 'all
noysome filth and uncleaness'.[3]

Newcastle upon Tyne was a city of some 10,000 people. It lay where
the main east-coast road to Scotland crossed the Tyne. Ships could berth
at the foot of its Castle Hill, and the quayside quarters lay within the walls
which enclosed about 150 acres, not all built over until the nineteenth
century. The sixteenth century had seen a great increase in the use of coal
in England generally, and the easily worked outcrops on the banks of the
Tyne were being rapidly exploited. Shipments of coal from Newcastle
increased fivefold between 1560 and 1600, and this activity was reflected
in the prosperity of the town. It was able to spend a large sum of money
to obtain a new and extended charter from the Crown in 1600. A bridge
across the Tyne joined the town to Gateshead, but this was a borough in
its own right, and attempts to annex it, during the sixteenth century, had
failed.[4]

[1] A. G. Dickens in P. M. Tillott (ed.), *V.C.H. The city of York* (1961), 121–2,
129–30.
[2] J. Speed, 47.   [3] W. Smith (1588, ed. 1879), 34.   [4] J. Speed, 89.

Exeter was a city of about 9,000 people in spite of severe epidemics in the 1590s. Its ancient walls enclosed only 93 acres, but ribbon-like suburbs had grown out along the main roads, especially towards London, and these may have housed as much as a quarter of its population. It had its own clothing industry, and was also the outlet for the products of the clothing centres around. Direct access to the sea had long ceased, but the growth of its outport at Topsham, four miles down the estuary of the Exe, reflected the prosperity of the city itself.[1]

Lesser provincial towns

Of the lesser provincial towns of the realm, some were losing ground like the old cloth-making centres of Coventry and Salisbury with about 6,000 inhabitants each.[2] Canterbury, with its cathedral, had about 5,000 people.[3] The majority of county towns rarely exceeded 3,000 to 4,000 inhabitants apiece. Some, like Worcester and Hereford, had long linear extensions along each of half a dozen highways leading from their town gates.[4] Others, like Northampton and Nottingham, had virtually no suburbs. Their layouts were very varied. Of the smaller towns some occupied strong points, like Durham and Buckingham, around their castles. Others were roughly star-shaped where roads converged, as at Hertford and Bedford. Yet others, with their houses, shops and public buildings were merely strung along highways, as at Huntingdon and Kendal.

Market towns of all sizes, with weekly markets, were widely distributed over the countryside, and they often specialised in one or more products such as grain, wool, leather or cheese. Annual fairs also specialised more particularly in cattle, sheep or horses, and these attracted people from 50 miles and more. A characteristic meeting place was not only the market or the fair but the inn. A census of inns and ale-houses of 1577 enumerated over 1,600 inns in 25 English counties; those at Romford in Essex were the Blue Boar, the Swan and the White Horse.[5] A number of inns offered

[1] R. Pickard, *The population and epidemics of Exeter in pre-Census times* (Exeter, 1947), 14–15; W. T. MacCaffrey, *Exeter, 1540–1640* (Cambridge, Mass., 1958), 12–13; W. G. Hoskins, 'The Elizabethan merchants of Exeter', in S. T. Bindoff et al. (eds.), *Elizabethan government and society* (London, 1961), 164.
[2] W. G. Hoskins, *Provincial England: essays in social and economic history* (London, 1963), 72; Anon., *The history and antiquities of the city of Coventry* (Coventry, 1810), 51, 55, 71.
[3] W. K. Jordan, *Social institutions in Kent, 1480–1660* (Kent Archaeol. Soc., 75, 1961), 3.   [4] M. D. Lobel, 'Hereford' in M. D. Lobel (ed.), 9.
[5] A. Everitt in J. Thirsk (1967), 559.

special facilities to traders and their wares, and became centres for private bargaining.

Some markets towns were very small. William Harrison, about 1580, thought that the common run of market towns rarely had more than 60 households or 200 to 300 communicants which would imply total populations of 500 or less; but within their limited spheres such small towns played a vital role.[1] Of quite a different character were those places with mineral springs. Bath, after long neglect, began once more to attract visitors, but its heyday was still a long way off in 1600. The springs at Buxton and Matlock, as Camden noted, were being increasingly frequented. But none of these English spas could compare with the more famous continental resorts.[2]

A number of the lesser provincial towns in 1600 were prospering as a result of recent industrial development. Colchester, with perhaps 5,000 people, had almost as many houses standing 'without the walls' as within. Dutch immigrants, who began to arrive about 1570, gave its cloth-making industry a new lease of life. By 1609 it had become 'so populous...that there was not one house to be had at any rate'; by this time, the immigrants numbered 1,300.[3] Nearby in East Anglia, other clothing-making towns were also flourishing – Hadleigh, Lavenham, Long Melford and Sudbury, together with a host of smaller centres. In the clothing area of the West Country there were also prosperous sizeable towns such as Tiverton with 4,000 people, Barnstaple with 3,500, Crediton and others.[4] In the north, Halifax, Leeds and Wakefield were beginning to stand out as centres of the clothing trade. Other industrial towns were also beginning to emerge: Sheffield had about 2,200 people, apart from the inhabitants of various villages in its large parish.

Along the coast there was another group of prospering towns. Plymouth had more than doubled its numbers in the last decades of the sixteenth century until in 1600 it included about 7,800 people. New streets were built all round Sutton Pool in response to the town's role as the principal naval base in the Spanish war.[5] Southampton, on the other hand,

---

[1] W. Harrison, 217; W. G. Hoskins, 'Provincial life', in A. Nicol (ed.), *Shakespeare in his own age* (Cambridge, 1964), 13–20.

[2] J. A. Patmore, 'The spa towns of Britain', in R. P. Beckinsale and J. M. Houston (eds.), *Urbanization and its problems* (Oxford, 1968), 51–2.

[3] J. Speed, 31; P. Morant, *The history and antiquities of the county of Essex*, I (London, 1768), 72, 75–8, 105–39.

[4] W. G. Hoskins (1954), 113, 455; W. Smith (1588, ed. 1879), 44.

[5] W. G. Hoskins, *Local history in England* (London, 1959), 143–4.

was in decay largely because it had lost its trade to London.[1] In the Thames estuary, Greenwich, Rochester and Chatham had about 3,000 inhabitants each, so had Sandwich in Kent.[2] In East Anglia, Ipswich with about 5,500 people, was 'one of the most famous towns in England at this present for traffic and other respects'. Speed's plan shows a town much expanded beyond its walls, with suburbs along all the roads leading from the town, and a quay on the Orwell stretching as far as Stoke Bridge; and he saw it as 'full of streets plenteously inhabited'.[3] Farther north along the east coast were Great Yarmouth, Lynn, Boston and Hull, all reflecting the agricultural and industrial prosperity of their hinterlands. In the north-west Chester was a flourishing city 'with very fair and large suburbs'. It was still the post-town for Ireland, but the Dee was silting up and the town had 'lost the advantage of a harbour'. It was relying upon outports, such as Neston and Heswall, and it was being supplanted by Liverpool, in whose grain market the Lancashire dealers stood on one side and the Cheshire dealers on the other.[4]

London

London stood far above all other urban areas, being about sixteen times or so the size of Norwich, the largest provincial city. By 1600, 'Greater London' (including Westminster and Southwark) probably had a population of about 250,000.[5] The rapid growth of the sixteenth century could have been sustained only by immigration from the rest of England. There was also a foreign element that added a cosmopolitan character to its life. A return for 1573 gave a total of 4,287 for the city – 3,160 Dutch, 440 French, 423 Burgundians, 137 Italians, 58 Spaniards and 32 Scots, to name the chief elements.[6] There were others in Westminster, Southwark and the suburbs. The numerical preponderance of the newly growing suburbs is clear, and became more pronounced because accommodation within the old walled city could be increased only 'by turning great houses into tenements, and by building upon a few gardens'.[7]

[1] A. A. Ruddock, *Italian merchants and shipping in Southampton, 1270–1600* (Southampton, 1960), 258–62.

[2] W. Smith (1588, ed. 1879), 7; W. K. Jordan, 4; J. W. Jessup, 100–104; C. W. Chalklin, *Seventeenth century Kent* (London, 1965), 23–6.

[3] W. G. Hoskins (1959), 177; J. Speed, 76.

[4] A. Everitt in J. Thirsk (1967), 481.

[5] N. G. Brett-James, *The growth of Stuart London* (London, 1935), 495–8.

[6] E. E. Rich, 263.

[7] N. G. Brett-James, 499, quoting John Graunt in 1662.

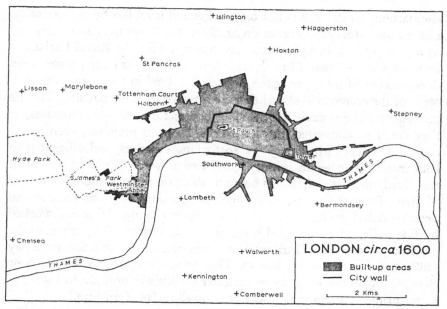

Fig. 63 London *circa* 1600
    Based on N. G. Brett-James, *The growth of Stuart London* (London, 1935),
    map opposite p. 78.

The outline of Greater London may be reconstructed from Norden's
plan of 1593 and Stow's unique description published in 1598 (Fig. 63).
Only London Bridge spanned the Thames, which was unembanked and
broader than it is today. At the southern end of the bridge lay the borough
of Southwark which had been annexed to the city in 1550, and a continuous line of houses ran along the south bank where there were theatres
and places for bull-baiting and bear-baiting. Below the bridge-head, sea-
going ships anchored in mid-river, in the 'Pool', and they were served by
barges, lighters and wherries which carried goods and passengers to the
riverside wharves, warehouses and landing-stages.[1]

As viewed from the river, the boldest feature in the city's profile was
the square tower of Old St Paul's, high above the steeples of the other
124 churches, and overlooking a warren of congested streets without
a central focus. For decades new houses, gardens, and yards had covered
any scrap of available land or made ground – in the city ditch, over
Walbrook and the Fleet, on London Bridge, within the churchyards,

[1] *Ibid.*, 31.

encroaching on the streets not only at ground level but by the extension of the upper storeys of houses on brackets. No fewer than 80 houses had to be demolished, in 1565–70, to provide a site for the Royal Exchange, between Cornhill and Threadneedle Street. There had been widespread development of the 23 religious houses dissolved in the 'Great Pillage'. Most of the conventual churches were retained to serve parishes, but the monastic buildings made fine dwelling-houses for nobles or merchants, or they could be converted into tenements, business premises, even tennis courts. This kept to the city the aristocrats, courtiers, and officials who later in the seventeenth century gravitated towards Westminster. With the greatly increased demands being made upon them, London's supplies of water from the rivers, wells, and springs diminished in volume and deteriorated in quality. After 1582 came the new 'artificial forciers' worked by tide-mills, which pumped water from the Thames along pipes to the conduits; it was then distributed by water-carriers or, in a few cases, by quills direct to rich men's houses. This system was eclipsed by the New River Company in 1613, when a regular supply was brought to Finsbury from springs over 38 miles away at Amwell in Hertfordshire.[1]

From the Tower around to Shoreditch in the east, mean ribbon-like suburbs already reached along the Thames frontage to Limehouse and down the main roads, in Stow's view 'no small blemish to so famous a city to have so unsavoury and unseemly an entrance'.[2] The northern and western suburbs were more compact and had houses of quality, as at Chancery Lane and Holborn Hill; here, too, Camden tells us were 'some Inns for the study of common law'. Farther west, the suburbs forged a first physical link between London and Westminster. This went along the Strand, with a long line of former bishops' palaces beside the river, now transformed into 'large and goodly houses'. After 1603 the palace and abbey of Westminster drew to themselves permanently the seats of court and government, although the image lingered of 'twin-sister cities as joined by one street'.[3]

There were many attempts to curb the growth of Greater London following the proclamation in 1580 that prohibited the building of new houses within three miles of the city, and sub-letting within it. But it was difficult to check the dispersal of industry to the suburbs, including Southwark, maintained as it was by journeymen and apprentices who had

---

[1] W. H. Chaloner and A. E. Musson, 26.
[2] N. G. Brett-James, 59.
[3] N. G. Brett-James, 61, quoting Thomas Heywood in 1635.

broken away from the craft gilds within the city; by industries seeking lower rents and others (like tanning or soap-boiling) which were obnoxious in a residential setting; and by foreign immigrants who settled in suburban tenements, plying their trades in competition with the gilds.[1] The suburbs continued to grow, and to a fastidious Venetian observer it seemed as if they were 'inhabited by an inept population of the lowest description'.[2] Even so, Camden rejoiced in what to many was the dismaying growth of London. To him it was 'the epitome of all Britain, the seat of the British Empire'.

[1] H. J. Dyos, *Victorian suburb: a study of the growth of Camberwell* (Leicester, 1961), 34; 'The growth of a pre-Victorian suburb: south London, 1580–1836', *Town Planning Rev.*, XXV (1954), 67.

[2] Horatio Busino, 1618, quoted by H. and P. Massingham, *The London anthology* (London, 1950), 455.

INDEX

Bignor, 63
Birmingham, 231, 242, 283
Black Death, 78, 80, 136, 137, 143, 144, 145,
146, 157, 167, 170, 172, 175, 177, 187,
188, 195, 197, 199, 200, 202, 203, 205,
206, 207, 219, 220, 229
Blackheath, 271
Bladon, 65
Blakeney, 270
Blast furnaces, 65, 228, 230–2, 234, 281, 282,
283, 284
Blencarn, 14
Bloomeries, 228, 229, 230, 231
Bodmin, 69
Bodmin Moor, 142, 261
Boldon Book, 112
Bolton, 280
Boroughs, 17, 22, 36, 66–74, 115, 116, 123–9,
131, 132, 137, 177–83
    Parliamentary, 129, 179
    Taxation, 81, 124, 127–9, 132, 137, 177–83
Boston (Lincs.), 109, 118, 122, 123, 127, 132,
134, 176, 177, 181, 183, 184, 185, 234,
239, 240, 243, 298
Box (Wilts.), 63, 111
Bradford on Avon, 69, 278
Bramley, 182
Bran Dyke, 9
Brass manufacture, 234, 276, 286
Breckland, 46, 49, 270
Bremhill, 182
Breweries, brewing, 74, 109, 171, 232, 284
Brickhill, Bow, Great and Little, 28
Bridge-building, bridges, 119, 134, 237,
294, 295, 299
Bridgnorth, 182, 185
Bridgwater, 182, 185, 240, 288
Bridlington, 70
Bridport, 69, 118, 175
Brigham, 286
Brightwell, 189
Brill, 14
Brindley (Yorks.), 30
Bristol, 70, 71, 73, 118, 127, 132, 134, 135,
163, 170, 173, 176, 177, 181, 184, 185,
188, 192, 222, 234, 236, 239, 240, 241,
243, 245–6, 287, 290, 293, 295
Bristol Channel, 12, 295
Broadcloth, 225, 226, 227, 228, 276, 277
Bruton, 69, 71

Brycheiniog, 8
Buckden, 40
Buckingham, 37, 69, 296
Buckinghamshire, 20, 50, 69, 81, 88, 141,
164–5, 179, 245, 255, 271
Building materials, 63, 109–12, 119, 149,
172–3, 175, 257–9, 285, 291, 294
Burghal Hidage, 17, 18, 36
Burghclere, 127
Burnham (Norf.), 28
Burton upon Trent, 129, 130
Bury St Edmunds, 70, 71, 73–4, 102, 128,
134, 182, 183, 184, 243
Butter, 173, 270, 271, 272, 291, 292
Buxton, 297
Bytham, Castle and Little, 62

Cabbages, 91
Caister (Norf.), 65, 134
Calamine, 234, 286
Calder, R., 223
Calne, 69
Cam, R. (Gloucs.), 277
Cambridge, 9, 20, 35, 37, 42, 52, 70, 71,
118, 132, 134, 181, 184, 243
Cambridgeshire, 7, 9, 44, 50, 57, 70, 76, 88,
140, 141, 142, 159, 163, 181, 189, 251
Campden, 181
Canals, 237, 291
Cannock Chase, 14, 100, 165, 265, 267
Canons Regular, 110
Canterbury, 38, 59, 69, 71, 73, 115, 118,
132, 134, 138, 181, 184, 203, 204, 237,
243, 272, 287, 288, 290, 296
Canterbury cathedral priory, 152–4, 155,
160–2, 168, 203, 204
Canvey Island, 162
Carlisle, 13, 20, 134, 174, 179, 288
Carrots, 272
Castle Acre, 268
Castle Combe, 167, 222, 225, 242
Castles, 71, 72, 110, 129, 147, 244, 294,
296
Castor (Northants), 182
Catraeth, Catterick, 5, 12
Cattle, 7, 27, 49, 50, 94, 97, 98, 120, 159,
161, 162, 167, 168, 211, 249, 259, 260,
261, 263, 265, 266, 267, 268, 269, 270,
271, 272, 276, 294, 296
    See also Livestock, Stock-breeding